PRACTICAL
RELIABILITY
ENGINEERING

'The concept of chance enters into the very first steps of scientific activity, by virtue of the fact that no observation is absolutely correct. I think chance is a more fundamental concept than causality, for whether in a concrete case a cause–effect relationship exists can only be judged by applying the laws of chance to the observations.'

Max Born,
Natural Philosophy of Cause and Chance

'A statistical relationship, however strong and however suggestive, can never establish a causal connection. Our ideas on causation must come from outside statistics, ultimately from some theory.'

Kendall & Stuart,
The Advanced Theory of Statistics

'Reliability is, after all, engineering in its most practical form.'

James R. Schlesinger
Former US Secretary of State for Defense

PRACTICAL RELIABILITY ENGINEERING

Second Edition

PATRICK D. T. O'CONNOR

British Aerospace Dynamics Group, Stevenage

A Wiley–Interscience Publication

JOHN WILEY & SONS

Chichester • New York • Brisbane • Toronto • Singapore

First Edition © 1981 by Heyden & Son Ltd.

Copyright © 1985 by John Wiley & Sons Ltd.

Reprinted with corrections December 1985
Reprinted with corrections April 1988
Reprinted June 1989

Library of Congress Cataloging in Publication Data:

O'Connor, Patrick D. T.
  Practical reliability engineering.

  Includes bibliographies and index.
  1. Reliability (Engineering)   I. Title.
TS173.029   1985   620'.00452   84-13101
ISBN 0 471 90551 8

British Library Cataloguing in Publication Data:

O'Conor, Patrick D. T.
  Practical reliability engineering. — 2nd ed.
  1. Reliability (Engineering)
  I. Title
  620'.00452   TS173

  ISBN 0 471 90551 8

Typeset by Mathematical Composition Setters, Ltd.
7 Ivy Street, Salisbury, Wiltshire
Printed and bound in Great Britain by
Courier International Ltd, Tiptree, Essex

ACKNOWLEDGEMENTS

I am deeply indebted to many people for their help and constructive criticism. In particular, Dr R. A. Evans provided invaluable comments, corrections and suggestions for improvement. My colleague Mr Norman Harris provided much valuable assistance and several new insights on the material of Chapters 2 and 3. Professor A. D. S. Carter of the Royal Military College of Science gave very helpful advice on the analysis methods in Chapter 4, and I am grateful for his permission to reproduce material from his published papers on load—strength interference which appear in the Bibliography at the end of that chapter. Dr Bev Littlewood of the City University, London, was very generous with suggestions and criticisms on Chapter 8 on software reliability. Mr Kenneth Blemel also provided many suggestions for improvement and corrections. I am also indebted to the many others who encouraged and helped me, either in dicussions, with comments on the text, or whose published work provided me with insights and guidance.

I acknowledge with thanks the permission I have been given to reproduce the following copyright material: Fig. 1.6, from British Standard, BS 5760, Part 1, British Standards Institution, London (1979; Table 2.4, from W. Volk, *Applied Statistics for Engineers*, McGraw-Hill, New York (1958); Fig. 2.7, from *Industrial Quality Control* (December 1965); Fig. 2.8, from H. F. Dodge and H. G. Romig, *Sampling Inspection Tables*, Bell Telephone Laboratories (1944); Figs 4.4 to 4.7, from A. D. S. Carter, 'Reliability reviewed', *Proc. Inst. Mech. Eng.*, **193** (4) (1979); Table 5.1, adapted from R. H. Myers, K. L. Wong and H. M. Gordy, *Reliability Engineering for Electronic Systems*, Wiley, New York (1964); Figs 7.16 and 7.17, and Table 7.7, from R. I. Anderson, RDH 376: *Reliability Design Handbook*, Reliability Analysis Center, IIT Research Institute (1976); Appendix 4 from A. Hald, *Statistical Tables and Formulas*, Wiley, New York (1952); probability and hazard plotting papers from Technical and Engineering Aids to Management Inc. and H. W. Peel and Co. Ltd.

Finally, I would like to thank Miss Linda Rendle for her speedy and efficient typing of the manuscript, and the management of British Aerospace Dynamics Group, Stevenage Division, for encouragement and assistance in writing the book.

ACKNOWLEDGEMENTS

CONTENTS

xii

PREFACE TO THE FIRST EDITION

This book is designed to provide an introduction to reliability engineering and management, both for students and for practising engineers and managers. The emphasis throughout is on practical applications, and the mathematical concepts described are accordingly limited to those necessary for solution of the types of problems covered. Practical approaches to problem-solving, such as the use of probability plotting techniques and computer programs, are stressed throughout. More advanced texts are cited for further reading on the mathematical and statistical aspects. The references given in the Bibliographies are limited to those considered to provide a direct continuation of the chapter material, with the emphasis on practical applications. Tables and charts are provided to complement the analytical methods described, and numerous worked examples are included.

The book describes and comments on the usage of the major national and government standards and specifications covering reliability engineering and management in the USA and the UK. It is considered that this is an important aspect of the practical approach, since so much engineering development work is now governed by such documents. The effects of current engineering, commercial and legislative developments, such as microelectronics, software-based systems, consumerism and product liability, are covered in some detail.

The requirements of the examination syllabi of the American Society for Quality Control, and the Institute of Quality Assurance (UK) in reliability engineering are covered, so the book will be suitable for use in courses leading to these qualifications. The emphasis on practical approaches to engineering and management, the comprehensive coverage of standards and specifications, and the overall layout of the book should make it equally as suitable as a general up to date reference for use in industry and in government agencies.

PREFACE TO THE SECOND EDITION

I have received much helpful criticism of the first edition of my book since it appeared in 1981. Whilst the reviews have generally not been unfavourable, critics have pointed out that, despite the title, the book was not quite practical enough in some areas. I have also come to realize this through my own work, particularly on the application of mathematical modelling and statistics to reliability problems. Consequently, much of the revision for the second edition has been to add to what I consider to be the practical aspects of management and engineering for reliability.

I have added to the sections on reliability prediction, demonstration and measurement, to explain and to stress the fundamental and considerable uncertainty associated with attempts to quantify and forecast a property of engineered products which is inherently non-deterministic. I believe that when people involved in reliability work manage to unshackle themselves from the tyranny of the 'numbers game' the way is cleared for the practical engineering and management approaches that are the only ways to achieve the highly reliable products demanded by the markets of today. I have not removed the descriptions of the methods for quantifying reliability, since I believe that, when these are applied with commonsense and understanding of their inherent limitations, they can help us to solve reliability problems and to design and make better products.

I have added three new chapters, all related to the practical aspects.

The first edition described how to analyse test data, but included little on how to test. I have therefore written a new chapter on reliability testing, covering environmental and stress testing and the integration of reliability and other development testing. I am indebted to Wayne Tustin for suggesting this and for his help and advice on this subject.

The quality of manufacture is obviously fundamental to achieving high reliability. This point was made in the first edition, but was not developed. I have added a complete chapter on quality assurance (QA), as well as new material on integrated management of reliability and QA programmes.

Maintenance also affects reliability, so I have added a new chapter on maintenance and maintainability, with the emphasis on how they affect reliability, how reliability affects maintenance planning and how both affect availability.

I have also added new material on the important topic of reliability analysis for repairable systems. Harry Ascher, of the US Naval Research Laboratory, has pointed out that the reliability literature, including the first edition of my book, has almost totally ignored this aspect, leading to confusion and analytical errors. How many reliability engineers and teachers know that Weibull analysis of repairable system reliability data can be quite misleading except under special, unrealistic conditions? Thanks to Harry Ascher, I know now, and I have tried to explain this in the new edition.

I have also brought other parts of the book up to date, particularly the sections on electronic and software reliability.

The third reprint of the first edition included many corrections, and more corrections are made in this edition.

I am extremely grateful to all those who have pointed out errors and have helped me to correct them. Paul Baird of Hewlett Packard, Palo Alto, was particularly generous. Colleagues at British Aerospace, particularly Brian Collett, Norman Harris, Chris Gilders and Gene Morgan, as well as many others, also provided help, advice and inspiration.

Finally, my thanks go to my wife Ina for much patience, support and typing.

NOTATION AND DEFINITIONS

LIST OF SYMBOLS

α	Producer's risk (in sample or sequential testing, s-probability that good lot or item will be rejected). Also Duane slope, also s-significance, s-confidence.
β	Consumer's risk (in sample or sequential testing, s-probability that bad lot or item will be accepted). Also shape parameter (slope) of Weibull distribution.
γ	Location parameter (minimum life) of Weibull distribution.
μ	Location parameter for a continuous distribution. Equal to mean value for s-normal distribution.
η	Characteristic life for Weibull distribution.
σ	Standard deviation (scale parameter) of s-normal distribution.
ν	Degrees of freedom.
θ	Mean time between failures (MTBF) for constant failure rate.
λ	Failure rate.
$\Phi(z)$	Cumulative value of standard s-normal variate, z.
$B(\cdot)$	B-life of Weibull distribution, e.g. B_{10} life.
C	s-Confidence.
d	Design ratio = design MTBF/low limit MTBF.
$E(\cdot)$	s-expected value. The s-expected value of x is denoted $\hat{x}$.
$f(\cdot)$	Function of. For a distribution, the probability density function (p.d.f.).
$F(\cdot)$	Cumulative distribution function (c.d.f.)
F	Variance ratio.
$h(\cdot)$	Hazard function.
$H(\cdot)$	Cumulative hazard function.
L	Load.
n	Sample size.
N	Population size.
$P(\cdot)$	s-Probability.
$P(\cdot \mid \cdot)$	Conditional s-Probability.
r	s-Correlation coefficient. Also rank value for median ranks.

R(·)   Reliability function.
S   Standard error. Also strength.
t   Time.
Var(·)   Variance (σ^2).
$\bar{x}$ Mean value of distributed variable, x.
z   Standard s-normal variate.

SELECTED DEFINITIONS

Reliability

Reliability. The ability of an item to perform a required function under stated conditions for a stated period of time (BS 4778).

(Note: the term reliability can also be denoted as a probability or as a success ratio.)

Redundancy. The existence of more than one means for accomplishing a given function. Each means of accomplishing the function need not necessarily be identical (MIL-STD-721B).

Redundancy, active. That redundancy wherein all redundant items are operating simultaneously rather than being switched on when needed (MIL-STD-721B).

Redundancy, standby. That redundancy wherein the alternative means of performing the function is inoperative until needed and is switched on upon failure of the primary means of performing the function (MIL-STD-721B).

Failure

Failure. The termination of the ability of an item to perform a required function (BS 4778).

Observed failure rate. For a stated period in the life of an item, the ratio of the total number of failures in a sample to the cumulative observed time on that sample. The observed failure rate is to be associated with particular and stated time intervals (or summation of intervals) in the life of the item, and under stated conditions (BS 4778).

Observed mean time between failures (MTBF). For a stated period in the life of an item, the mean value of the length of time between consecutive failures computed as the ratio of the cumulative observed time to the number of failures under stated conditions (BS 4778).

Observed mean time to failure (MTTF) (for non-repairable items). For a stated period in the life of an item, the ratio of the cumulative time for a sample to the total number of failures in the sample during the period under stated conditions (BS 4778).

Observed B-Percentile life. The length of observed time at which stated proportion ($B\%$) of a sample of items has failed (BS 4778).

Maintainability

Maintainability. The ability of an item, under stated conditions of use, to be retained in, or restored to, a state in which it can perform its required functions, when maintenance is performed under stated conditions and using prescribed procedures and resources (BS 4778).

Mean time to repair (MTTR). The total corrective maintenance time divided by the total number of corrective maintenance actions during a given period of time (MIL-STD-721B).

Maintenance, corrective. The actions performed, as a result of failure, to restore an item to a specified condition (MIL-STD-721B).

Maintenance, preventive. The actions performed in an attempt to retain an item in a specified condition by providing systematic inspection, detection and prevention of incipient failure (MIL-STD-721B).

Availability

Availability. The ability of an item (under combined aspects of its reliability, maintainability and maintenance support) to perform its required function at a stated instant of time or over a stated period of time (BS 4778).

Chapter 1

INTRODUCTION TO RELIABILITY ENGINEERING

WHY TEACH RELIABILITY?

Engineering education is traditionally concerned with teaching how manufactured products work. The ways in which products fail, the effects of failure and aspects of design which affect the likelihood of failure are not usually taught, mainly because it is necessary to understand how a product works before considering ways in which it might fail. For many products the tendency to approach the failed state is analogous to entropy. The engineer's tasks are to design and maintain the product so that the failed state is deferred. In these tasks he faces the problems inherent in the variability of engineering materials, processes and applications. Engineering education is basically deterministic, and does not usually pay sufficient attention to variability. Yet variability and chance play a vital role in determining the reliability of most products. Basic parameters like mass, dimensions, friction coefficients, strengths and stresses are never absolute, but are in practice subject to variability due to process and materials variations, human factors and applications. Some parameters also vary with time. Understanding the laws of chance and the causes and effects of variability is therefore necessary for the creation of reliable products and for the solution of problems of unreliability.

However, there are practical problems in applying statistical knowledge to engineering problems. These problems have probably deterred engineers in the past from using statistical methods, and texts on reliability engineering and mathematics have generally stressed the theoretical aspects without providing guidance on their practical application. To be helpful a theoretical basis must be credible, and statistical methods which work well for insurance actuaries, market researchers or agricultural experimenters may not work as well for engineers. This is not because the theory is wrong, but because engineers usually have to cope with much greater degrees of uncertainty, mainly due to human factors in production and use.

Some highly reliable products are produced by design and manufacturing teams who practise the traditional virtues of reliance on experience and maintenance of high quality. They do not see reliability engineering as a subject requiring specialist consideration, and a book such as this would teach

1

them little that they did not already practise in creating their reliable products. Engineers and managers might therefore regard a specialist reliability discipline with scepticism. However, many pressures now challenge the effectiveness of the traditional approaches. Competition, the pressure of schedules and deadlines, the cost of failures, the rapid evolution of new materials, methods and complex systems, the need to reduce product costs and safety considerations all increase the risks of product development. Figure 1.1 shows the pressures that lead to the overall perception of risk. Reliability engineering has developed in response to the need to control these risks. Engineers and managers involved in high risk developments generally accept the need for interdisciplinary liaison and control. However, these controls must not stifle progress and flair, and the control must not become an end in itself.

Later chapters will show how reliability engineering methods can be applied to design, development and management to control the level of risk. The extent to which the methods are applicable must be decided for each project and for each design area. They must not replace normal good practice, such as safe design for components subject to cyclic loading, or application guidelines for electronic components. They should be used to supplement good practice. However, there are times when new risks are being taken, and the normal rules and guidelines are inadequate or do not apply. Sometimes we take risks unwittingly, when we assume that we can extrapolate safely from our present knowledge. Designers and managers are often overoptimistic or are reluctant to point out risks about which they are unsure.

It is for these reasons that an understanding of reliability principles and methods is now an essential ingredient of modern engineering.

WHAT IS RELIABILITY?

No one disputes the need for articles to be reliable. The average consumer is acutely aware of the problem of less than perfect reliability in domestic pro-

Figure 1.1   Perception of risk

ducts such as TV sets and automobiles. Organizations such as airlines, the military and public utilities are aware of the costs of unreliability. Manufacturers often suffer high costs of failure under warranty. Argument and misunderstanding begin when we try to quantify reliability values, or try to put financial or other benefit values to levels of reliability.

The simplest, purely producer-oriented or inspectors' view of reliability is that in which a product is assessed against a specification or set of attributes, and when passed is delivered to the customer. The customer, having accepted the product, accepts that it might fail at some future time. This simple approach is often coupled with a warranty, or the customer may have some protection in law, so that he may claim redress for failures occurring within a stated or reasonable time. However, this approach provides no measure of quality over a period of time, particularly outside a warranty period. Even within a warranty period, the customer usually has no grounds for further action if the product fails once, twice or several times, provided that the manufacturer repairs the product as promised each time. If it fails often, the manufacturer will suffer high warranty costs, and the customers will suffer inconvenience. Outside the warranty period, only the customer suffers. In any case, the manufacturer will also probably incur a loss of reputation, possibly affecting future business.

We therefore come to the need for a time-based concept of quality. The inspectors' concept is not time-dependent. The product either passes a given test or it fails. On the other hand, reliability is usually concerned with failures in the time domain. This distinction marks the difference between traditional quality control and the modern approach to reliability.

Quality control (QC) of manufacturing processes obviously makes an essential contribution to the reliability of a product. In the chapters on manufacturing (Chapter 10) and reliability management (Chapter 14), the need to consider QC as an integral part of an overall reliability programme will be emphasized.

Reliability is, then, generally concerned with failures during the life of a product. (The reliability, or success ratio, of 'one-shot' items will be covered separately.) We therefore need to understand why items fail. A stressed item will fail only when the applied load exceeds the strength at the time of application. (The reader will think of failures which are not caused by the application of loads; this is an aspect of reliability which we will cover later.) This is easy to appreciate for most mechanical items, but it can be taken as an axiom for all engineered products. For example, a transistor will fail if the current through it exceeds its ability to conduct without overheating to the point of failure of the substrate or of a wire bond. A bearing may seize if it has degraded to the point that the load causes local break-down of the lubricating film.

Reliability is therefore an aspect of engineering uncertainty. Whether an item works for a particular period is a question which can be answered as a probability. This results in the usual engineering definition of reliability (for a non-repaired item) as:

The probability that an item will perform a required function without failure under stated conditions for a stated period of time.

PROBABILISTIC RELIABILITY

The concept of reliability as a probability means that any attempt to quantify it must involve the use of statistical methods. An understanding of statistics as applicable to reliability enegineering is therefore a necessary basis for progress, except for the special cases when reliability is perfect (we know the item will never fail) or it is zero (the item will never work). For practical purposes a hammer might be 100 per cent reliable if used for driving tacks, but would have a zero reliability if used to stop a train. If used to break rocks, its reliability for a given period might have some value between 0 and 100 per cent. In engineering we try to ensure 100 per cent reliability, but our experience tells us that we do not always succeed. Therefore reliability statistics are usually concerned with probability values which are very high (or very low: the probability that a failure does occur, which is 1 − reliability). Quantifying such numbers brings increased uncertainty, since we need correspondingly more information. Other sources of uncertainty are introduced because reliability is often about people who make and people who use the product, and because of the widely varying environments in which typical products might operate.

Further uncertainty, often of a subjective nature, is introduced when engineers begin to discuss failures. Should a failure be counted if it was due to an error that it is hoped will not be repeated? If design action is taken to reduce the risk of one type of failure, how can we quantify our trust in the designer's success? Was the machine under test typical of the population of machines?

Reliability is quantified in other ways. We can specify a reliability as the mean number of failures in a given time (failure rate), or as the *mean time between failures* (MTBF) for items which are repaired and returned to use, or as the *mean time to failure* (MTTF) for items which are not repaired. For repaired items, it is often assumed that failures occur at a constant rate, in which case the failure rate $\lambda = (\text{MTBF})^{-1}$. However, this is only a special case, valuable because it is often true and because it is easy to understand.

The application and interpretation of statistics in reliability are less straightforward than in, say, public opinion polls or measurement of human variations such as IQ or height. In these applications, most interest is centred around the behaviour of the larger part of the population or sample, variation is not very large and data are plentiful. In reliability we are concerned with the behaviour of unlikely combinations of load and strength, variability is often hard to quantify and data are expensive.

Chapter 2 provides the statistical basis of reliability engineering, but it must always be remembered that quality and reliability data contain many sources of uncertainty and variability which cannot be rigorously quantified. Nevertheless, it is necessary to derive values for decision-making so the mathematics are essential. The important point is that the reliability engineer or manager is not, like an insurance actuary, a powerless observer of his statistics. Statistical derivations of reliability are not a guarantee of results, and these results can be significantly affected by actions taken by quality and reliability engineers and managers.

LOAD AND STRENGTH

Obviously the relationship between load and strength is important in reliability engineering. Generally, failure will occur if load exceeds strength. The probability of failure due to the load exceeding the strength can be modelled by the rules of statistics, since the loads and strengths are themselves often variable.

Figure 1.2 shows a simple situation, in which an item has a constant strength and is subjected to a fixed stress which is less than the strength. Such an item, under such an application, will not fail. Figure 1.3 shows a more typical case,

Figure 1.2 Load–strength—discrete values

Figure 1.3 Load–strength—distributed values

Figure 1.4 Load–strength—interfering distribution

in which both load and strength are variable, distributed about a mean value, i.e. there is a finite probability that either value exceeds or is less than the mean value. If there is no overlap of the distributed values, again failures cannot occur. Figure 1.4, however, shows an overlap of the load and strength distributions. In such a population of items, any item in the left-hand 'tail' of the strength distribution which is subjected to a load in the overlapping right-hand 'tail' of the load distribution will fail at that time. This is a situation in which there is *interference* between the load and strength distributions. Strength does not always have a constant average value, e.g. when material fatigue or corrosion occurs, and the strength distribution then varies with time or with repeated load applications. Load–strength interference is a very important aspect of reliability engineering, and the ways in which the load and strength are distributed affect the pattern of failures as well as the probability of failure. Because of its importance, load–strength interference is covered in detail in Chapter 4.

REPAIRABLE AND NON-REPAIRABLE ITEMS

It is important to distinguish between repairable and non-repairable items when predicting or measuring reliability.

For a non-repairable item such as a light bulb, a transistor, a rocket motor or an unmanned spacecraft, reliability is the survival probability over the item's expected life, or for a period during its life, *when only one failure can occur*. During the item's life the instantaneous probability of the first and only failure is called the *hazard rate*. Life values such as the mean life or *mean time to failure* (MTTF), or the expected life by which a certain percentage might have failed (say 10 per cent.), are other reliability characteristics that can be used. Note that non-repairable items may be individual parts (light bulbs, transistors) or systems comprised of many parts (spacecraft, microprocessors). When a part fails in a non-repairable system, the system fails (usually) and system reliability is, therefore, a function of the time to the first part failure.

For items which are repaired when they fail, reliability is the probability that failure will not occur in the period of interest, when *more than one failure can occur*. It can also be expressed as the *failure rate* or the *rate of occurrence of failures* (ROCOF). However, the failure rate expresses the instantaneous probability of failure per unit time, when several failures can occur in a time continuum. It is unfortunate that the expression *failure rate* has been applied loosely in the past to non-repairable items. For example, US Military Handbook 217—*Reliability Prediction for Electronic Systems*—gives 'failure rates' for electronic parts. What is really meant is that, in a repairable system which contains a part type, the part will contribute by that amount to the system failure rate. The part, being non-repairable, cannot have a failure rate. This distinction does not always have practical significance, but it can, as will be shown later. The expression ROCOF is sometimes used to stress this point.

Repairable system reliability can also be characterized by the *mean time between failures* (MTBF), but only under the particular condition of a constant failure rate. We are also concerned with the *availability* of repairable items, since repair takes time. Availability is affected by the rate of occurrence of failures (failure rate) and by maintenance time. Maintenance can be corrective (i.e. repair) or preventive (to reduce the likelihood of failure, e.g. lubrication). We therefore need to understand the relationship between reliability and maintenance, and how both reliability and maintainability can affect availability.

Sometimes an item may be considered as both repairable and non-repairable. For example, a missile is a repairable system whilst it is in store and subjected to scheduled tests, but it becomes a non-repairable system when it is launched. Reliability analysis of such systems must take account of these separate states. Repairability might also be determined by other considerations. For example, whether a TV electronics board is treated as a repairable item or not will depend upon the cost of repair. An engine or vehicle will be treated as repairable only up to a certain age.

Repairable system reliability data analysis is covered in Chapter 3 and availability and maintainability in Chapter 9.

THE PATTERN OF FAILURES WITH TIME
(NON-REPAIRABLE ITEMS)

There are three basic ways in which the pattern of failures can change with time. The hazard rate may be decreasing, increasing or constant. We can tell much about the causes of failure and about the reliability of the item by appreciating the way the hazard rate behaves in time.

A constant hazard rate is characteristic of failures which are caused by the application of loads in excess of the design strength, at a constant average rate. For example, overstress failures due to accidental or transient circuit overload, or maintenance-induced failures of mechanical equipment, typically occur randomly and at a generally constant rate.

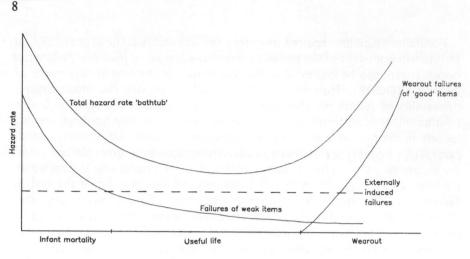

Figure 1.5 The 'bathtub' curve

Material fatigue brought about by strength deterioration due to cyclic loading is a failure mode which does not occur for a finite time, and then exhibits an increasing probability of occurrence.

Decreasing hazard rates are observed in items which become less likely to fail as their survival time increases. This is often observed in electronic equipment and parts. 'Burn-in' of electronic parts is a good example of the way in which knowledge of a decreasing hazard rate is used to generate an improvement in reliability. The parts are operated under failure-provoking stress conditions for a time before delivery. As substandard parts fail and are rejected the hazard rate decreases and the surviving parts are more reliable.

The combined effect generates the so-called *bathtub curve* (Fig. 1.5). This shows an initial decreasing hazard rate or infant mortality period, an intermediate useful life period and a final wearout period. Death is a good analogy to failure of a non-repairable system, and the bathtub curve model is similar to actuarial statistical models.

THE PATTERN OF FAILURES WITH TIME (REPAIRABLE ITEMS)

The failure rates (or ROCOF) of repairable items can also vary with time, and important implications can be derived from these trends.

A constant failure rate (CFR) is indicative of externally induced failures, as in the constant hazard rate situation for non-repairable items. A CFR is also typical of complex systems subject to repair and overhaul, where different parts exhibit different patterns of failure with time and parts have different ages since repair or replacement. Repairable systems can show a decreasing failure rate (DFR) when reliability is improved by progressive repair, as defective parts which fail relatively early are replaced by good parts. 'Burn-in' is applied to electronic systems, as well as to parts, for this purpose.

An increasing failure rate (IFR) occurs in repairable systems when wearout failure modes of parts begin to predominate.

The pattern of failures with time of repairable systems can also be illustrated by use of the bathtub curve (Fig. 1.5), but with the failure rate (ROCOF) plotted against age instead of the hazard rate.

The statistical treatment of failure data is covered in Chapters 2 and 3.

THE DEVELOPMENT OF RELIABILITY ENGINEERING

Reliability engineering, as a separate engineering discipline, originated in the United States during the 1950s. The increasing complexity of military electronic systems was generating failure rates which resulted in greatly reduced availability and increased costs. Solid state electronics technology offered long term hope, but conversely miniaturization was to lead to proportionately greater complexity, which offset the reliability improvements expected. The gathering pace of electronic device technology meant that the developers of new military systems were making increasing use of large numbers of new component types, involving new manufacturing processes, with the inevitable consequences of low reliability. The users of such equipment were also finding that the problems of diagnosing and repairing the new complex equipment were seriously affecting its availability for use, and the costs of spares, training and other logistics support were becoming excessive. Against this background the US Department of Defense and the electronics industry jointly set up the Advisory Group on Reliability of Electronic Equipment (AGREE) in 1952. The AGREE report concluded that, to break out of the spiral of increasing development and ownership costs due to low reliability, disciplines must be laid down as integral activities in the development cycle for electronic equipment. The report laid particular stress on the need for new equipments to be tested for several thousand hours in high stress cyclical environments including high and low temperatures, vibration and switching, in order to discover the majority of weak areas in a design at an early enough stage to enable them to be corrected before production commenced. Until that time, environmental tests of tens of hours duration had been considered adequate to prove the suitability of a design. The report also recommended that formal demonstrations of reliability, in terms of statistical confidence that a specified MTBF had been exceeded, be instituted as a condition for acceptance of equipment by the procuring agency. A large part of the report was devoted to providing detailed test plans for various levels of statistical confidence and environmental conditions.

The AGREE report was accepted by the Department of Defense, and AGREE testing quickly became a standard procedure. Companies which invested in the expensive environmental test equipment necessary soon found that they could attain levels of reliability far higher than by traditional methods. It was evident that designers, particularly those working at the fringes of advanced technology, could not be expected to produce highly reliable equipment without it being subjected to a test regime which would show up weaknesses. Complex systems and the components used in them included too many variables and interactions for the human designer to cope with infallibly, and even the most careful design reviews and disciplines could not provide sufficient protection. Consequently it was necessary to make the product speak for itself, by causing it to fail, and then to eliminate the weaknesses that caused the failures. The Department of Defense (DOD) reissued the AGREE report on testing as US Military Standard (MIL-STD) 781, *Reliability Qualification and Production Approval Tests*.

Meanwhile the revolution in electronic device technology continued, led by integrated microcircuitry. Increased emphasis was now placed on improving the quality of devices fitted to production equipments. Screening techniques, in which devices are temperature cycled, vibrated, centrifuged, operated at electrical overstress and otherwise abused, were introduced in place of the traditional sampling techniques. With component populations on even single printed circuit boards becoming so large, sampling no longer provided sufficient protection against the production of defective equipment. These techniques were formalized in military standards covering the full range of electronic components. Components produced to these standards were called 'Hi-rel' components.

Engineering reliability effort in the United States developed quickly, and the AGREE and reliability programme concepts were adopted by NASA and many other major suppliers and purchasers of high technology equipment. In 1965 the DOD issued MIL-STD-785—*Reliability Programs for Systems and Equipment*. This document made mandatory the integration of a programme of reliability engineering activities with the traditional engineering activities of design, development and production, as it was by then realized that such an integrated programme was the only way to ensure that potential reliability problems would be detected and eliminated at the earliest, and therefore the cheapest, stage in the development cycle. Much written work appeared on the cost—benefit of higher reliability, to show that effort and resources expended during early development and during production testing, plus the imposition of demonstrations of specified levels of reliability to MIL-STD-781, led to reductions in in-service costs which more than repaid the reliability programme expenditure. The concept of life cycle costs (LCC), or whole life costs, was introduced.

In the United Kingdom, Defence Standard 00–40, *The Management of Reliability and Maintainability* was issued in 1981. The British Standards Institution has issued BS 5760—*Guide on Reliability of Systems, Equipments and Components*.

Specifications and test systems for electronic components, based upon the US Military Standards, have been developed in the United Kingdom and in continental Europe. Electronic component standards including test and quality aspects are being harmonized internationally through the International Electrotechnical Commission (IEC).

COURSES, CONFERENCES AND LITERATURE

Reliability engineering and management are now taught in engineering courses at a large number of universities, colleges and polytechnics, and at many specialist short courses.

Conferences on general and specific reliability engineering and management topics have been held regularly in the United States since the 1960s and in Europe and elsewhere since the 1970s. The best known is the annual US Reliability and Maintainability Symposium (RAMS), sponsored by most of the important engineering associations and institutions in the United States. These conference proceedings contain much useful information and are often cited.

Journals on reliability have also appeared; some are referenced at the end of this chapter. Several books have been published on the subjects of reliability engineering and management; some of these are referenced at the end of this and other chapters.

Much of the reliability literature has tended to emphasize the mathematical and analytical aspects of the subject, with the result that reliability engineering is often considered by designers and others to be a rather esoteric subject. This is unfortunate, since it creates barriers to communication. More recently there has been a trend to emphasize the more practical aspects and to integrate reliability work into the overall management and engineering process. These aspects are covered in later chapters.

ORGANIZATIONS INVOLVED IN RELIABILITY WORK

Several organizations have been created to develop policies and methods in reliability engineering and to undertake research and training. A summary of these is given at Appendix 8.

RELIABILITY AS AN EFFECTIVENESS PARAMETER

With the increasing cost and complexity of many modern systems, the importance of reliability as an effectiveness parameter, which should be specified and paid for, has become apparent. For example, a radar station, a process plant or an airliner must be available when required, and the cost of non-availability, particularly if it is unscheduled, can be very high. In the weapons field, if an anti-aircraft missile has a less than 100 per cent probability of functioning correctly throughout its engagement sequence, operational planners must consider deploying the appropriate extra quantity to provide the required level of defence. The Apollo project second stage rocket was powered by six rocket motors; any five would have provided sufficient impulse, but an additional motor was specified to cater for a possible failure of one. As it happened there were no failures, and every launch utilized an 'unnecessary' motor. These considerations apply equally to less complex systems, e.g.

vending and copying machines, even if the failure costs are less dramatic in absolute terms.

As an effectiveness parameter, reliability can be 'traded off' against other parameters. Reliability generally affects availability, and in this context maintainability is also relevant. Reliability and maintainability are often related to availability by the formula:

$$\text{Availability} = \frac{\text{MTBF}}{\text{MTBF} + \text{MTTR}}$$

where MTTR is the mean time to repair. This is the simplest steady-state situation. It is clear that availability improvements can be achieved by improving either MTBF or MTTR. For example, automatic built-in test equipment can greatly reduce diagnostic times for electronic equipment, at a cost of a slight reduction in overall reliability and an increase in unit costs. Many other parameters can be considered in trade-offs, such as weight, redundancy, cost of materials, parts and processes, or reduction in performance.

The greatest difficulty in estimating relationships for reliability trade-offs derives from the fact that, whereas it is possible to estimate quite accurately such factors as the cost and weight penalties of built-in test equipment, the cost of materials and components, or the worth of a measurable performance parameter, the effect on reliability cannot generally be forecast accurately, and reliability measurements can at best be made only within statistical limits imposed by the number of data available. Even for quite simple systems, quantifying the separate contributions to a measured change in reliability due to different causes is a very uncertain exercise. Consequently, while there is considerable literature on the effectiveness of various reliability methods, few generally applicable estimating relationships are available. Selection of trade-offs must therefore be very much a matter of experience of similar projects in the knowledge that wide margins of error can exist.

RELIABILITY PROGRAMME ACTIVITIES

What, then, are the actions that managers and engineers can take to influence reliability? One obvious activity already mentioned is quality control, the whole range of functions designed to ensure that delivered products are compliant with the design. For many products, QC is sufficient to ensure high reliability, and we would not expect a match factory or a company mass-producing simple diecastings to employ reliability staff. In such cases the designs are simple and well proven, the environments in which the products will operate are well understood and the very occasional failure has no significant financial or operational effect. QC, together with craftsmanship, can provide adequate assurance for simple products or when the risks are known to be very low. Risks are low when safety margins can be made very large, as in most stuctural engineering. Reliability engineering disciplines may justifiably be absent in many types of product development and manufacture. QC disciplines are, however, essential elements of any integrated reliability programme.

A formal reliability programme is necessary whenever the risks or costs of failure are not low. We have already seen how reliability engineering developed as a result of the high costs of unreliability of military equipment. In many respects reliability engineering has achieved the highest status, and the most dramatic results, in the space programmes, particularly in the United States. However, the same disciplines are used to great effect when unreliability affects business, and the high reliability, often taken for granted, of many everyday products owes much to the reliability engineering methods developed in these programmes. Good examples are provided by domestic electronic equipment and office copying machines, where competition and new technology force an awareness of the need for a disciplined approach to reliability. Risks of failure usually increase in proportion to the number of components in a system, so reliability programmes are required for any product whose complexity leads to an appreciable risk.

An effective reliability programme should be based on the conventional wisdom of responsibility and authority being vested in one person. Let us call him or her the reliability programme manager. The responsibility must relate to a defined objective, which may be a maximum warranty cost figure, an MTBF to be demonstrated or a requirement that failure will not occur. Having an objective and the authority, how does the reliability programme manager set about his or her task, faced as he or she is with a responsibility based on statistical uncertainties? This question will be answered in detail in subsequent chapters, but a brief outline is given below.

The reliability programme must begin at the earliest, conceptual phase of the project. It is at this stage that fundamental decisions involving trade-offs between reliability, on one hand, and factors such as complexity, performance and price, on the other, are made. The shorter the development time-scale, the more important is this need, particularly if there will be few opportunities for an iterative approach. The activities appropriate to this phase are an involvement in the assessment of these trade-offs and the generation of reliability objectives. The reliability staff can perform these functions effectively only if they are competent to contribute to the give-and-take inherent in the trade-off negotiations, which may be conducted between designers, production staff, marketing staff, customer representatives and finance staff.

As development proceeds from initial study to detail design,the reliability risks are controlled by a formal, documented, approach to the review of design and to the imposition of design rules relating to components, materials and process selection, de-rating policy, tolerancing, etc. The objectives at this stage are to ensure that known good practices are applied, that deviations are detected and corrected, and that areas of uncertainty are highlighted for further action. The programme continues through the initial hardware manufacturing and test stages, by planning and executing tests to generate confidence in the design and by collecting, analysing and acting upon test data. During production, QC

Figure 1.6 Reliability programme (Extracts from BS 5760: Part 1 are reproduced by permission of the British Standards Institution. Complete copies of the document can be obtained from BSI at Linford Wood, Milton Keynes, MK14 6LE)

activities ensure that the proven design is repeated, and further testing may be applied to eliminate weak items and to maintain confidence. The data collection, analysis and action process continues through the production and in-use phases. Throughout the product life cycle, therefore, the reliability is assessed, first by initial predictions based upon past experience in order to determine feasibility and set objectives, then by refining the predictions as detail design proceeds and subsequently by recording performance during the test, production and in-use phases. This performance is fed back to generate corrective action, and to provide data and guidelines for future products.

The elements of a reliability programme are outlined in documents such as MIL-STD-785, Defence Standard 00–40 and BS 5760. Figure 1.6 (from BS 5760) indicates the cyclical nature of an effective programme and shows the range of activities involved. The activities are described fully in subsequent chapters.

RELIABILITY ECONOMICS

Obviously the reliability programme activities described can be expensive. For any product there is a theoretical optimum expenditure on reliability in relation to the subsequent benefit. Figure 1.7 shows how the total life cycle cost can be minimized by selecting this optimum expenditure. However, there is no way of forecasting precisely how much the reliability will be affected by a given expenditure on a particular activity. Estimating relationships have been derived, but there is much scope for variation and uncertainty. A badly run reliability programme may be expensive and ineffective (this is often the case when reliability is viewed as a parallel activity, not directly associated with other project development activities: the 'lip service' case). On the other hand, a tightly knit, integrated team of high competence may develop a product which is right first time and requires little testing to demonstrate compliance. Obviously the aim should be to create products in this way. Therefore the

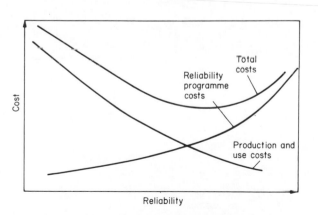

Figure 1.7   Reliability and life cycle costs

reliability programme manager should be concerned with project organization and team competence. Also, since downstream activities such as test and corrective action on completed designs become progressively more expensive, he or she should aim at expenditure on those early stage activities which will minimize the need for progressively higher expenditure as development proceeds. Typically, the discovery and correction of a potential weakness at an early design review may cost a hundred times less than if it were found during testing of production standard hardware.

Guidance on the costs and benefits of reliability programme elements is covered in Chapter 14.

BIBLIOGRAPHY

These are general texts which cover a range of reliability topics.
1. D. J. Smith, *Reliability and Maintainability in Perspective*, Macmillan (2nd edn., 1985).
2. D. K. Lloyd and M. Lipow, *Reliability Management, Methods, Mathematics*, 2nd edn. Published by the authors (1977).
3. A. E. Green and A. J. Bourne, *Reliability Technology*. Wiley, New York (1972).
4. W. G. Ireson, *Reliability Handbook*. McGraw-Hill, New York (1966).
5. *Quality and Reliability Engineering International*. Wiley (published quarterly).
6. *IEEE Trans. Reliab.* IEEE (published quarterly).
7. *Proceedings of the US Reliability and Maintainability Symposia*. American Society for Quality Control and IEEE (published annually).
8. US MIL-STD-721: *Definitions of Effectiveness Terms for Reliability, Maintainability, Human Factors and Safety*. Available from the National Technical Information Service, Springfield, Virgina.
9. British Standard, BS 4778: *Glossary of Terms Used in Quality Assurance* (includes reliability and maintainability). British Standards Institution, London.
10. British Standard, BS 5760: *Reliability of Systems, Equipments and Components*. British Standards Institution, London.

Chapter 2

RELIABILITY MATHEMATICS

INTRODUCTION

The methods used to quantify reliability are the mathematics of probability and statistics. In reliability work we are dealing with uncertainty. In fact this is the case in much of modern engineering, and the probabilistic as opposed to the deterministic approach to engineering problems is becoming more widely applied. As an example, data may show that a certain type of transistor fails at a constant average rate of once per 10^7 h. If an equipment contains 1,000 such transistors, and we operate it for 100 h, we cannot say with certainty whether it will fail in that time or not. We can, however, make a statement about the *probability* of failure. We can go further and state that, within specified statistical *confidence limits*, the probability of failure lies between certain values above and below this probability. If a sample of such equipments is tested, we obtain data which are called *statistics*.

Reliability statistics can be broadly divided into the treatment of *discrete functions*, *continuous functions* and *point processes*. For example, a switch may either work or not work when selected or a pressure vessel may pass or fail a test — these situations are described by discrete functions. In reliability we are often concerned with two-state discrete systems, since equipment is in either an operational or a failed state. Continuous functions describe those reliability situations which are governed by a continuous variable, such as time or distance travelled. The electronic equipment mentioned above would have a reliability function in this class. The distinction between discrete and continuous functions is one of how the problem is treated, and not necessarily of the physics or mechanics of the situation. For example, whether or not a pressure vessel fails a test may be a function of its age, and its reliability could therefore be treated as a continuous function. The statistics of point processes are used in relation to repairable systems, when more than one failure can occur in a time continuum. The choice of method will depend upon the problem and on the type of data available.

Statistical concepts such as confidence are often written as *s*-confidence to distinguish them from the non-statistical meanings of the terms. This convention is adopted in the remainder of this book.

PROBABILITY CONCEPTS

Any event has a probability of occurrence, which can be in the range 0–1. A zero probability means that the event will not occur; a probability of 1 means that it will occur. A coin has a 0.5 (even) probability of landing heads, and a die has a 1/6 probability of giving any one of the six numbers. Such events are *s-independent*, i.e. the coin and the die logically have no memory, so whatever has been thrown in the past cannot affect the probability of the next throw. No 'system' can beat the statistics of these situations; waiting for a run of blacks at roulette and then betting on reds only appears to work because the gamblers who won this way talk about it, whilst those who lost do not.

With coins, dice and roulette wheels we can predict the probability of the outcome from the nominal nature of the system. A coin has two sides, a die six faces, a roulette wheel equal numbers of reds and blacks. Assuming that the coin, die and wheel are fair, these outcomes are also *unbiased*, i.e. they are all equally probable. In other words, they occur *randomly*.

With many systems, such as the sampling of items from a production batch, the probabilities can only be determined from the statistics of previous experience.

We can define probability in two ways:

1. If an event can occur in N equally likely ways, and if the event with attribute A can happen in n of these ways, then the probability of A occurring is

$$P(A) = \frac{n}{N}$$

2. If, in an experiment, an event with attribute A occurs n times out of N experiments, then as N becomes large, the probability of event A approaches n/N, i.e.

$$P(A) = \lim_{n \to \infty} \left(\frac{n}{N} \right)$$

The first definition covers the cases described earlier, i.e. equally likely *s*-independent events such as rolling dice. The second definition covers typical cases in quality control and reliability. If we test 100 items and find that 30 are defective, we may feel justified in saying that the probability of finding a defective item in our next test is 0.30, or 30 per cent.

However, we must be careful in making this type of assertion. The probability of 0.30 of finding a defective item in our next test may be considered as our *degree of belief*, limited by the size of the sample, in this outcome. This leads to a third, subjective, definition of probability. If, in our tests of 100 items, seven of the defectives had occurred in a particular batch of ten and we had taken corrective action to improve the process so that such a problem batch was less likely to occur in future, we might assign some lower probability to the next item being defective. This subjective approach is quite valid, and is very often necessary in quality control and reliability work. Whilst it is important to have an understanding of the rules of probability, there are usually so many variables which can affect the properties of manufactured items that we must always keep an open mind about statistically derived values. We must ensure that the sample from which statistics have been derived represents the new sample, or the overall population, about which we plan to make an assertion based upon our sample statistics.

A sample represents a population if all the members of the population have an equal chance of being sampled. This can be achieved if the sample is selected so that this condition is fulfilled. Of course in engineering this is not always practicable; e.g. in reliability engineering we often need to make an assertion about items that have not yet been produced, based upon statistics from prototypes.

To the extent that the sample is not representative, we will alter our assertions. Of course, subjective assertions can lead to argument, and it might be necessary to perform additional tests to obtain more data to use in support of our assertions. If we do perform more tests, we need to have a method of interpreting the new data in relation to the previous data: we will cover this aspect later.

The assertions we can make based on sample statistics can be made with a degree of s-confidence which depends upon the size of the sample. If we had decided to test ten items after introducing a change to the process, and found one defective, we might be tempted to assert that we have improved the process, from 30 per cent. defectives being produced to only 10 per cent. However, since the sample is now much smaller, we cannot make this assertion

Figure 2.1 Samples with defectives (black squares)

with as high s-confidence as when we used a sample of 100. In fact, the true probability of any item being defective might still be 30 per cent. i.e. the population might still contain 30 per cent. defectives.

Figure 2.1 shows the situation as it might have occurred, over the first 100 tests. The black squares indicate defectives, of which there are 30 in our batch of 100. If these are randomly distributed, it is possible to pick a sample batch of ten which contains fewer (or more) than three defectives. In fact, the smaller the sample, the greater will be the sample-to-sample variation about the population average, and the s-confidence associated with any estimate of the population average will be accordingly lower. The derivation of s-confidence limits is covered later in this chapter.

RULES OF PROBABILITY

In order to utilize the statistical methods used in reliability engineering, it is necessary to understand the basic notation and rules of probability. These are:

1. The probability of obtaining an outcome A is denoted by $P(A)$, and so on for other outcomes.
2. The *joint* probability that A *and* B occur is denoted by $P(AB)$.
3. The probability that A *or* B occurs is denoted by $P(A + B)$.
4. The *conditional* probability of obtaining outcome A, *given that B has occurred*, is denoted by $P(A \mid B)$.
5. The probability of the complement, i.e. of A *not* occurring, is $P(\bar{A}) = 1 - P(A)$.
6. If (and only if) events A and B are *s-independent*, then

$$P(A \mid B) = P(A \mid \bar{B}) = P(A)$$

and

$$P(B \mid A) = P(B \mid \bar{A}) = P(B) \tag{2.1}$$

i.e. $P(A)$ is unrelated to whether or not B occurs, and vice versa.
7. The joint probability of the occurrence of two s-independent events A and B is the product of the individual probabilities:

$$P(AB) = P(A)P(B) \tag{2.2}$$

This is also called the *product rule* or *series rule*. It can be extended to cover any number of s-independent events. For example, in rolling a die, the probability of obtaining any given sequence of numbers in three throws is

$$\tfrac{1}{6} \times \tfrac{1}{6} \times \tfrac{1}{6} = \tfrac{1}{216}$$

8. If events A and B are *s-dependent*, then

$$P(AB) = P(A)P(B \mid A) = P(B)P(A \mid B) \tag{2.3}$$

i.e. the probability of A occurring times the probability of B occurring given that A has already occurred, or vice versa.

If $P(A) \neq 0$, Eqn (2.3) can be rearranged to

$$P(B \mid A) = \frac{P(AB)}{P(A)} \qquad (2.4)$$

9. The probability of any one of two events A or B occurring is

$$P(A + B) = P(A) + P(B) - P(AB) \qquad (2.5)$$

10. The probability of A or B occurring, if A and B are s-independent, is

$$P(A + B) = P(A) + P(B) - P(A)P(B) \qquad (2.6)$$

Figure 2.2 Dual redundant system

The derivation of this equation can be shown by considering the system shown in Fig. 2.2, in which either A or B, or A and B, must work for the system to work. If we denote the system success probability as P_s, then the failure probability, $P_f = 1 - P_s$. The system failure probability is the joint probability of A and B failing, i.e.

$$P_f = [1 - P(A)] [1 - P(B)]$$
$$= 1 - P(A) - P(B) + P(A)P(B)$$
$$P_s = 1 - P_f = P(A + B) = P(A) + P(B) - P(A)P(B)$$

11. If events A and B are *mutually exclusive*, i.e. A and B cannot occur simultaneously, then

$$P(AB) = 0$$

and

$$P(A + B) = P(A) + P(B) \qquad (2.7)$$

12. If multiple, mutually exclusive probabilities of outcomes B_i jointly give a probability of outcome A, then

$$P(A) = \sum_i P(AB_i) = \sum_i P(A \mid B_i)P(B_i) \qquad (2.8)$$

13. Rearranging Eqn (2.3)

$$P(AB) = P(A)P(B \mid A) = P(B)P(A \mid B)$$

we obtain

$$P(A \mid B) = \frac{P(A)P(B \mid A)}{P(B)} \qquad (2.9)$$

This is a simple form of *Bayes' theorem*. A more general expression is

$$P(A \mid B) = \frac{P(A)P(B \mid A)}{\sum_i P(B \mid E_i)P(E_i)} \qquad (2.10)$$

where E_i is the ith event.

Example 2.1

The reliability of a missile is 0.85. If a salvo of two missiles is fired, what is the probability of at least one hit? (Assume s-independence of missile hits.)

Let A be the event 'first missile hits' and B the event 'second missile hits'. Then

$$P(A) = P(B) = 0.85$$
$$P(\bar{A}) = P(\bar{B}) = 0.15$$

There are four possible, mutually exclusive outcomes, AB, $\bar{A}$B, A$\bar{B}$, $\bar{A}\bar{B}$. The probability of both missing, from Eqn (2.2), is

$$P(\bar{A})P(\bar{B}) = P(\bar{A}\bar{B})$$
$$= 0.15^2 = 0.0225$$

Therefore the probability of at least one hit is

$$P_s = 1 - 0.0225 = 0.9775$$

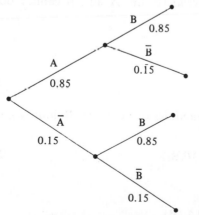

$P(AB) = P(A)\,P(B) = 0.85 \times 0.85 = 0.7225$

$P(A\bar{B}) = P(A)\,P(\bar{B}) = 0.85 \times 0.15 = 0.1275$

$P(\bar{A}B) = P(\bar{A})\,P(B) = 0.15 \times 0.85 = 0.1275$

$P(\bar{A}\bar{B}) = P(\bar{A})\,P(\bar{B}) = 0.15 \times 0.15 = 0.0225$

We can derive the same result by using Eqn (2.6):

$$P(A + B) = P(A) + P(B) - P(A)P(B)$$
$$= 0.85 + 0.85 - 0.85^2 = 0.9775$$

Another way of deriving this result is by using the *sequence tree diagram* on page 22.

The probability of a hit is then derived by summing the products of each path which leads to at least one hit. We can do this since the events defined by each path are mutually exclusive.

$$P(AB) + P(A\bar{B}) + P(\bar{A}B) = 0.9775$$

(Note that the sum of all the probabilities is unity.)

Example 2.2

In Example 2.1 the missile hits are not *s*-independent, but are *s*-dependent, so that if the first missile fails the probability that the second will also fail is 0.2. However, if the first missile hits, the hit probability of the second missile is unchanged at 0.85. What is the probability of at least one hit?

$$P(A) = 0.85$$
$$P(B \mid A) = 0.85$$
$$P(B \mid A) = 0.15$$
$$P(\bar{B} \mid \bar{A}) = 0.2$$
$$P(B \mid \bar{A}) = 0.8$$

The probability of at least one hit is

$$P(AB) + P(\bar{A}B) + P(\bar{B}A)$$

Since A, B and A$\bar{B}$ are *s*-independent,

$$P(AB) = P(A)P(B)$$
$$= 0.85 \times 0.85 = 0.7225$$

and

$$P(A\bar{B}) = P(A)P(\bar{B})$$
$$= 0.85 \times 0.15 = 0.1275$$

Since $\bar{A}$ and B are *s*-dependent, from Eqn (2.3),

$$P(\bar{A}B) = P(\bar{A})P(B \mid \bar{A})$$
$$= 0.15 \times 0.8 = 0.12$$

and the sum of these probabilities is 0.97.
This result can also be derived by using a sequence tree diagram:

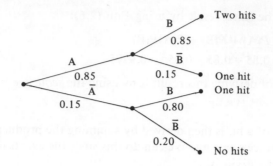

As in Example 2.1, the probability of at least one hit is calculated by adding the products of each path leading to at least one hit, i.e.

$$P(A)P(B) + P(A)P(\bar{B}) + P(\bar{A})P(B)$$
$$= (0.85 \times 0.85) + (0.85 \times 0.15) + (0.15 \times 0.80) = 0.97$$

Example 2.3

In the circuit shown, the probability of any switch being closed is 0.8 and all events are s-independent. (a) What is the probability that a circuit will exist? (b) Given that a circuit exists, what is the probability that switches a and b are closed?

Let the events that a, b, c and d are closed be A, B, C and D. Let X denote the event that the circuit exists.

(a) $X = AB + (C + D)$

$$P(X) - P(AB) + P(C + D) - P(AB)P(C + D)$$
$$P(AB) = P(A)P(B)$$
$$= 0.8 \times 0.8 = 0.64$$
$$P(C + D) = P(C) + P(D) - P(C)P(D)$$
$$= 0.8 + 0.8 - 0.64 = 0.96$$

Therefore

$$P(X) = 0.64 + 0.96 - (0.96 \times 0.64) = 0.9856$$

(b) From Eqn (2.4),

$$P(AB \mid X) = \frac{P(ABX)}{P(X)}$$

A and B jointly give X. Therefore, from Eqn (2.8),

$$P(ABX) = P(AB)$$

So

$$P(AB \mid X) = \frac{P(AB)}{P(X)} = \frac{P(A)P(B)}{P(X)}$$

$$= \frac{0.8 \times 0.8}{0.9856} = 0.6494$$

Example 2.4

A test set has a 98 per cent. probability of correctly classifying a faulty item as defective and a 4 per cent. probability of classifying a good item as defective. If in a batch of items tested 3 per cent. are actually defective, what is the probability that when an item is classified as defective, it is truly defective?

Let D represent the event that an item is defective and C represent the event that an item is classified defective. Then

$$P(D) = 0.03$$
$$P(C \mid D) = 0.98$$
$$P(C \mid \bar{D}) = 0.04$$

We need to determine $P(D \mid C)$. Using Eqn (2.10),

$$P(D \mid C) = \frac{P(D)P(C \mid D)}{P(C \mid D)P(D) + P(C \mid \bar{D})P(\bar{D})}$$

$$= \frac{(0.03)(0.98)}{(0.98)(0.03) + (0.04)(0.97)} = 0.43$$

This indicates the importance of a test equipment having a high probability of correctly classifying good items as well as bad items.

PROBABILITY DISTRIBUTIONS

If we plot measured values which can vary about an average (e.g. the diameter of machined parts or the gain of a transistor) as a histogram, for a given sample we may obtain a representation such as Fig. 2.3(a).

Figure 2.3 (a) Frequency histogram of a random sample, (b) frequency histogram of another random sample from the same population, (c) data of many samples shown with measurement intervals of 0.5

In this case 30 items have been measured and the frequencies of occurrence of the measured values are as shown. The values range from 2 to 9, with most items having values between 5 and 7. Another random sample of 30 from the same population will usually generate a different histogram, but the general shape is likely to be similar, e.g. Fig. 2.3(b). If we plot a single histogram showing the combined data of many such samples, but this time show the values in measurement intervals of 0.5, we get Fig. 2.3(c). Note that now we have used a percentage frequency scale. We now have a rather better picture of the distribution of values, as we have more information from the larger sample. If we proceed to measure a large number and we further reduce the measurement interval, the histogram tends to a curve which describes the population *probability density function* (p.d.f.) or simply the *distribution* of values. Figure 2.4 shows a general *unimodal* probability distribution, f(x) being the probability density of occurrence, related to the variable x. The value of x at which the distribution peaks is called the *mode*. Multimodal distributions are encountered in reliability work as well as unimodal distributions. However, we will deal only with the statistics of unimodal distributions in this book, since multimodal distributions are usually generated by the combined effects of separate unimodal distributions.

The area under the curve is equal to unity, since it describes the total probability of all possible values of x, as we have defined a probability which is a

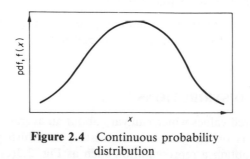

Figure 2.4 Continuous probability distribution

certainty as being a probability of one. Therefore

$$\int_{-\infty}^{\infty} f(x)\, dx = 1 \qquad (2.11)$$

The probability of a value falling between any two values x_1 and x_2 is the area bounded by this interval, i.e.

$$P(x_1 < x < x_2) = \int_{x_1}^{x_2} f(x)\, dx \qquad (2.12)$$

To describe a p.d.f. we normally consider four aspects:

1. The *central tendency*, about which the distribution is grouped.
2. The *spread*, indicating the extent of variation about the central tendency.
3. The *skewness*, indicating the lack of symmetry about the central tendency.
4. The *kurtosis*, indicating the 'peakedness' of the p.d.f.

Measures of central tendency

For a sample containing n items the sample *mean* is denoted by $\bar{x}$.

$$\bar{x} = \sum_{i=1}^{n} \frac{x_i}{n} \qquad (2.13)$$

The sample mean can be used to *estimate* the population mean, which is the average of all possible outcomes. For a continuous distribution, the mean is derived by extending this idea to cover the range $-\infty$ to $+\infty$.

The mean of a distribution is usually denoted by μ. The mean is also referred to as the *location parameter*, *average value* or *s-expected value*, $E(x)$. In reliability work θ is used to denote the mean in certain special cases (see page 38).

$$\mu = \int_{-\infty}^{\infty} x f(x)\, dx \qquad (2.14)$$

This is analogous to the centre of gravity of the p.d.f. The *estimate* of a population mean from sample data is denoted by $\hat{\mu}$.

Other measures of central tendency are the *median*, which is the mid-point

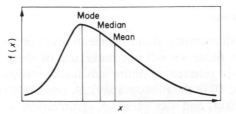

Figure 2.5 Measures of central tendency

of the distribution, i.e. the point at which half the measured values fall to either side, and the *mode*, which is the value (or values) at which the distribution peaks. The relationship between the mean, median and mode for a right-skewed distribution is shown in Fig. 2.5. For a symmetrical distribution, the three values are the same, whilst for a left-skewed distribution the order of values is reversed.

Spread of a distribution

The spread, or dispersion, i.e. the extent to which the values which make up the distribution vary, is measured by its *variance*. For a sample size n the variance, $\text{Var}(x)$ or $E(x - \bar{x})^2$, is given by

$$\text{Var}(x) = E(x - \bar{x})^2 = \frac{\sum_{i=1}^{n} (x_i - \bar{x})^2}{n} \tag{2.15}$$

Where sample variance is used to estimate the population variance, we use $(n - 1)$ in the denominator of Eqn (2.15) instead of n, as it can be shown to provide a better estimate. The *estimate of population variance* from a sample is denoted $\hat{\sigma}^2$ where

$$\hat{\sigma}^2 = \sum_{i=1}^{n} \frac{(x_i - \bar{x})^2}{n - 1} \tag{2.16}$$

The *population* variance σ^2, for a finite population N, is given by

$$\sigma^2 = \frac{\sum_{i=1}^{N} (x_i - \mu)^2}{N} \tag{2.17}$$

For a continuous distribution.

$$\sigma^2 = \int_{-\infty}^{\infty} (x - \mu)^2 f(x) \, dx \tag{2.18}$$

σ is called the *standard deviation* (SD) and is frequently used in practice instead of the variance. It is also referred to as the *scale parameter*. σ^2 is the second moment about the mean and is analogous to a radius of gyration.

Skewness and kurtosis

The third and fourth moments about the mean give the skewness and kurtosis of the distribution. Since we will not make use of these parameters in this book, the reader is referred to more advanced statistical texts for their derivation (e.g. No. 10 in the Bibliography). A positively skewed distribution is skewed to the right, and vice versa. A symmetrical distribution has zero skew.

Figure 2.6 Typical cumulative distribution function (c.d.f.)

The cumulative distribution function

The cumulative distribution function (c.d.f.), $F(x)$, gives the probability that a measured value will fall between $-\infty$ and x:

$$F(x) = \int_{-\infty}^{x} f(x)\, dx \tag{2.19}$$

Figure 2.6 shows the typical ogive form of the c.d.f. with $F(x) \to 1$ as $x \to \infty$.

Reliability and hazard functions

In reliability engineering we are concerned with the probability that an item will survive for a stated interval (e.g. time, cycles, distance, etc.), i.e. that there is no failure in the interval (0 to x). This is the *s-reliability*, and it is given by the reliability function $R(x)$. From this definition, it follows that

$$R(x) = 1 - F(x) = \int_{x}^{\infty} f(x)\, dx = 1 - \int_{-\infty}^{x} f(x)\, dx \tag{2.20}$$

The *hazard function* or *hazard rate* $h(x)$ is the conditional probability of failure in the interval x to $(x + dx)$, given that there was no failure by x:

$$h(x) = \frac{f(x)}{R(x)} = \frac{f(x)}{1 - F(x)} \tag{2.21}$$

The *cumulative hazard function* $H(x)$ is given by

$$H(x) = \int_{-\infty}^{x} h(x)\, dx = \int_{-\infty}^{x} \frac{f(x)}{1 - F(x)}\, dx \tag{2.22}$$

Range of definite integrals

In engineering we do not usually encounter measured values below zero and the lower limit of the definite integral is then 0.

DISCRETE DISTRIBUTIONS

The binomial distribution

The binomial distribution describes a situation in which there are only two outcomes, such as pass or fail, and the probability remains the same for all trials. (Trials which give such results are called *Bernoulli trials*.) Therefore, it is obviously very useful in QA and reliability work. The p.d.f. for the binomial distribution is

$$f(x) = \frac{n!}{x!(n-x)!}\, p^x q^{(n-x)} \qquad (2.23)$$

$$\frac{n!}{x!(n-x)!} \quad \text{may be written} \quad \binom{n}{x}$$

This is the probability of obtaining x good items and $(n-x)$ bad items, in a sample of n items, when the probability of selecting a good item is p and of selecting a bad item is q. The mean of the binomial distribution (from Eqn 2.13 is given by

$$\mu = np \qquad (2.24)$$

and the SD from (Eqn 2.17)

$$\sigma = (npq)^{1/2} \qquad (2.25)$$

The binomial distribution can only have values at points where x is an integer. The c.d.f. of the binomial distribution (i.e. the probability of obtaining r or fewer successes in n trials) is given by

$$F(r) = \sum_{x=0}^{r} \binom{n}{x} p^x q^{(n-x)} \qquad (2.26)$$

Example 2.5

A frequent application of the cumulative binomial distribution is in QC acceptance sampling. For example, if the acceptance criterion for a production line is that not more than 4 defectives may be found in a sample of 20, we can determine the probability of acceptance of a lot if the production process yields 10 per cent. defectives.

For Eqn (2.26),

$$F(4) = \sum_{x=0}^{4} \binom{20}{x} 0.1^x 0.9^{(20-x)}$$

$$= 0.957$$

When approximate values will suffice the nomogram in Fig. 2.7 can be used.

Figure 2.7 Nomogram of the cumulative binomial distribution

Example 2.6

An aircraft landing gear has 4 tyres. Experience shows that tyre bursts occur on average on 1 landing in 1,200. Assuming that tyre bursts occur s-independently of one another, and that a safe landing can be made if not more than 2 tyres burst, what is the probability of an unsafe landing?

If n is the number of tyres and p is the probability of a tyre bursting,

$$n = 4$$

$$p = \frac{1}{1,200} = 0.000\ 83$$

$$q = (1 - p) = 0.999\ 17$$

The probability of a safe landing is the probability that not more than 2 tyres burst.

$$F(2) = \binom{4}{2}(0.000\ 83)^2(0.999\ 17)^2 + \binom{4}{1}(0.000\ 83)^1(0.999\ 17)^3$$

$$+ \binom{4}{0}(0.000\ 83)^0(0.999\ 17)^4$$

$$= 0.000\ 004\ 159\ 7 + 0.003\ 325\ 006\ 9 + 0.996\ 670\ 831$$

$$= 0.999\ 999\ 997\ 7$$

Therefore the probability of an unsafe landing is

$$1 - 0.999\ 999\ 997\ 7 = 2.3 \times 10^{-9}$$

The Poisson distribution

If events are Poisson-distributed they occur at a *constant average rate*, with only one of the two outcomes countable, e.g. the number of failures in a given time or defects in a length of wire:

$$f(x) = \frac{\mu^x}{x!} \exp(-\mu) \qquad (x = 0, 1, 2, \ldots) \tag{2.27}$$

where μ is the mean rate of occurrence. The Poisson distribution can also be considered as an extension of the binomial distribution, in which n is considered infinite.

Figure 2.8 provides a chart for the cumulative Poisson distribution. For example, if we need to know the probability of not more than three failures occurring in 1,000 h of operation of a system, when the mean rate of failures is 1 per 1,000 h, we can read the answer from Fig. 2.8 as 0.98 ($\mu = 1/1,000$, $x = 3$).

Since the Poisson distribution can represent the limiting case of the binomial distribution it gives a good approximation to the binomial distribution, when p or q are small and n is large. This is useful in sampling work where the proportion of defectives is low (i.e. $p < 0.1$).

The Poisson approximation is

$$f(x) = \frac{(np)^x}{x!} \exp(-np) \tag{2.28}$$

$$[\mu = np; \ \sigma = (np)^{1/2} = \mu^{1/2}]$$

This approximation allows us to use Poisson tables or charts in appropriate cases and also simplifies calculations.

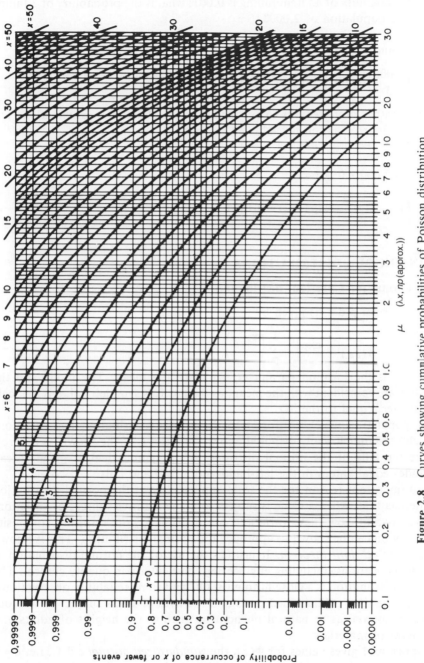

Figure 2.8 Curves showing cumulative probabilities of Poisson distribution

Example 2.7

If the probability of an item failing is 0.001, what is the probability of 3 failing out of a population of 2,000?

The binomial solution is

$$\binom{2,000}{3} 0.999^{1997} \, 0.001^3 = 0.1805$$

This is rather more tedious to evaluate than the Poisson approximation. The Poisson approximation is evaluated as follows:

$$\mu = np$$
$$= 2,000 \times 0.001 = 2$$

$$P(x = 3) = \frac{2^3}{3!} \exp(-2) = 0.1804$$

CONTINUOUS DISTRIBUTIONS

The normal (or Gaussian) distribution

The s-normal p.d.f. is given by

$$f(x) = \frac{1}{\sigma(2\pi)^{1/2}} \exp\left[-\frac{1}{2} \left(\frac{x - \mu}{\sigma} \right)^2 \right] \tag{2.29}$$

where μ is the location parameter, equal to the mean. The mode and the median are coincident with the mean, as the p.d.f. is symmetrical. σ is the scale parameter, equal to the SD.

A population which conforms to the normal distribution has variations which are symmetrically disposed about the mean (i.e. the skewness is zero). Since the tails of the normal distribution are symmetrical, a given spread includes equal values in the left-hand and right-hand tails.

An important reason for the wide applicability of the normal distribution is the fact that, when a value is subject to many additive sources of variation, irrespective of how these variations are distributed, the resulting composite distribution can be shown to approach the normal distribution. This is known as the *central limit theorem.* It justifies the use of the normal distribution in many applications, including engineering, particularly in quality control. The normal distribution is a close fit to most QC and some reliability observations, such as the sizes of machined parts and the lives of items subject to wearout failures, as well as to natural phenomena such as the heights of adults and strengths of materials.

Appendix 1 gives values for $\Phi(z)$, the *standardized normal* c.d.f., i.e. $\mu = 0$ and $\sigma = 1$. z represents the number of SDs displacement from the mean. Any normal distribution can be evaluated from the standardized normal distribution by calculating the standardized normal variate z, where

$$z = \frac{x - \mu}{\sigma}$$

and finding the appropriate value of $\Phi(z)$.

Example 2.8

The life of an incandescent lamp is normally distributed, with mean 1,200 h and SD 200 h. What is the probability that a lamp will last (a) at least 800 h? (b) at least 1,600 h?

(a) $z = (x - \mu)/\sigma$, i.e. the distance of x from μ expressed as a number of SDs. Then

$$z = \frac{800 - 1,200}{200} = -2 \text{ SD}$$

Appendix 1 shows that the probability of a value not exceeding 2 SD is 0.977. Figure 2.9(a) shows this graphically, on the p.d.f. (the shaded area).

(b) The probability of a lamp surviving more than 1,600 h is derived similarly:

$$z = \frac{1,600 - 1,200}{200} = 2 \text{ SD}$$

This represents the area under the p.d.f. curve beyond the +2 SD point [Fig. 2.9(a)] or 1 − (area under the curve to the left of +2 SD) on the c.d.f. [Fig. 2.9(b)]. Therefore the probability of surviving beyond 1,600 h is (1 − 0.977) = 0.023.

As the normal distribution represents a limiting case of the binomial and Poisson distributions, it can be used to provide a good approximation to these distributions. For example, it can be used when $0.1 > p > 0.9$ and n is large.

Figure 2.9 (a) The p.d.f. [f(x)] versus x; (b) the c.d.f. [F(x)] versus x (see Example 2.8)

Then

$$\mu = np$$
$$\sigma = (npq)^{1/2}$$

Example 2.9

What is the probability of having not more than 20 failures if $n = 100$, $p = 0.14$?

Using the binomial distribution,

$$P_{20} = 0.9640$$

Using the normal approximation

$$\mu = np = 14$$
$$\sigma = (npq)^{1/2} = 3.470$$
$$z = \frac{20 - 14}{3.47} = 1.73$$

Referring to Appendix 1, $P_{20} = 0.9582$.

As $p \to 0.5$, the approximation improves, and we can then use it with smaller values of n. Typically, if $p = 0.4$, we can use the approximation with $n = 50$.

The lognormal distribution

The lognormal distribution is a more versatile distribution than the normal as it has a range of shapes, and therefore is often a better fit to reliability data, such as for populations with wearout characteristics. Also, it does not have the normal distribution's disadvantage of extending below zero to $-\infty$. The lognormal p.d.f. is

$$f(x) = \begin{cases} \dfrac{1}{\sigma x (2\pi)^{1/2}} \exp\left[-\dfrac{1}{2} \left(\dfrac{\ln x - \mu}{\sigma} \right)^2 \right] & \text{(for } x \geqslant 0) \\ 0 & \text{(for } x < 0) \end{cases} \tag{2.30}$$

In other words, it is the normal distribution with $\ln x$ as the variate. The mean and SD of the lognormal distribution are given by

$$\text{Mean} = \exp\left(\mu + \frac{\sigma^2}{2} \right)$$
$$\text{SD} = [\exp(2\mu + 2\sigma^2) - \exp(2\mu + \sigma^2)]^{1/2}$$

where μ and σ are the mean and SD of the ln data.

When $\mu \gg \sigma$, the lognormal distribution approximates to the normal distribution. The normal distribution has a hazard rate which increases as x increases. The lognormal distribution describes reliability situations in which the hazard rate increases from $x = 0$ to a maximum and then decreases.

The exponential distribution

The exponential distribution describes the situation wherein the hazard rate is constant. A Poisson process generates a constant hazard rate. The p.d.f. is

$$f(x) = \begin{cases} a \exp(-ax) & \text{(for } x \geqslant 0) \\ 0 & \text{(for } x < 0) \end{cases} \tag{2.31}$$

This is an important distribution in reliability work, as it has the same central limiting relationship to life statistics as the normal distribution has to non-life statistics. It describes the constant hazard rate situation. As the hazard rate is often a function of time, we will denote the independent variable by t instead of x. The constant hazard rate is denoted by λ. The mean life, or mean time to failure (MTTF), is $1/\lambda$. The p.d.f. is then written as

$$f(t) = \lambda \exp(-\lambda t) \tag{2.32}$$

The probability of no failures occurring before time t is obtained by integrating Eqn (2.32) between 0 and t and subtracting from 1:

$$R(t) = 1 - \int_0^t f(t) \, dt = \exp(-\lambda t) \tag{2.33}$$

$R(t)$ is the *reliability function* (or survival probability). For example, the reliability of an item with an MTTF of 500 h over a 24 h period is

$$R(24) = \exp\left(\frac{-24}{500}\right) = 0.953$$

Values of $\exp(-x)$ are tabulated in Appendix 2. Note that for small values of $\lambda t (< 0.1)$, the first two terms of the negative exponential series $(1 - \lambda t + \ldots)$ can be used to give a good approximation for $R(t)$.

$R(t)$ can also be derived as follows. If there are N items surviving at time t, then in a small period of time δt the number failing is δN. Then

$$\frac{\delta N}{\delta t} = -\lambda N$$

i.e. constant proportion to number surviving

$$\frac{\delta N}{N} = -\lambda \delta t$$

$$\ln N = -\lambda t - \ln C$$

$$N = C \exp(-\lambda t)$$

If N_0 is the number of items at $t = 0$ then $C = N_0$:

$$N = N_0 \exp(-\lambda t)$$

$$R(t) = \frac{N}{N_0} = \exp(-\lambda t)$$

For items which are repaired, λ is called the *failure rate*, and $1/\lambda$ is called the *mean time between failures* (MTBF) (θ). Note that 63.2 per cent. of items will have failed by $t = \theta$.

If times to failure are exponentially distributed, the probability of x failures is Poisson-distributed. For example, if the MTBF is 100 h, the probability of having more than 15 failures in 1,000 h is derived as:

$$\text{Expected number of failures} = \frac{1,000}{100} = 10$$

Referring to Fig. 2.8, for $\lambda(\mu) = 10$, $x = 15$.

$$P = 0.965$$

This is the probability of having up to 15 failures. Therefore the probability of more than 15 failures is

$$1 - 0.965 = 0.035$$

The gamma distribution

The gamma distribution describes, in reliability terms, the situation when partial failures can exist, i.e. when a given number of partial failure events must occur before an item fails, or the time to the ath failure when time to failure is exponentially distributed. The p.d.f. is

$$f(x) = \begin{cases} \dfrac{\lambda}{\Gamma(a)} (\lambda x)^{a-1} \exp(-\lambda x) & \text{(for } x \geqslant 0) \\ 0 & \text{(for } x < 0) \end{cases} \tag{2.34}$$

$$\mu = \frac{a}{\lambda}$$

$$\sigma = \frac{a^{1/2}}{\lambda}$$

where λ is the failure rate (complete failures) and a the number of partial failures per complete failure, or events to generate a failure. $\Gamma(a)$ is the *gamma function*:

$$\Gamma(a) = \int_0^\infty x^{a-1} \exp(-x) \, dx \tag{2.35}$$

When $(a - 1)$ is a positive integer, $\Gamma(a) = (a - 1)!$ This is the case in the partial failure situation. The exponential distribution is a special case of the gamma distribution, when $a = 1$, i.e.

$$f(x) = \lambda \exp(-\lambda x)$$

The gamma distribution can also be used to describe a decreasing or increasing hazard rate. When $a < 1$, $h(x)$ will decrease whilst for $a > 1$, $h(x)$ increases.

The χ^2 distribution

The χ^2 (chi-square) distribution is a special case of the gamma distribution, where $\lambda = \frac{1}{2}$, and $v = 2a$, where v is called the number of *degrees of freedom* and must be a positive integer. This permits the use of the χ^2 distribution for evaluating reliability situations, since the number of failures, or events to failure, will always be positive integers. The χ^2 distribution is really a family of distributions, which range in shape from that of the exponential to that of the normal distribution. Each distribution is identified by the degrees of freedom.

In statistical theory, the χ^2 distribution is very important, as it is the distribution of the sums of squares of n, s-independent, s-normal variates. This allows it to be used for statistical testing, goodness-of-fit tests and evaluating s-confidence. These applications are covered later. The c.d.f. for the χ^2 distribution is tabulated for a range of degrees of freedom in Appendix 3.

The Weibull distribution

The Weibull distribution has the great advantage in reliability work that by adjusting the distribution parameters it can be made to fit many life distributions. The Weibull p.d.f. is (in terms of time t)

$$f(t) = \begin{cases} \dfrac{\beta}{\eta^\beta} t^{\beta-1} \exp\left[-\left(\dfrac{t}{\eta}\right)^\beta \right] & \text{(for } t \geqslant 0) \\ 0 & \text{(for } t < 0) \end{cases} \tag{2.36}$$

The corresponding reliability function is

$$R(t) = \exp\left[-\left(\frac{t}{\eta}\right)^\beta \right] \tag{2.37}$$

The hazard rate is

$$\frac{\beta}{\eta^\beta} t^{\beta-1}$$

β is the *shape parameter* and η is the *scale parameter*, or *characteristic life* — it is the life at which 63.2 per cent. of the population will have failed.

When $\beta = 1$, the exponential reliability function (constant hazard rate) results, with

η = mean life $(1/\lambda)$.

When $\beta < 1$, we get a *decreasing* hazard rate reliability function.
When $\beta > 1$, we get an *increasing* hazard rate reliability function.

When $\beta = 3.5$, for example, the distribution approximates to the normal distribution. Thus the Weibull distribution can be used to model a wide range of life distributions characteristic of engineered products.

So far we have dealt with the two-parameter Weibull distribution. If, however, failures do not start at $t = 0$, but only after a finite time γ, then the Weibull reliability function takes the form

$$R(t) = \exp\left[-\left(\frac{t - \gamma}{\eta}\right)^{\beta} \right] \quad \text{(for } t \geq \gamma; \text{ R}(t) = 1 \text{ for } 0 \leq t \leq \gamma) \tag{2.38}$$

i.e. a three-parameter distribution. γ is called the *failure free time*, *location parameter* or *minimum life*. It is sometimes denoted as t_0.

The extreme value distributions

In reliability work we are often concerned not with the distribution of variables which describe the bulk of the population but only with the extreme values which can lead to failure. For example, the mechanical properties of a semiconductor wire bond are such that under normal operating conditions good wire bonds will not fracture or overheat. However, extreme high values of electrical load or extreme low values of bond strength can result in failure. In other words, we are concerned with the implications of the tails of the distributions in load–strength interactions. However, we often cannot assume

Table 2.1 Sample data taken randomly from a common population.

Sample	Data							
1	30	31	$\overline{41}$	$\underline{29}$	39	36	$\underline{38}$	30
2	31	34	$\underline{23}$	27	29	32	$\overline{35}$	35
3	$\underline{26}$	33	$\overline{35}$	32	34	29	30	34
4	$\underline{27}$	33	30	31	31	36	28	$\overline{40}$
5	$\underline{18}$	$\overline{39}$	25	32	31	34	27	37
6	$\underline{22}$	36	$\overline{42}$	27	33	27	31	31
7	$\overline{39}$	35	32	39	32	$\underline{27}$	28	32
8	33	34	32	30	34	$\overline{35}$	33	$\underline{28}$
9	32	32	$\overline{37}$	25	33	35	35	$\underline{19}$
10	28	32	36	37	$\underline{17}$	31	$\overline{42}$	32
11	26	$\underline{22}$	32	23	33	$\overline{36}$	36	31
12	36	31	$\overline{45}$	$\underline{24}$	30	27	24	27

that, because a measured value appears to be, say, normally distributed, that this distribution necessarily is a good model for the extremes. Also, few measurements are likely to have been made at these extremes. Extreme value statistics are capable of describing these situations asymptotically.

Extreme value statistics are derived by considering the lowest or highest values in each of a series of equal samples. For example, consider the sample data in Table 2.1, taken randomly from a common population. The overall data can be plotted as shown in Fig. 2.10 as f(x). However, if we plot separately the lowest values and the highest values in each sample, they will appear as $g_L(x)$ and $g_H(x)$. $g_L(x)$ is the extreme value distribution of the lowest extreme whilst $g_H(x)$ is the extreme value distribution of the highest extreme in each sample. For many distributions the distribution of the extremes will be one of three types:

Type I — also known as the *extreme value* or *Gumbel distribution*.
Type II — also known as the *log extreme value distribution*.
Type III — for the lowest extreme values. This is the Weibull distribution.

Figure 2.10 Extreme value distributions

Extreme value type I

The type I extreme value distributions for maximum and minimum values are the limiting models for the right and left tails of the exponential types of distribution, where this is defined as any distribution whose cumulative probability approaches unity at a rate which is equal to or greater than that for the exponential distribution. This includes most reliability distributions, such as the normal, lognormal and exponential distributions.

The probability density functions for maximum and minimum values, respectively, are

$$f(x) = \frac{1}{\sigma} \exp\left\{ -\frac{1}{\sigma}(x-\mu) - \exp\left[-\frac{1}{\sigma}(x-\mu) \right] \right\} \qquad (2.39)$$

$$f(x) = \frac{1}{\sigma} \exp\left\{ \frac{1}{\sigma}(x-\mu) - \exp\left[\frac{1}{\sigma}(x-\mu) \right] \right\} \qquad (2.40)$$

The *reduced variate* is given by

$$y = \frac{x-\mu}{\sigma}$$

Substituting in Eqns (2.39) and (2.40), we can derive the c.d.f. in terms of the reduced variate y.

For maximum values:

$$F(y) = \int_{-\infty}^{y} \exp\{ -[x + \exp(-x)] \} \, dx = \exp[-\exp(-y)] \qquad (2.41)$$

For minimum values:

$$F(y) = 1 - \exp[-\exp(-y)] \qquad (2.42)$$

The distribution of maximum values is right-skewed and the distribution of minimum values is left-skewed. The hazard function of maximum values approaches unity with increasing x, whilst that for minimum values increases exponentially.

σ and μ are scale and location parameters, respectively, but they do not represent the mean and standard deviation. In fact μ is the mode of the distribution, and it can be-shown that the means are:

Maximum value:

$$\mu + 0.577\sigma$$

Minimum value:

$$\mu - 0.577\sigma$$

The standard deviation is 1.283σ in both cases.

Extreme value type II

The extreme type II distribution does not play an important role in reliability work. If the logarithms of the variables are extreme value distributed, then the variable is described by the extreme value type II distribution. Thus its relationship to the type I extreme value distribution is analogous to that of the lognormal to the normal distribution.

Type III extreme value distribution

The type III extreme value distribution for minimum values is the limiting model for the left-hand tail for distributions which are bounded at the left. In fact, the Weibull distribution is the type III extreme value distribution for minimum values, and although it was initially derived empirically, its use for describing the strength distribution of materials has been justified using extreme value theory.

The extreme value distributions related to load and strength

The type I extreme value distribution for maximum values is often an appropriate model for the occurrence of load events, when these are not bounded to the right, i.e. when there is no limiting value.

It is well known that engineering materials possess strengths well below their theoretical capacity, mainly due to the existence of imperfections which give rise to non-uniform stresses under load. In fact, the strength will be related to the effect of the imperfection which creates the greatest reduction in strength, and hence the extreme value distribution for minimum values suggests itself as an appropriate model for strength.

The strength, and hence the time to failure, of many types of product can be considered to be dependent upon imperfections whose extent is bounded, since only very small imperfections will escape detection by inspection or process control, justifying use of a type III (Weibull) model. On the other hand, a type I model might be more representative of the strength of an item which is mass-produced and not 100 per cent. inspected, or in which defects can exist whose extent is not bounded, but which are not detected, e.g. a long wire, whose strength will then be a function of length.

For a system consisting of many components in series, where the system hazard rate is decreasing from $t = 0$ (i.e. bounded) a type III (Weibull) distribution will be a good model for the system time to failure.

SUMMARY OF CONTINUOUS STATISTICAL DISTRIBUTIONS

Figure 2.11 (overleaf) is a summary of the continuous distributions described above.

Type of distribution	Parameters	Probability density function, f(t)	Reliability function, R(t) = 1 − F(t)	Hazard function (instantaneous failure rate), $h(t) = \dfrac{f(t)}{R(t)}$
Normal	Mean, μ Standard deviation, σ	$f(t) = \dfrac{1}{\sigma(2\pi)^{1/2}} \exp\left[-\dfrac{(t-\mu)^2}{2\sigma^2}\right]$	$R(t) = \int_t^\infty f(t)\,dt$	$h(t) = \dfrac{f(t)}{R(t)}$ (general expression)
Lognormal	Mean, μ Standard deviation, σ	$f(t) = \dfrac{1}{\sigma t(2\pi)^{1/2}} \exp\left[-\dfrac{\ln(t-\mu)^2}{2\sigma^2}\right]$	$R(t) = \int_t^\infty f(t)\,dt$	$h(t) = \dfrac{f(t)}{R(t)}$ (general expression)
Exponential	Failure rate, λ MTBF (=SD), θ $\theta = \lambda^{-1}$	$f(t) = \lambda \exp(-\lambda t)$	$R(t) = \exp(-\lambda t)$	$h(t) = \lambda = \theta^{-1}$

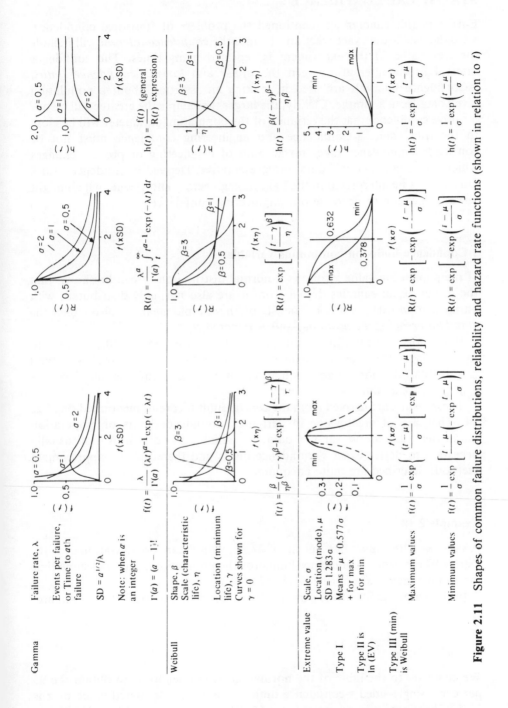

Figure 2.11 Shapes of common failure distributions, reliability and hazard rate functions (shown in relation to t)

STATISTICAL CONFIDENCE

Earlier in this chapter we mentioned the problem of statistical confidence. *s*-Confidence is the exact fraction of times the *confidence interval* will include the true value, if the experiment is repeated many times. The confidence interval is the interval between the *upper* and *lower s-confidence limits*. *s*-Confidence intervals are used in making an assertion about a population given data from a sample. Clearly, the larger the sample the greater will be our intuitive confidence that the estimate of the population parameter will be close to the true value. *s*-Confidence and engineering confidence must not be confused; *s*-confidence takes no account of engineering or process changes which might make sample data unrepresentative. Derived *s*-confidence values must always be interpreted in the light of engineering knowledge, which might serve to increase or decrease our engineering confidence.

s-Confidence limits on continuous variables

If the population value x follows a normal distribution, it can be shown that the means, $\bar{x}$, of samples drawn from it are also normally distributed, with variance σ^2/n (SD $= \sigma/n^{1/2}$). The SD of the sample means is also called the *standard error of the estimate*, and is denoted S_x.

If x is not normally distributed, provided that n is large (> 30), $\bar{x}$ will tend to a normal distribution. If the distribution of x is not excessively skewed (and is unimodal) the normal approximation for $\bar{x}$ at values of n as small as 6 or 7 may be acceptable.

These results are derived from the central limit theorem, mentioned on page 34. They are of great value in deriving *s*-confidence limits on population parameters, based on sample data. In reliability work it is not usually necessary to derive exact *s*-confidence limits and therefore the approximate methods described are quite adequate.

Example 2.10

A sample of 100 values has a mean of 27.56, with a standard deviation of 1.10. Derive 95 per cent. *s*-confidence limits for the population mean. (Assume that the sample means are normally distributed.)

In this case, the SD of the sample means, or standard error of the estimate, is

$$\frac{\sigma}{n^{1/2}} = \frac{1.1}{(100)^{1/2}} = 0.11$$

We can refer to the table of the normal c.d.f. (Appendix 1) to obtain the 95 per cent. single-sided *s*-confidence limits. The closest tabulated value of z is 1.65. Therefore, approximately ± 1.65 SDs are enclosed within the 95 per cent. single-sided s-confidence limits. Since the normal distribution is symmetrical, the 90 per cent. *double-sided* s-confidence interval will exclude 5 per cent. of values at either limit.

In the example, 1.65 SDs = 0.18. Therefore the 95 per cent. s-confidence limits on the population mean are 27.56 ± 0.18, and the 90 per cent. s-confidence interval is $(27.56 - 0.18)$ to $(27.56 + 0.18)$.

As a guide in s-confidence calculations, assuming a normal distribution:

± 1.65 SDs enclose approximately 90 per cent. s-confidence limits (i.e. 5 per cent. lie in each tail).
± 2.0 SDs enclose approximately 95 per cent s-confidence limits (i.e. 2.5 per cent lie in each tail).
± 2.5 SDs enclose approximately 99 per cent. s-confidence limits (i.e. 0.5 per cent lie in each tail).

In many reliability situations we assume a CFR, in which case the failure data will be exponentially distributed and thus highly skewed. Therefore, the normal approximation for s-confidence limit estimation is inappropriate (unless n is large). We use the χ^2 distribution for estimating the confidence limits in these circumstances.

It can be shown that the upper and lower (one-sided) s-confidence limits for data which are generated by a Poisson process, such as failure data when the failure rate is constant, are given by

$$\theta_1 = \frac{2T}{\chi_\alpha^2(\nu = 2n)}$$

$$\theta_u = \frac{2T}{\chi_{1-\alpha}^2(\nu = 2n)}$$

(2.43)

where θ_1 and θ_u are the lower and upper confidence limits on MTBF, T the total test time, α the confidence level and n the number of failures. Equation (2.43) is correct when the test is stopped at the nth failure, i.e. a *failure truncated test*. For *time truncated tests*, use $\nu = 2n + 2$ degrees of freedom for evaluating the lower confidence limit. Values of χ^2 for different values of ν are given in Appendix 3.

Example 2.11

Ten units were tested for a total of 1,000 h and 3 failures occurred. The test was then failure-truncated. Assuming a CFR, what is the 90 per cent. lower confidence limit (LCL) on MTBF?

$$\hat{\theta} = \frac{1,000}{3} = 333 \text{ h}$$

48

From Appendix 3, for $\nu = 6$ and $\alpha = 0.9$, $\chi^2 = 10.6$. Therefore,

$$\theta_1 = \frac{2 \times 1,000}{10.6} = 189 \text{ h}$$

Figure 2.12 may also be used for this estimation.

s-Confidence limits for discrete data

The normal and Poisson approximations to the binomial distribution may not be acceptable for evaluating s-confidence limits with sample sizes below 100.

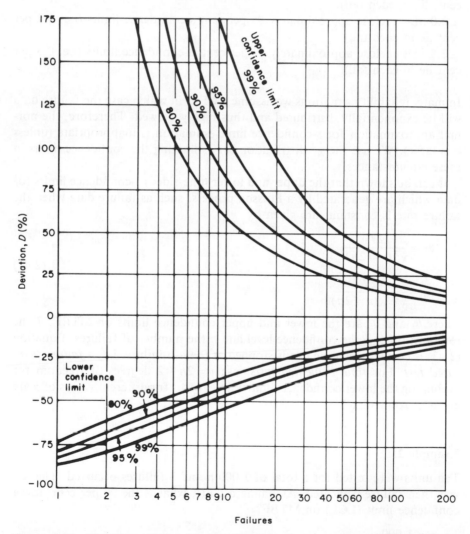

Figure 2.12 Confidence limits (double-sided) for measurement of MTBF for time truncated tests (see Example 2.11). Key: compute confidence limits as $\hat{\theta}(1 + D/100)$, where $\hat{\theta}$ is MTBF estimate

In these situations we can conveniently use the nomogram in Fig. 2.7 to derive the lower confidence limit. For example, if four failures occur in 100 tests, the 95 per cent. LCL on reliability of the population is given by drawing the line from the 0.05 point on the P scale through the 100/4 point and reading the value on the p scale. In this case $p = 0.09$, i.e. the 95 per cent. LCL on reliability is 0.91.

STATISTICAL HYPOTHESIS TESTING

It is often necessary to determine whether observed differences between the statistics of a sample and prior knowledge of a population, or between two sets of sample statistics, are s-significant or due merely to chance. The variation inherent in sampling makes this distinction itself subject to chance. We need, therefore, to have methods for carrying out such tests. Statistical hypothesis testing is similar to s-confidence estimation, but instead of asking the question *How s-confident are we that the population parameter value is within the given limits?* (on the assumption that the sample and the population come from the same distribution), we ask *How s-significant is the deviation of the sample?*

In statistical hypothesis testing, we set up a *null hypothesis*, i.e. that the two sets of information are derived from the same distribution. We then derive the *s-significance* to which this inference is tenable. As in s-confidence estimation, the s-significance we can attach to the inference will depend upon the size of the sample. Many s-signficance test techniques have been developed for dealing with the many types of situation which can be encountered.

In this section we will cover a few of the simpler methods commonly used in reliability work. However, the reader should be aware that the methods described and the more advanced techniques are readily accessible on modern calculators and as computer programs. The texts listed in the Bibliography should be used to identify appropriate tests and tables for special cases.

Tests for differences in means (z test)

A very common s-significance test is for the hypothesis that the mean of a set of data is the same as that of an assumed s-normal population, with known μ and σ. This is the z *test*. The z-*statistic* is given by

$$z = \frac{|\mu - \bar{x}|}{S_{\bar{x}}} = \frac{|\mu - \bar{x}|}{\sigma n^{-1/2}} \tag{2.44}$$

where n is the sample size, μ the population mean, $\bar{x}$ the sample mean and σ the population SD. We then derive the s-significance level from the normal c.d.f. table.

Example 2.12

A type of roller bearing has a time to failure which is normally distributed, with a mean of 6,000 h and an SD of 450 h. A sample of 9, using a changed lubricant, gave a mean life of 6,400 h. Has the new lubricant resulted in an s-significant change in mean life?

$$z = \frac{|6,000 - 6,400|}{450 \times 9^{-1/2}} = 2.67$$

From Appendix 1, $z = 2.67$ indicates a cumulative probability of 0.996. This indicates that there is only 0.004 probability of observing this change purely by chance, i.e. the change is s-significant at the 0.4 per cent. level. Thus we reject the null hypothesis that the sample data are derived from the same normal distribution as the population, and infer that the new lubricant does provide an increased life.

s-Significance is denoted by α. In engineering, an s-significance level of less than 5 per cent. can usually be considered to be sufficient evidence upon which to reject a null hypothesis. An s-significance of greater than 10 per cent. would not normally constitute sufficient evidence, and we might either reject the null hypothesis or perform further trials to obtain more data. The s-significance level considered sufficient will depend upon the importance of the decision to be made based on the evidence. As with s-confidence, s-significance should also be assessed in the light of engineering knowledge.

Instead of testing a sample against a population, we may need to determine whether there is an s-significant difference between the means of two samples. The SD of the distribution of the difference in the means of the samples is

$$S_{(\bar{x}_1 - \bar{x}_2)} = \sqrt{\frac{\sigma_1^2}{n_1} + \frac{\sigma_2^2}{n_2}} \tag{2.45}$$

The SD of the distribution of the difference of the sampling means is called the *standard error of the difference*. This test assumes that the SDs are the population SDs. Then

$$z = \frac{\text{difference in sample means}}{\text{standard error of the difference}}$$

Example 2.13

In Example 2.12, if the mean value of 6,000 and SD of 450 were in fact derived from a sample of 60, does the mean of 6,400, with an SD of 380 from a sample of 9 represent an s-significant difference?

The difference in the means is

$$6,400 - 6,000 = 400$$

The standard error of the difference is

$$S_d = \sqrt{\frac{\sigma_1{}^2 + \sigma_2{}^2}{n_1 + n_2}}$$

$$= \sqrt{\frac{450^2}{60} + \frac{380^2}{9}} = 139$$

$$z = \frac{400}{139} = 2.88$$

$$\alpha = 1 - \Phi(z) = 0.002 \ (0.2 \text{ per cent.})$$

We can therefore say that the difference is highly s-significant, a similar result to that of Example 2.12.

Use of the z test for binomial trials

We can also use the z test for testing the s-significance of binomial data. Since in such cases we are concerned with both extremes of the distribution, we use a two-sided test, i.e. we use 2α instead of α.

Example 2.14

Two sets of tests give the results in Table 2.2. We need to know if the differences in test results are s-significant.

Table 2.2 Results for tests in Example 2.14.

Test	Number tested, n	Number failed
1	217	16
2	310	14

The null hypothesis that the tests are without difference is examined by combining the test results:

$$P = \frac{\text{total failed}}{\text{total tested}} = \frac{30}{527} = 0.057$$

The standard error of the difference in proportions is

$$S_d = \left[pq \left(\frac{1}{n_1} + \frac{1}{n_2} \right) \right]^{1/2}$$

$$= \left[0.057 \times 0.943 \left(\frac{1}{217} + \frac{1}{310} \right) \right]^{1/2}$$

$$= 0.02$$

The proportion failed in test 1 is $16/217 = 0.074$. The proportion failed in test 2 is $14/310 = 0.045$. The difference in proportions is $0.074 - 0.045 = 0.029$. Therefore $z = 0.029/0.020 = 1.45$, giving

$$\alpha = 1 - \Phi(z) = 7.35 \text{ per cent}$$
$$2\alpha = 14.7 \text{ per cent}$$

With such a result, we might accept the null hypothesis and infer that the difference between the tests is not very s-significant.

χ^2 Test for significance

The χ^2 test for the s-significance of differences is used when we can make no assumptions about the underlying distributions. The value of the χ^2 statistic is calculated by summing the terms

$$\frac{(x_i - E_i)^2}{E_i}$$

where x_i and E_i are the ith observed and expected values, respectively. This value is compared with the χ^2 value appropriate to the required s-significance level.

Example 2.15

Using the data of Example 2.14, the χ^2 test is set up as follows:

Test	Failure		Success		Totals
1	16	12.35	201	204.65	217
2	14	17.65	296	292.35	310
Totals	30		497		527

The first number in each column is the observed value and the second number is the expected value based upon the totals of the observations (e.g. expected failures in test $1 = 30/527 \times 217 = 12.35$).

$$\chi^2 = \frac{(16 - 12.35)^2}{12.35} + \frac{(201 - 204.65)^2}{204.65} + \frac{(14 - 17.65)^2}{17.65} + \frac{(296 - 292.35)^2}{292.35}$$

$$= 1.94$$

The number of DF is one less than the number of different possibilities which could exist. In this case there is only one DF, since there are two possibilities — pass and fail. The value of χ^2 of 1.94 for 1 DF (from Appendix 3) occurs with a cumulative probability of between 80 and 90 per cent. Therefore α (two-sided) is between 10 and 20 per cent. The difference between the observed data sets is therefore not s-significant. This inference is the same as that derived in Example 2.14.

Tests for differences in variances

Variance ratio test (F test)

The s-significance tests for differences in means described above have been based on the assumption in the null hypothesis that the samples came from the same s-normal distribution, and therefore should have a common mean. We can also perform s-significance tests on the differences of variances. The *variance ratio*, F, is defined as

$$F = \frac{\text{greater estimate of population variance}}{\text{lesser estimate of population variance}}$$

Values of the F distribution are tabulated in Appendix 4 against the number of degrees of freedom in the two variance estimates (for a sample size n, DF $= n - 1$). The use of the F test is illustrated by Example 2.16.

Example 2.16

Life test data on two items give the results in Table 2.3.

Table 2.3 Life test data on two items

	Sample size, n	Sample standard deviation, σ	Sample variance, σ^2
Item 1	20	37	1,369
Item 2	10	31	961

$$F = \frac{1,369}{961} = 1.42$$

Entering the tables of F values at 19 DF for the greater variance estimate and 9 DF for the lesser variance estimate, we see that at the 5 per cent. level our value for F is less than the tabulated value. Therefore the difference in the variances is not s-significant at the 5 per cent. level.

NON-PARAMETRIC INFERENTIAL METHODS

Methods have been developed for measuring and comparing statistical variables when no assumption is made as to the form of the underlying distributions. These are called non-parametric (or distribution-free) statistical methods. They are only slightly less powerful than parametric methods in terms of the accuracy of the inferences derived for assumed s-normal distributions. However, they are more powerful when the distributions are not s-normal. They are also simple to use. Therefore they can be very useful in reliability work provided that the data being analysed are independently and identically distributed (IID). The implications of data not independently and identically distributed are covered on page 61 and in the next chapter.

Comparison of median values

The sign test

If a null hypothesis states that the median values of two samples are the same, then about half the values of each sample should lie on either side of the median. Therefore about half the values of $(x_i - \bar{x})$ should be positive and half negative. If the null hypothesis is true and r is the number of differences with one sign, then r has a binomial distribution with parameters n and $p = \frac{1}{2}$. We can therefore use the binomial distribution to determine critical values of r to test whether there is an s-significant difference between the median values. Table 2.4 gives critical values for r for the sign test where r is the number of less frequent signs. If the value of r is equal to or less than the tabulated value the null hypothesis is rejected.

Table 2.4 Critical values of r for the sign test.

	Significance level per cent		
n	10	5	1
8	1	0	0
10	1	1	0
12	2	2	1
14	3	2	1
16	4	3	2
18	5	4	3
20	5	5	3
25	7	7	5
30	10	9	7
35	12	11	9
40	14	13	11
45	16	15	13
50	18	17	15
55	20	19	17
60	23	21	19
75	29	28	25
100	41	39	36

Example 2.17

Ten items are tested to failure, with lives

98, 125, 141, 72, 119, 88, 64, 187, 92, 114

Do these results indicate an s-significant change from the previous median life of 125?

The sign test result is

$$- 0 + - - - - + - -$$

i.e. $r = 2$, $n = 9$ (since one difference $= 0$, we discard this item).

Table 2.4 shows that r is greater than the critical value for $n = 9$ at the 10 per cent. s-significant level, and therefore the difference in median values is not significant at this level.

The weighted sign test

We can use the sign test to determine the likely magnitude of differences between samples when differences in medians are significant. The amount by which the samples are believed to differ are added to (or subtracted from) the values of one of the samples, and the sign test is then performed as described above. The test then indicates whether the two samples differ s-significantly by the weighted value.

Tests for variance

Non-parametric tests for analysis of variance are given in Chapter 13.

Reliability estimates

Non-parametric methods for estimating reliability values are given in Chapter 12.

GOODNESS OF FIT

In analysing statistical data we need to determine how well the data fit an assumed distribution. The goodness of fit can be tested statistically, to provide a level of s-significance that the null hypothesis (i.e. that the data do fit the assumed distribution) is rejected. Goodness-of-fit testing is an extension of s-significance testing in which the sample c.d.f. is compared with the assumed true c.d.f.

A number of methods are available to test how closely a set of data fits an assumed distribution. As with s-significance testing, the power of these tests in rejecting incorrect hypotheses varies with the number and type of data available, and with the assumption being tested.

The χ^2 goodness-of-fit test

A commonly used and versatile test is the χ^2 goodness-of-fit test, since it is equally applicable to any assumed distribution, provided that a reasonably large number of data points is available. For accuracy, it is desirable to have at least three data classes, or *cells*, with at least five data points in each cell.

The justification for the χ^2 goodness-of-fit test is the assumption that, if a sample is divided into n cells (i.e. we have ν degrees of freedom where $\nu = n - 1$), then the values within each cell would be normally distributed about the expected value, if the assumed distribution is correct, i.e. if x_i and E_i are the observed and expected values for cell i:

$$\sum_{i}^{n} \frac{(x_i - E_i)^2}{E_i} = \chi^2 \quad \text{(with } n - 1 \text{ degrees of freedom)}$$

High values of χ^2 cast doubt on the null hypothesis. The null hypothesis is usually rejected when the value of χ^2 falls outside the 90th percentile. If χ^2 falls inside the 10th percentile the null hypothesis can be accepted. Intermediate values indicate statistical uncertainty.

The application can be described by use of an example.

Example 2.18

Failure data of transistors are given in Table 2.5.

If an assumed distribution gave expected values of 20, 15, 12, 10, 9, then

$$\chi^2 = \frac{(18 - 20)^2}{20} + \frac{(14 - 15)^2}{15} + \frac{(10 - 12)^2}{12} + \frac{(12 - 10)^2}{10} + \frac{(8 - 9)^2}{9}$$

$$= 1.11$$

Referring to Appendix 3 for values of x^2 with $(n - 1) = 4$ degrees of freedom, $x^2 = 1.11$ lies close to the 10th percentile. Therefore we accept the null hypothesis of the assumed hazard rate distribution at the 90 per cent. level.

Table 2.5 Data from an overstress life test of transistors.

Cell (h)	Number in cell	Cell (h)	Number in cell
0–999	18	3,000–3,999	12
1,000–1,999	14	4,000–4,999	6
2,000–2,999	10		

The Kolmogorov–Smirnov test

Another goodness-of-fit test commonly used in reliability work is the Kolmogorov–Smirnov (K–S) test. It is rather simpler to use than the χ^2 test and can give better results with small numbers of data points. It is also convenient to use in conjunction with probability plots (see Chapter 3), since it is based upon cumulative ranked data, i.e. the sample c.d.f. The procedure is:

1. Tabulate the ranked failure data. Calculate the values of $|x_i - E_i|$ where x_i is the ith cumulative rank value and E_i the expected cumulative rank value for the assumed distribution.
2. Determine the highest single value.
3. Compare this value with the appropriate K–S value.

Example 2.19

Table 2.6 shows failure data with the ranked values of x_i. We wish to test the null hypothesis that the data do not fit a normal distribution with parameters which give the tabulated cumulative values of E_i. Therefore, in the E_i column we list the expected value of proportion failed at each failure time.

The largest value of $|x_i - E_i|$ is 0.08 (shown by *). The Kolmogorov–Smirnov table (Appendix 5) shows that, for $n = 12$, the critical value of $|x_i - E_i|$ is 0.338 at the 10 per cent. s-significance level. Therefore the null hypothesis is not rejected at this level, and we can accept the data as coming from the hypothesized normal distribution.

Example 2.19 shows quite a large difference between the critical K–S value and the largest value of $|x_i - E_i|$. When the parameters of the assumed c.d.f. are being estimated from the sample data, as in this example, the critical K–S

Table 2.6 Failure data with ranked values of x_i.

Event	Time to failure (h)	x_i	E_i	$|x_i - E_i|$
1	12.2	0.056	0.035	0.021
2	13.1	0.136	0.115	0.021
3	14.0	0.217	0.29	0.073
4	14.1	0.298	0.32	0.022
5	14.6	0.379	0.44	0.061
6	14.7	0.459	0.46	0.001
7	14.7	0.54	0.46	0.08*
8	15.1	0.621	0.58	0.041
9	15.7	0.702	0.73	0.028
10	15.8	0.783	0.75	0.033
11	16.3	0.864	0.85	0.014
12	16.9	0.94	0.95	0.006

values are too large and give lower s-significance levels than are appropriate in the circumstances. In order to correct for this, the critical values should be multiplied by the factors:

0.70 $(\beta > 3.0)$

0.75 $(3.0 > \beta > 1.5)$

0.80 $(\beta < 1.5)$

where β is the Weibull shape parameter. Therefore, in Example 2.19, since the Weibull β value appropriate to the normal distribution is >3, the corrected K$-$S critical value is $0.338 \times 0.70 = 0.237$.

Least squares test

The least squares goodness-of-fit test is used to measure the *linear correlation* of data with the equation of the straight line which best fits the plotted data. The line which most nearly fits a set of plotted data is called the *regression line*, and the goodness of fit as derived by the least squares method is called the *correlation coefficient*.

The correlation coefficient is

$$r = \frac{S_{xy}}{S_x S_y}$$

where

$$\left. \begin{array}{l} S_x^2 = \dfrac{1}{n} \sum_{i=1}^{n} (x_i - \bar{x})^2 \\[12pt] S_y^2 = \dfrac{1}{n} \sum_{i=1}^{n} (y_i - \bar{y})^2 \end{array} \right\} \quad \text{(i.e. the sample variances)} \qquad (2.46)$$

$$S_{xy} = \frac{1}{n} \sum_{i=1}^{n} (x_i - \bar{x})(y_i - \bar{y}) \quad \text{(the } \textit{covariance}\text{)} \qquad (2.47)$$

x_i, y_i are the coordinates of the plotted data. The regression line of the plotted data is given by

$$y - \bar{y} = \frac{S_{xy}}{S_x^2} (x - \bar{x}) \qquad (2.48)$$

If r is positive, it indicates that the data are positively correlated, i.e. the regression line slope is positive, and vice versa; if $|r| = 1$ we have perfect correlation, with all plotted points lying on the regression line; if $r = 0$ the variates are not linearly correlated.

r^2 is often used instead of r to indicate correlation, since it provides a more sensitive indication, particularly with probability plots (see Chapter 3). It is called the *sample coefficient of determination*.

Linear regression analysis can be used for data that are not linearly correlated if the axes are transformed to linearize the equation. Therefore, the method can be used for evaluating the goodness of fit of data plotted on probability papers (Chapter 3).

SERIES OF EVENTS (POINT PROCESSES)

Situations in which discrete events occur randomly in a continuum (e.g. time) cannot be truly represented by a single continuous distribution function. Failures occurring in repairable systems, aircraft accidents and vehicle traffic flow past a point are examples of series of discrete events. These situations are called *stochastic point processes*. They can be analysed using the statistics of *event series*.

The Poisson distribution function (Eqn 2.27) describes the situation in which events occur randomly and at a constant average rate. This situation is described by a *homogeneous Poisson process* (HPP). An HPP is a *stationary* point process, since the distribution of the number of events in an interval of fixed length does not vary, regardless of when (where) the interval is sampled.

The Poisson distribution function is (from Eqn 2.27)

$$f(x) = \frac{(\lambda x)^x}{x!} \exp(-\lambda x) \qquad \text{(for } n = 0, 1, 2, \ldots) \tag{2.49}$$

where λ is the mean rate of occurrence, so that λx is the expected number of events in $(0, x)$. Fig. 2.8 can be used for estimating the probability of a given number of events in a given interval (see p. 33).

In a *non-homogeneous* Poisson process (NHPP) the point process is non-stationary, so that the distribution of the number of events in an interval of fixed length changes as x increases. Typically, the discrete events (e.g. failures) might occur at an increasing or decreasing rate.

Note that an essential condition of any homogeneous Poisson process is that the probabilities of events occuring in any period are independent of what has occurred in preceding periods. An HPP describes a sequence of independently and identically exponentially distributed (IIED) random variables. A NHPP describes a sequence of random variables which is neither independently nor identically distributed.

Trend analysis

When analysing data from a stochastic point process it is important to determine whether the process has a trend, i.e. to know whether a failure rate is increasing, decreasing or constant. We can test for trends by analysing the *arrival values* of the event series. The arrival values $x_1, x_2, \ldots, x_n$ are the values of the independent variables (e.g. time) from $x = 0$ at which each event occurs. The *interarrival values* $X_1, X_2, \ldots, X_n$ are the intervals between

Figure 2.13 Arrival and interarrival values

successive events 1, 2, ..., n, from $x = 0$. Figure 2.13 shows the distinction between arrival and interarrival values.

If x_0 is the period of observation, then the test statistic for trend is

$$U = \frac{\sum x_i / n - x_0 / 2}{x_0 \sqrt{1/(12n)}}$$ (2.50)

This is called the *centroid test* or the *Laplace test*. It compares the centroid of the observed arrival values with the mid-point of the period of observation. If $U = 0$ there is no trend, i.e. the process is stationary. If $U < 0$ the trend is decreasing, i.e. the interarrival values are tending to become larger. Conversely, when $U > 0$ the trend is increasing, i.e. interarrival values are tending to become progressively smaller.

If the period of observation ends at an event, use $(n - 1)$ instead of n and exclude the time to the last event from the summation $\sum x_i$.

We can test the null hypothesis that there is no trend in the chronologically ordered data by testing the value of U against the values of the standard normal variate, z. For example, using Appendix 1, if $U = 1.65$, for $z = 1.65$, $\Phi(z) = 0.95$. Therefore we can reject the null hypothesis at the 5 per cent. *s*-significance level.

The centroid test is theoretically adequate if $n \geqslant 4$, when the observation interval ends with an event, and if $n \geqslant 3$, when the interval is terminated at a predetermined time.

Example 2.20

Arrival values (x_i) and interarrival values (X_i) between 12 successive failures of a component are as follows (observation ends at the last failure):

x_i	X_i
175	175
196	21
304	108
415	111
504	89
516	12
618	102
641	23
679	38
726	47
740	14
791	51

$$\sum x_i = 5{,}514 \qquad \text{(excluding 791)}$$

$$n - 1 = 11$$

$$\frac{\sum x_i}{n - 1} = 501.3$$

$$\frac{x_0}{2} = 395.5$$

$$U = \frac{501.3 - 395.5}{791\sqrt{1/(12 \times 11)}} = 1.54 \qquad \begin{array}{l}\text{Referring to Appendix 1, for} \\ z = 1.54, \ \Phi(z) = 0.94\end{array}$$

Therefore we can reject the null hypothesis that there is no trend at the 6 per cent. s-significance level. The interarrival times are becoming shorter, i.e. the failure rate is increasing.

The existence of a trend in the data, as in Example 2.20, indicates that the interarrival values *are not independently and identically distributed* (IID). This is a very important point to consider in the analysis of failure data, as will be explained in the next chapter.

Superimposed processes

If a number of separate stochastic point processes combine to form an overall process, e.g. failure processes of individual components (or sockets) in a system, these are called *superimposed processes*. If the individual random variables are IID exponential then the overall process variable is also IID exponential and the process is HPP.

If the individual variables are IID non-exponential, the overall process will tend to an HPP. Such a process is called a *renewal process*. Figure 2.14 shows these processes.

Figure 2.14 Rate of occurrence for superimposed processes

BILIOGRAPHY

Helpful introductory sources

1. M. J. Moroney, *Facts from Figures*. Penguin, Harmondsworth (1965).
2. R. Langley, *Practical Statistics Simply Explained*, 2nd Edn. Pan, London (1979).
3. W. W. Hines and D. C. Montgomery, *Probability and Statistics in Engineering and Management Science* (2nd Edn.). Wiley, New York (1980).
4. L. Mann, *Applied Engineering Statistics*. Barnes & Noble, New York (1970).
5. C. Chatfield, *Statistics for Technology*, 2nd Edn. Chapman & Hall, London (1978).
6. D. R. Cox and P. A. W. Lewis, *The Statistical Analysis of Series of Events*. Chapman and Hall, London (1966).
7. W. J. Conover, *Practical Non-parametric Statistics*. Wiley, New York (1971).
8. British Standard, BS 5760: *Reliability of Systems, Equipments and Components*, Part 2. British Standards Institution, London.
9. Handbooks for statistical and programmable calculators, e.g. Commodore S61, Texas Instruments T159, Hewlett-Packard HP41C.

More advanced works

10. A. J. Duncan, *Quality Control and Industrial Statistics*, 4th Edn. Irwin, Homewood, Illinois (1974).
11. A. H. Bowker and G. J. Lieberman, *Engineering Statistics*, 2nd Edn. Prentice-Hall, Englewood Cliffs (1972).
12. G. J. Hahn and S. S. Shapiro, *Statistical Models in Engineering*. Wiley, New York (1967).
13. N. L. Johnson and S. Kotz, *Distributions in Statistics*, 4 Vols. Wiley, New York (1970).
14. M. L. Shooman, *Probabilistic Reliability — An Engineering Approach*. McGraw-Hill, New York (1968).

15. E. S. Pearson and H. O. Hartley, *Biometrika Tables for Statisticians*. Biometrika Trust, London.
16. R. E. Barlow and F. Proschan, *Mathematical Theory of Reliability*. Wiley, New York (1965).
17. R. E. Barlow and F. Proschan, *Statistical Theory of Reliability and Life Testing*, Holt, Rinehart & Winston, New York (1975).
18. N. Mann, R. E. Schafer and N. D. Singpurwalla, *Methods for Statistical Analysis of Reliability and Life Data*. Wiley, New York (1974).
19. M. Hollander and D. A. Wolfe, *Non-parametric Statistical Methods*. Wiley, New York (1973).
20. H. Goldberg, *Extending the Limits of Reliability Theory*. Wiley, New York (1981).
21. H. F. Martz and R. A. Waller, *Bayesian Reliability Analysis*. Wiley, New York (1982).

Chapter 3

ANALYSING RELIABILITY DATA

PROBABILITY PLOTTING

It is frequently useful in reliability engineering to determine which distribution best fits a set of failure data and to derive interval estimates of the distribution parameters. The mathematical basis for the approaches to these problems was covered in Chapter 2. Graphical estimation methods can greatly ease this task, and probability plotting papers have been developed for the purpose. These are based upon the cumulative distribution function (c.d.f.) [$F(x)$] of the distribution concerned. The axes of probability plotting papers are transformed in such a way that the true c.d.f. plots as a straight line. Therefore if the plotted data can be fitted by a straight line, the data fit the appropriate distribution. Further constructions permit the distribution parameters to be estimated. Thus reliability data can be evaluated quickly, without a detailed knowledge of the statistical mathematics being necessary. This facilitates analysis and presentation of data.

Note that probability plotting methods to derive distribution parameters are only applicable when times to failure are independently and identically distributed (IID). This is usually the case for non-repairable components and systems but may *not* be the case with failure data from repairable systems. Reliability data analysis for repairable systems will be covered in later sections.

RANKING OF DATA

Mean ranking (normal probability plots)

Probability graph papers are based upon plots of the variable of interest against the cumulative percentage probability. The data therefore need to be ordered and the cumulative probability calculated. For example, consider the data on times to failure of 12 items (Table 3.1). For the first failure, the cumulative proportion is 1/12 or or 8.3 per cent. For the second, the cumulative proportion is 2/12 or 16.7%, and so on to 12/12 or 100 per cent. for the twelfth failure. However, for probability plotting, it is better to make an adjustment to allow for the fact that each failure represents a point on a distribution. Thus considering the whole population of which the 12 items represent a sample, the times by which 8.3, 16.7 ,..., 100 per cent. will have

Table 3.1 Data on times to failure of 12 items.

Order no.	Time to failure (h)	Cumulative per cent (c.d.f.)	Mean rank per cent (c.d.f.)	Median rank per cent (c.d.f.)
1	12.2	8.3 (1/12)	7.7	5.6
2	13.1	16.7 (2/12)	15.4	13.6
3	14.0	25.0 (3/12)	23.1	21.7
4	14.1	33.3 (4/12)	30.8	29.8
5	14.6	41.7 (5/12)	38.5	37.9
6	14.7	50.0 (6/12)	46.2	45.9
7	14.7	58.3 (7/12)	53.8	54.0
8	15.1	66.7 (8/12)	61.5	62.1
9	15.7	75.0 (9/12)	69.2	70.2
10	15.8	83.3 (10/12)	76.9	78.3
11	16.3	91.7 (11/12)	84.6	86.4
12	16.9	100.0 (12/12)	92.3	94.4

failed in several samples of 12 will all be randomly distributed. However, the data in Table 3.1 show a bias, in that the first failure is shown much further from the zero cumulative percentage point than is the last from 100 per cent. (in fact it coincides). To overcome this, and thus to improve the accuracy of the estimation, *mean* or *median* ranking of cumulative percentages is used for probability plotting.

The usual method for mean ranking is to use $(n + 1)$ in the denominator, instead of n, when calculating the cumulative percentage position. Thus in Table 3.1 the cumulative percentages (mean ranks) would be

$$\frac{100}{12 + 1} = 7.7\%$$

$$\frac{200}{12 + 1} = 15.4\%$$

$$\vdots$$

$$\frac{1,200}{12 + 1} = 92.3\%$$

These data are shown plotted on normal probability paper in Fig. 3.1 (circles). The plotted points show a reasonably close fit to the straight line drawn 'by eye'. Therefore, we can say that the data appear to fit the cumulative normal distribution represented by the line.

Median ranking

Mean ranking is the appropriate method for a symmetrical distribution, such as the normal. However, for skewed distributions median ranking provides a

Figure 3.1. Normal probability plot

 80 Equal divisions x probability

better correction. The most common approximation used for median ranking is that due to Bénard. The ith rank value is given by

$$r_i = \frac{i - 0.3}{n + 0.4}$$

where i is the ith order value and n is the sample size.

Median ranking is the method most frequently used in probability plotting, particularly if the data are known not to be normally distributed. Appendix 6 gives tables of median ranks for samples up to 50. Figure 3.1 shows the data of Table 3.1 plotted using median ranks (triangles). The effect is to alter the slope slightly, but the linearity is not affected.

PROBABILITY PLOTTING TECHNIQUES

Fitting the line

Having ranked and plotted the data, the question that often arises is *What is the best straight line fit to the data?* (assuming, of course, that there is a reasonable straight line fit). There can be a certain amount of subjectivity or even a temptation to adjust the line a little to fit a preconception. Two basic considerations apply:

1. Since the plotted data are cumulative, the points at the high cumulative proportion end of the plot are more important than the early points.
2. If the conclusions to be derived can be influenced by subjective differences in drawing the line through the data, then either there are insufficient data or there is not a good enough fit to the assumed distribution. (The same argument applies to choices of which ranking method should be used.)

Normally, a line which gives a good 'eyeball fit' to the plotted data is satisfactory, and more refined methods will give results which differ by amounts well within the 5 per cent. s-significance level. However, a simple and accurate procedure to use, if rather more objectivity is desired, is to place a transparent rule on the last point and draw a line through this point such that an equal number of points lie to either side of the line.

Statistical goodness-of-fit tests can be applied, as described in Chapter 2. It is important to realize that cumulative probability plots are to a large extent self-aligning, since succeeding points can only continue upwards and to the right. Goodness-of-fit tests will nearly always indicate good correlation with any straight line drawn through such points. Therefore, it is sometimes prudent to test the data against the assumed distribution using the methods described. Engineering knowledge can also be of value in determining the likely best distribution.

s-Confidence limits on plotted data

s-Confidence limits can be evaluated for the rank values of sample data, as shown in Chapter 2. Appendix 6 includes 5 per cent. and 95 per cent. median ranks, which can be used to plot the upper and lower 95 per cent. s-confidence limits on the c.d.f. These are shown plotted for the data of Table 3.1 in Fig. 3.1.

Censored data

When analysing field or test failure data, we may have to consider samples in which not all items have failed. A proportion of items may have survived to the end of the test, or some may have been withdrawn prior to the end of the test. Data on such a sample are called *censored* data. Items which have survived or which have been removed are called *suspended items*. The derivation of median ranks for censored data is carried out as follows:

1. List order number (i) of failed items.
2. List increasing ordered sequence of life values (t_i) of failed items.
3. Against each failed item, list the number of items which have survived to a time between that of the previous failure and this failure (or between $t = 0$ and the first failure).
4. For each failed item, calculate the *mean order number* i_{t_i} using the formula

$$i_{t_i} = i_{t_{i-1}} + N_{t_i} \tag{3.1}$$

where

$$N_{t_i} = \frac{(n + 1) - i_{t_{i-1}}}{1 + (n - \text{number of preceding items})} \tag{3.2}$$

in which n is sample size.

5. Calculate median rank for each failed item, using the formula

$$r_{t_i} = \frac{i_{t_i} - 0.3}{n + 0.4}\% \tag{3.3}$$

Example 3.1

Of a sample of 50 items, the running times (h) for the failed and surviving items are as shown in Table 3.2 (12 failures, 38 survivors).

From Eqns (3.1) to (3.3),

$$N_1 = \frac{50 + 1 - 0}{1 + (50 - 2)} = 1.04$$

$$i_1 = i_0 + N_1 = 0 + 1.04 = 1.04$$

$$N_2 = \frac{50 + 1 - 1.04}{1 + (50 - 3)} = 1.04$$

$$\vdots$$

$$i_2 = i_1 + N_2 = 1.04 + 1.04 = 2.08$$

$$\vdots$$

$$r_1 = \frac{i_1 - 0.3}{n + 0.4} = \frac{1.04 - 0.3}{50 + 0.4} = 1.46 \text{ per cent}$$

$$r_2 = \frac{i_2 - 0.3}{n + 0.4} = \frac{2.08 - 0.3}{50 + 0.4} = 3.53 \text{ per cent}$$

$$\vdots$$

These data (Table 3.3) are shown plotted in Weibull probability paper in Fig. 3.4.

'Sudden death' testing

'Sudden death' testing is a useful technique for obtaining test data quickly on a relatively large sample. The sample is divided into a number of equal groups (n groups, each of k items) by random selection. All items are then tested until one fails in each group. These failures are then used as the data points on the life plot. However, the line obtained represents the distribution of times to the first failure, in a sample in which all the non-failed items are suspended. The times to failure of the failed items are ranked and plotted, giving a sudden

Table 3.2 Running times for failed and surviving items for a sample of 50 items.

Item[a]	Running time (h)	Item[a]	Running time (h)	Item[a]	Running time (h)
S1	40	S10	141	S25	165
S2	51	S11	147	S26	165
F1	54	S12	147	S27	166
F2	70	S13	150	S28	166
S3	73	F9	153	S29	166
S4	73	S14	153	S30	168
S5	80	S15	153	S31	168
F3	85	S16	154	S32	171
S6	90	S17	156	F11	173
F4	96	S18	156	S33	177
S7	102	S19	156	S34	181
F5	108	S20	158	S35	185
F6	118	S21	158	S36	188
S8	128	S22	158	F12	200
S9	128	F10	161	S37	202
F7	132	S23	162	S38	205
F8	141	S24	162		

[a]S is survivor, F is failure.

Table 3.3 Derivation of median ranks for Example 3.1.

Failure order no.	Ordered life values, t_i(h)	No. of intervening survivors, $S(t_i)$	Mean order, i_{t_i}	Median rank, r_{t_i} per cent
1	54	2	1.04	1.46
2	70	0	2.08	3.53
3	85	3	3.19	5.74
4	96	1	4.33	7.99
5	108	1	5.50	10.31
6	118	0	6.66	12.62
7	132	2	7.89	15.07
8	141	0	8.71	16.69
9	153	4	10.12	19.47
10	161	9	12.15	23.51
11	173	10	16.47	32.08
12	200	4	25.10	49.21

death line. For example, Fig. 3.2 shows a plot of the times to first failure of ten samples of six each, with the rank values based on n (10). The first failure in a sample of six has a median rank of 10.91 per cent. (Appendix 6). To derive the population median line, it is necessary to drop a vertical from the point where the sudden death line intersects the 50 per cent. failed ordinate. A line drawn through the point where this intersects the 10.91 per cent ordinate, parallel to the sudden death line, will represent the population life distribution.

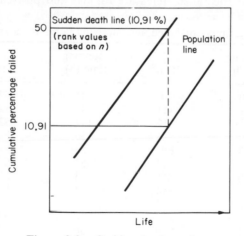

Figure 3.2. Sudden death testing

Mixed distributions

A plot of failure data may relate to one failure mode or to multiple failure modes within an item. If a straight line does not fit the failure date, particularly if an obvious change of slope is apparent, the causes of failure should be investigated to ensure that not more than one failure mode is included. For example, after a certain length of time on test, a second failure mode may become apparent, or an item may have two superimposed failure modes. In such cases each failure mode must be isolated and analysed separately, after the data on each have been appropriately censored (i.e. each A-type failure is a survival as far as B-type failures are concerned, and vice versa). However, such separation is appropriate only if the failure processes are s-independent, i.e. there is no interaction.

Screened samples

In QC and reliability work we often deal with samples which have been screened in some way. For example, machined parts will have been inspected, oversize and undersize parts removed, and electronic parts may have been screened so that no zero-life ('dead-on-arrival') parts exist. Screening can show up on probability plots, as a curvature in the tails. For example, a plot of time to failure of a fatigue specimen will normally be curved since quality control will have removed items of very low strength. In other words, there will be a positive minimum life. We will cover this aspect in more detail in the section on Weibull plotting of life data.

Estimation of parameters

We can also estimate the distribution parameters from the plot. In the example of the normal probability plot given above (Fig. 3.1) the mean life can be seen to be 14.8 h. The standard deviation can be derived by measuring the life interval covered by 34 per cent. either side of the 50 per cent. point on the probability scale. In this case it is 1.55 h. The 95 per cent. s-confidence limits on the mean are 14 h and 15.7 h. The estimation of the parameters of other distributions will be covered in later sections of this chapter, dealing with the Weibull and extreme value distributions.

LOGNORMAL PROBABILITY PLOTS

Lognormal probability paper is used in exactly the same way as normal probability paper, except that the ordinate scale is constructed logarithmically.

WEIBULL PROBABILITY PLOTS

In reliability engineering Weibull probability analysis is the most widely used of the technques described, due to the flexibility of the Weibull distribution in describing a number of failure patterns.

Weibull transformation

The axes of Weibull probability paper are derived by performing the transformation

$$R(t) = 1 - F(t) = \exp\left[-\left(\frac{t}{\eta}\right)^{\beta}\right]$$

assuming γ(failure-free time) is 0 (i.e. we use the two-parameter Weibull distribution). Therefore

$$\frac{1}{1 - F(t)} = \exp\left(\frac{t}{\eta}\right)^{\beta}$$

taking double logarithms,

$$\ln \ln \frac{1}{1 - F(t)} = \beta \ln t - \beta \ln \eta$$

This is a straight line of the form $y = ax + b$.

Thus the Weibull paper is constructed by having a log log reciprocal ordinate (y) scale, representing cumulative probability of failure (or failure percentage), and the abscissa (x) scale is a log scale representing the life value. The slope of a straight line plotted on this paper will be β, the shape parameter.

Examples of Weibull probability paper are shown in Figs 3.3 and 3.4, as produced by Technical and Engineering Aids to Management — TEAM (USA) and Chartwell (UK). On the TEAM paper the slope β is measured by constructing a line parallel to the line of plotted data, through the estimation point, and reading the value of β at the intersection with the β scale (1.6 in the example shown). The value of β is estimated on the Chartwell paper on the β scale by the intersection of the perpendicular from the estimation point to the line of the plotted data. η is given by the life value at the intersection with the η estimator line (63.2 per cent. cumulative failure). The Chartwell paper also includes a scale for estimating the mean life μ. This is derived by reading the survival probability from the P scale, where the β estimator line intersects, and then reading the life value relative to the probability on the cumulative percentage failure scale.

Another Weibull parameter often used is the B life. This is the life by which a given percentage of the population will have failed. For example, the B_{10} life is the life at which 10 per cent. of the population will have failed. It is estimated by reading the appropriate life value against the intersection of the line of the plotted data with the 10 per cent. cumulative failure line. For example, the B_{10} life in Fig. 3.4 is 110 h. ($\hat{B}_{10}$ indicates that this is an estimate from the sample data.)

Probability papers are available covering various decade ranges. For example, the two papers shown in Figs 3.3 and 3.4 cover 3 and 2 decades, respectively. If the correct decade range is not available, the data may still be

73

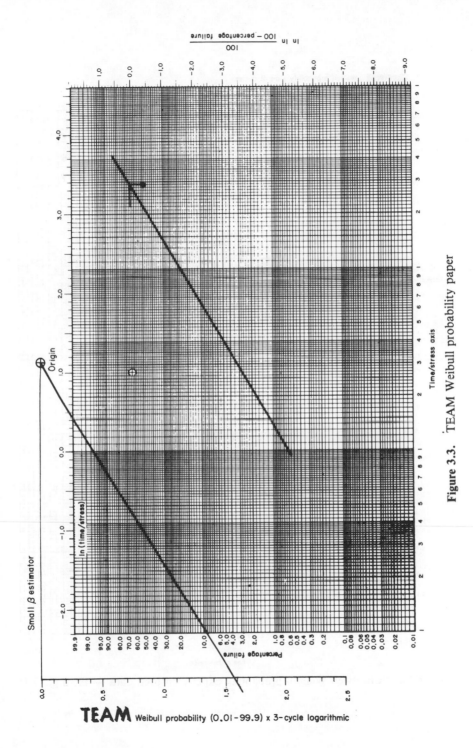

Figure 3.3. TEAM Weibull probability paper

74

Figure 3.4. Chartwell Weibull probability paper (censored data — Example 3.1)

plotted by allowing the plot to reappear on the left-hand side of the paper or by attaching more decade ranges to the right-hand side.

Failure-free life

The examples given so far all relate to cases where γ is 0, i.e. there is no failure-free time. Of course this is not necessarily the case in reliability work. If an item has a finite (positive) failure-free time under test, e.g. a fatigue test specimen, the failure data will plot as a curve, seen convex from above, since the transformation to achieve the Weibull scales assumes that the data fit a two-parameter distribution. The effect of a finite life is to shift the age-at-failure scale to the left. It is possible to have an apparent negative value for γ, for example, if the items under test had accumulated unrecorded operating time before the start of the test. In this case the curve will appear concave from above.

In order to transform such data into a straight line for analysis of the Weibull parameters it is necessary to transform the three-parameter distribution into a two-parameter one. To achieve this, equally spaced horizontals are drawn as shown in Fig. 3.5 with the middle line intersecting the mid-point of the plotted data. The corresponding values of t_1, t_2 and t_3 are then read off the life scale. Then

$$\hat{\gamma} = t_2 - \frac{(t_3 - t_2)(t_2 - t_1)}{(t_3 - t_2) - (t_2 - t_1)}$$

The data are then replotted, with the value of $\hat{\gamma}$ so derived subtracted from each life value. The life parameters estimated from the plot must then have the value of $\hat{\gamma}$ added to give the true life values.

Discretion must be used in interpreting data that do not plot as a straight line, since the cause of the non-linearity may be due to the existence of mixed distributions or because the data do not fit the Weibull distribution. It is quite

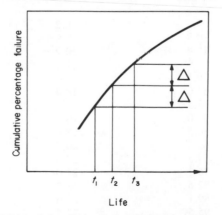

Figure 3.5. Life plot: failure-free time

likely to be due simply to the randomness in the sample. The failure mechanisms must be studied, and engineering judgement used, to ensure that the correct interpretations are made. It is a common error to assume that, because a straight line provides a reasonably good fit to the data, that there is no failure-free life. However, in most cases wearout failure modes do exhibit a finite failure-free life. Therefore a value for γ can sometimes be estimated from knowledge of the product and its application. Alternatively, the time to first failure is often a satisfactory estimate of γ. In these cases the procedure described above is not necessary. Generally, data on several failure modes in a system are likely to fit a two-parameter distribution ($\gamma = 0$), but single wearout failure modes are more likely to have positive values of γ.

Example 3.2

Failure data on 15 electrical relays tested to failure, where the criterion for failure is a predetermined level of contact resistance, are shown in Table 3.4.

The cycles to failure are plotted in Fig. 3.6. The best fit appears to be the curve shown indicating a finite minimum life. This conclusion is accepted as being valid from knowledge of the relay contact behaviour.

The geometrical mid-point on the percentage failure scale gives $t_2 = 13 \times 10^5$. t_1 and t_3 are 6.2×10^5 and 34.4×10^5, respectively.

$$\hat{\gamma} = t_2 - \frac{(t_3 - t_2)(t_2 - t_1)}{(t_3 - t_2) - (t_2 - t_1)}$$

$$= \left[13 - \frac{(34.4 - 13)(13 - 6.7)}{(34.4 - 13) - (13 - 6.7)} \right] \times 10^5 = 4.07 \times 10^5$$

Table 3.4. Failure data on 15 electrical relays tested to failure.

Order no.	Median rank per cent	Cycles to failure ($\times 10^5$)	Cycles to failure, adjusted ($\times 10^5$)
1	4.5	6.2	2.1
2	10.9	9.0	4.9
3	17.4	10.2	6.1
4	23.9	12.1	8.0
5	30.5	12.6	8.5
6	37.0	14.4	10.3
7	43.5	14.7	10.6
8	50.0	16.1	12.0
9	56.5	18.6	14.5
10	63.0	20.5	16.4
11	69.5	20.6	16.5
12	76.1	23.0	18.9
13	82.6	26.7	22.6
14	89.1	27.6	23.5
15	95.5	34.4	30.3

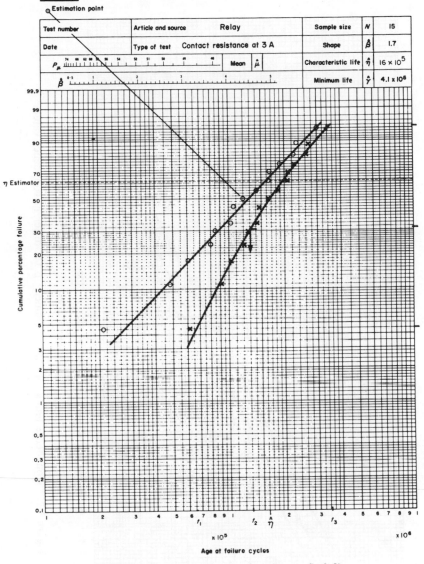

Figure 3.6. Relay life plot (see Example 3.2)

Subtracting 4.1 from each of the values in the third column gives the results in the fourth column. When these are plotted in Fig. 3.6 the adjusted data can be fitted by the straight line shown. From this we can read the other Weibull distribution parameters: $\hat{\beta} = 1.7$, $\hat{\eta} = 16 \times 10^5$ cycles.

s-Confidence limits on the shape parameter

It is important to derive s-confidence limits on the shape parameter β, since

78

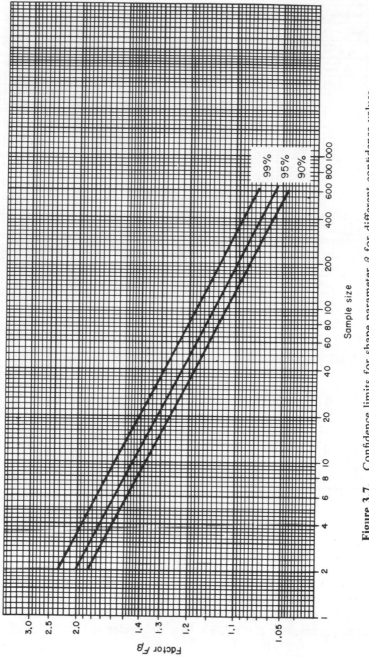

Figure 3.7. Confidence limits for shape parameter β for different confidence values

decisions may be based upon this value. Figure 3.7 gives factors F_β against sample size for different s-confidence levels (99, 95, 90 per cent) on β. The upper and lower s-confidence limits are then

$$\beta_u = \hat{\beta} F_\beta$$

$$\beta_1 = \hat{\beta} \, \frac{1}{F_\beta}$$

Example 3.3

Derive the upper and lower confidence limits if $n = 10, \beta = 1.6$ for $C = 90$ per cent. (double-sided).

From Fig. 3.7, $F_\beta = 1.37$; therefore,

$$\beta_u = 1.6 \times 1.37 = 2.19$$

$$\beta_1 = \frac{1.6}{1.37} = 1.17$$

i.e. we have a 90 per cent. s-confidence that $2.19 \geqslant \beta \geqslant 1.17$.

EXTREME VALUE PROBABILITY PLOTTING

Extreme value statistics are described in Chapter 2. TEAM extreme value probability paper is shown in Fig. 3.8. It uses a linear scale for the random variable and a cumulative probability scale. Two auxiliary scales give the *reduced variate, y,* and the *return period.* The return period is

where p is the cumulative probability that a future observed value will not exceed the previous value. Thus the return period represents the period over which there is a 50 per cent. probability of the event occurring. Both scales are read by drawing the vertical through the point of intersection of the value of the variable with the line of the plotted data.

Example 3.4

The breaking strength of a long wire was tested, using a sample of 15 equal lengths. Since the strength of a wire can be considered to depend upon the existence of imperfections, the extreme value distribution of minimum values might be an appropriate fit. The results are shown in Table 3.5 and plotted in Fig. 3.8.

80

TEAM Extreme value probability x 100 divisions

Figure 3.8. TEAM extreme value probability paper (see also Example 3.4)

Table 3.5 Breaking strengths of 15 samples of wire of equal length.

Rank order	Cumulative probability per cent (median ranks, c.d.f)	Breaking strength (N)
1	4.5	76
2	10.9	75
3	17.4	75
4	23.9	72.5
5	30.4	72
6	36.9	69
7	43.4	69
8	50.0	65
9	56.5	64
10	63.0	63
11	69.5	62
12	76.0	61
13	82.5	58
14	89.0	52
15	95.4	48

Since this would be a distribution of minimum value it will be left-skewed, and the data are therefore arranged in descending order of magnitude. Plotting the data the other way around would generate a convex curve, viewed from above. A plot of an extreme value distribution of maximum values would be made with the data in ascending order. In this case the probability scale represents the cumulative probability that the breaking strength will be greater than the value indicated, so that there is a 95 per cent. probability that the strength of a wire *of this length* will be greater than 46 N. If the wire is longer it will be likely to be weaker, since the probability of its containing extreme value imperfections will be higher.

The mode $\hat{\mu}$ is estimated directly from the plot. In this case $\hat{\mu} = 69.5$ N. $\hat{\sigma}$, the measure of variability, can be derived from the expression (see page 42):

$$\text{Mean} = \hat{\mu} - 0.577\hat{\sigma}$$

$$\hat{\sigma} = \frac{\hat{\mu} - \text{mean}}{0.577}$$

$$= \frac{69.5 - 65}{0.577} = 7.8 \text{ N}$$

The value of $1/\sigma$ is sometimes referred to as the *Gumbel slope*. The return period $1/(1 - p)$ represents the average value of x (e.g. number of times, or time) between recurrences of a greater (or lesser) value than represented by the return period. In Example 3.4 there is a 50–50 chance that a length of wire of breaking strength less than 45 N will occur in a batch of 26 lengths. The return period is used in forecasting the likelihood of extreme events such as floods and high winds, but it is not often referred to in reliability work.

HAZARD PLOTTING

Instead of plotting the cumulative proportion of items failed, we can plot the cumulative hazard function, using hazard papers. This technique has particular advantages when dealing with censored data, as will be shown.

The theory underlying hazard plotting is as follows. From Chapter 2, we remember that the hazard rate $h(x)$ is given by

$$h(x) = \frac{f(x)}{1 - F(x)} \tag{2.21}$$

$$F(x) = \int_0^x f(x) \, dx \tag{2.19}$$

The cumulative hazard function

$$
\begin{aligned}
H(x) &= \int_0^x h(x) \, dx \\
&= \int_0^x \frac{f(x)}{1 - F(x)} \, dx \\
&= - \ln \, [1 - F(x)]
\end{aligned}
\tag{2.22}
$$

This relationship allows the derivation of cumulative hazard plotting paper. For example, for the Weibull distribution,

$$h(x) = \frac{\beta}{\eta^\beta} x^{\beta - 1}$$

Therefore,

$$H(x) = \left(\frac{x}{\eta}\right)^\beta$$

If H is the cumulative hazard value,

$$\log x = \frac{1}{\beta} \log H + \log \eta$$

Thus, Weibull hazard paper is a log–log paper. The nominal slope of the plotted data would be $1/\beta$. When $H = 1$, $x = \eta$. Figure 3.9 is an example of Weibull hazard paper. Hazard papers for normal, lognormal, exponential and type I extreme value distributions are also available.

The hazard plotting procedure is:

1. Tabulate the times to failure in order.

Figure 3.9. TEAM Weibull hazard paper (see also Example 3.5)

2. For each failure, calculate the hazard interval

$$\Delta H_i = \frac{1}{\text{number of items remaining after previous failures/censoring}}$$

3. For each failure, calculate the cumulative hazard function

$$H = \Delta H_1 + \Delta H_2 + \cdots + \Delta H_n$$

4. Plot the cumulative hazard against life value on the chosen hazard paper.

As with Weibull probability paper, Weibull hazard paper is based upon the two-parameter Weibull distribution, and a non-zero value of failure-free life will result in a curved cumulative hazard line. A value for $\hat{\gamma}$ can be derived as described on page 75, except that equal length divisions are measured off on the abscissa instead of on the ordinate if the geometrical construction method is being used.

Hazard plotting should be used in preference to cumulative probability plotting when dealing with censored data, or when the data include multiple failure modes and we wish to analyse the overall failure distribution, as well as individual failure modes. (This is, of course, a type of censoring.) One tabulation may then be used, rather than a separate tabulation for each failure mode (see Example 3.5).

Example 3.5

Table 3.6 gives distance to failure on a vehicle shock absorber, taken from fleet records. The data are tabulated and treated as described above. F1 and F2 relate to the two failure modes encountered. Note that, compared with the method of dealing with censored data described earlier (page 68) the calculation of plotting positions is much easier. The data are shown plotted in Fig. 3.9.

In this example, for the combined failure modes, $\hat{\beta} = 2.6$, $\hat{\eta} = 29,000$ km (the life equivalent to 100 per cent. cumulative hazard). (This is the 63.2 per cent. c.d.f. on the probability scale.) For failure mode F1, $\hat{\beta} = 3.5$, $\hat{\eta} = 39,500$ km and $\hat{\gamma} = 6,800$ km.

CHOOSING THE DISTRIBUTION AND ASSESSING THE RESULTS

The types of failure pattern and modes of failure which can be described by the continuous statistical distributions covered in Chapter 2 and in this chapter have been mentioned in the relevant sections. The distribution which is likely to provide the best fit to a set of data is not always readily apparent, and it can be useful to follow guidelines which should lead to the most revealing presentation.

Table 3.6. Distance to failure on a vehicle shock absorber.

No.	Distance (km)	Δ Hazard per cent	Cumulative hazard — F1 and F2 per cent	Cumulative hazard — F1 only per cent
1 (38)	6 700 (F1)	2.63 (1/38)	2.63	2.63
2 (37)	6 950			
3 (36)	7 820			
4 (35)	8 790			
5 (34)	9 120 (F2)	2.94 (1/34)	5.57	
6 (33)	9 660			
7 (32)	9 820			
8 (31)	11 310			
9 (30)	11 690			
10 (29)	11 850			
11 (28)	11 880			
12 (27)	12 140			
13 (26)	12 200 (F1)	3.85	9.42	6.48
14 (25)	12 870			
15 (24)	13 150 (F2)	4.17	13.59	
16 (23)	13 330			
17 (22)	13 470			
18 (21)	14 040			
19 (20)	14 300 (F1)	5.00	18.59	11.48
20 (19)	17 520 (F1)	5.26	23.85	16.74
21 (18)	17 540			
22 (17)	17 890			
23 (16)	18 450			
24 (15)	18 960			
25 (14)	18 980			
26 (13)	19 410			
27 (12)	20 100 (F2)	8.33	32.18	
28 (11)	20 100			
29 (10)	20 150			
30 (9)	20 320			
31 (8)	20 900 (F2)	12.50	44.68	
32 (7)	22 700 (F1)	14.29	58.96	31.03
33 (6)	23 490			
34 (5)	26 510 (F1)	20.0	78.96	51.03
35 (4)	27 410			
36 (3)	27 490 (F1)	33.3	112.29	84.36
37 (2)	27 890			
38 (1)	28 100			

The steps to be taken are:

1. Consider the situation. What do we expect the failure pattern to be? For example, do we expect an increasing or decreasing hazard rate, or a failure-free life?
2. Plot the ranked data on the appropriate Weibull paper (probability or hazard).

3. Draw the line of best fit after estimating the failure-free life if appropriate.
4. If there is a good fit to a reasonably large sample ($\geqslant 10$ data points), estimate the parameters and determine the 5 per cent. and 95 per cent. confidence limits.
5. If the estimated value of β is near 3.5, try normal paper. If β is between 2.5 and 3.5, try lognormal paper, but only if the sample size is large (> 20).
6. If a good Weibull fit cannot be obtained, and the conditions make a fit to an extreme value type I or type II distribution likely, try extreme value paper. If extreme value is unlikely, look for multiple failure modes.

Statistical goodness-of-fit tests should be applied to test the fit to the assumed underlying distributions, when finer judgements are to be made and when there is a large database. The refinement inherent in such tests is not of much value for sample sizes less than about 20.

If a constant hazard rate ($\beta = 1$) is apparent, this can be an indication that multiple failure modes exist or that the time-to-failure data are suspect. This is often the case with systems in which different parts have different ages and individual part operating times are not available. A constant hazard rate can also indicate that failures are due to external events, such as faulty use or maintenance.

Whilst wearout failure modes are characterized by $\beta > 1$, situations can arise in which wearout failures occur after a finite failure-free time, and a β value of $\leqslant 1$ is obtained. This can happen when only a proportion of the sample is defective, the defect leading to failures after a finite time. The Weibull parameters of the wearout failure mode can be derived if it is possible to identify the defective items and analyse their life and failure data separately from that of the non-defective items.

It is very important that the physical nature of failures be investigated as part of any evaluation of failure data, if the results are to be used as a basis for engineering action, as opposed to logistic action. For example, if an assembly has several failure modes, but the failure data are being used to determine requirements for spares and repairs, then we do not need to investigate each mode and the parameters applicable to the mixed distributions are sufficient. If, however, we wish to understand the failure modes in order to make improvements, we must investigate all failures and analyse the distributions separately, since two or more failure modes may have markedly different parameters, yet generate overall failure data which are fitted by a distribution which is different to that of any of the underlying ones. Example 3.5 illustrates this sort of situation.

RELIABILITY ANALYSIS OF REPAIRABLE SYSTEMS

In earlier sections we described methods for analysing data related to the time to first failure. The distribution function of times to first failure are obviously important when we need to understand failure processes of parts which are replaced on failure, or when we are concerned with survival probability, for example, for missiles, spacecraft or underwater telephone repeaters.

However, for repairable systems, which really represent the great majority of everyday reliability experience, the distribution of times to first failures are much less important than is the *failure rate* or *rate of occurrence of failures* (ROCOF) of the system.

Any repairable system may be considered as an assembly of parts, the parts being replaced when they fail. The system can be thought of as comprising 'sockets' into which non-repairable parts are fitted. We are concerned with the pattern of successive failures of the 'sockets'. Some parts are repaired (e.g. adjusted, lubricated, tightened, etc.) to correct system failures, but we will consider first the case where the system consists only of parts that are replaced on failure (e.g. most electronic systems). Therefore, as each part fails a new part takes its place in the 'socket'. If we ignore replacement (repair) times, which are usually small in comparison with standby or operating times, and if we assume that the time to failure of any part is independent of any repair actions, then we can use the methods of event series analysis in Chapter 2 to analyse the system reliability.

Consider the data of Example 2.20. The interarrival and (chronologically ordered) arrival values between successive component failures were as shown in columns 1 and 2:

1 X_i	2 Chronological x_i	3 Ranked X_l
175	175	12
21	196	14
108	304	21
111	415	23
89	504	38
12	516	47
102	618	51
23	641	89
38	679	102
47	726	108
14	740	111
51	791	175

Example 2.20 showed that the failure rate was increasing, the interarrival values tending to become shorter. In other words, the interarrival values are not IID. If, however, we had not performed the centroid test and assumed that the data were IID, we might order the data in rank order (column 3) and plot on probability paper. These are shown plotted on Weibull paper in Fig. 3.10.

Figure 3.10. Plotted Data of Example 2.20

The plot shows an apparently exponential component life distribution. This is obviously a misleading result, since there is clearly an increasing failure rate trend for the 'socket' when the data are studied chronologically.

This example shows how important it is for failure data to be analysed correctly, depending on whether we need to understand the reliabililty of a non-repairable part or of a repairable system consisting of 'sockets' into which parts are fitted. The presence of a trend when the data are ordered *chronologically* shows that times to failure are not IID, and ordering by magnitude, which implies IID, will therefore give misleading results. *Whenever failure data* are reordered all trend information is ignored.

We can derive the system reliability over a period by plotting the cumulative times to failure in chronological order (column 2) rather than in rank order. This is shown in Fig. 3.10. It shows the progressively increasing failure rate (though the 'socket' times to failure are not Weibull-distributed).

Multisocket Systems

Now we will consider a more typical system, comprised of several parts which exhibit independent failure patterns. Each part fills a 'socket'. The failure pattern of such a system, comprising six 'sockets', is shown in Fig. 3.11.

Socket 1 generates a high, constant rate of system failures. Socket 2 generates an increasing rate of system failures as the system ages, and so on. The combined failure rate can be seen on the bottom line. The estimate of U for each part and for the system is shown. When U is negative, it denotes a 'happy' socket, with an increasing interarrival time between failures (DFR). A positive value of U indicates a 'sad' socket (IFR).

If there are no perturbations (which will be discussed below) the failure rate will tend to a constant value after most parts have been replaced at least once, regardless of the failure trends of the sockets (see page 61). This is one of the main reasons why the CFR assumption has become so widely used for systems, and why part hazard function has been confused with failure rate. However, the time by which most parts have been replaced in a system is usually very long, well beyond the expected life of most systems.

If part times to failure (in a series system, see Chapter 5) are independently and identically exponentially distributed (IID exponential) the system will have a CFR which will be the sum of the reciprocals of the part mean times to failure, i.e.

$$\lambda_s = \sum_1^n \frac{1}{x_i}$$

The assumption of IID exponential for part times to failure within their sockets in a repairable system can be very misleading. The reasons for this are (adapted from Reference 4, with permission):

1. The most important failure modes of systems are usually caused by parts which have failure probabilities which increase with time (wearout failures).
2. Failure and repair of one part may cause damage to other parts. Therefore times between successive failures *are not necessarily independent*.
3. Repairs often do not 'renew' the system. Repairs are often imperfect or they introduce other defects leading to failures of other parts.
4. Repairs might be made by adjustment, lubrication, etc., of parts which are wearing out, thus providing a new lease of life, but not 'renewal', i.e. the system is not made as good as new.
5. Replacement parts, if they have a decreasing hazard rate, can make subsequent failure initially more likely to occur.

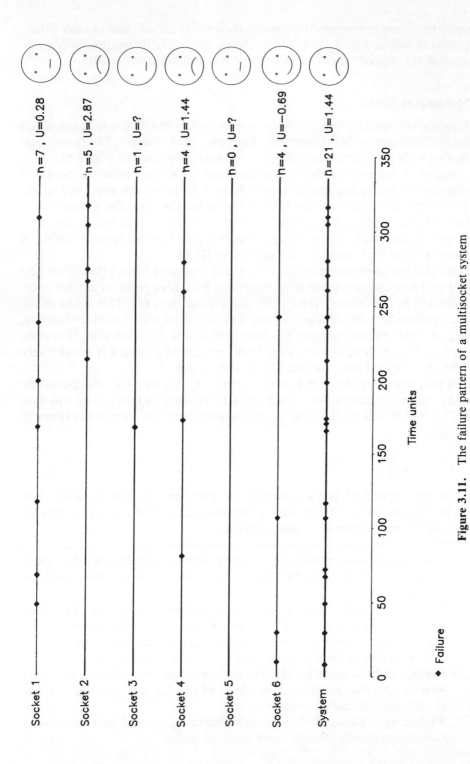

Figure 3.11. The failure pattern of a multisocket system

6. Repair personnel learn by experience, so diagnostic ability (i.e. the probability that the repair action is correct) improves with time. Generally, changes of personnel can lead to reduced diagnostic ability and therefore more reported failures.
7. Not all part failures will cause system failures.
8. Factors such as on–off cycling, different modes of use, different system operating environments or different maintenance practices are often more important than operating times in generating failure-inducing stress.
9. Reported failures are nearly always subject to human bias and emotion. What an operator or maintainer will tolerate in one situation might be reported as a failure in another, and perception of failure is conditioned by past experience, whether repair is covered by warranty, etc. Wholly objective failure data recording is very rare.
10. Failure probability is affected by scheduled maintenance or overhaul. Systems which are overhauled often display higher failure rates shortly after overhaul, due to disturbance of parts which would otherwise not have failed. If there is a post-overhaul test period before the system is returned to service, many of these failures might be repaired then. The failure data might or might not include these failures.
11. Replacement parts are not necessarily drawn from the same population as the original parts — they may be better or worse.
12. System failures might be caused by parts which individually operate within specification (i.e. do not fail) but whose combined tolerances cause the system to fail.
13. Many reported failures are not caused by part failures at all, but by events such as intermittent connections, improper use, maintainers using opportunities to replace 'suspect' parts, etc.
14. Within a system not all parts operate to the overall system cycle.
    Any practical person could add to this list from his or her own experience. The factors listed above often predominate in systems to be modelled and in collected reliability data. Large data-collection systems, in which failure reports might be coded and analysed remotely from the work locations, are usually most at fault in perpetrating the analytical errors described. Such data systems might generate 'MTBFs' for systems and for parts by merely counting total reported failures and dividing into total operating time. For example, MTBFs in flying hours are quoted for aircraft electronic equipment, when the equipment only operates for part of the flight, or MTBFs in hours are quoted for valves, ignoring whether they are normally closed, normally open or how often they are opened and closed. These data are often used for reliability predictions for new systems (see Chapter 5), thus adding insult to injury.

A CFR is often a practicable and measurable first-order assumption, particularly when data are not sufficient to allow more detailed analysis.

The effect of successive repairs on the reliability of an ageing system are shown vividly in the next example (from Ref. 4 in the Bibliography).

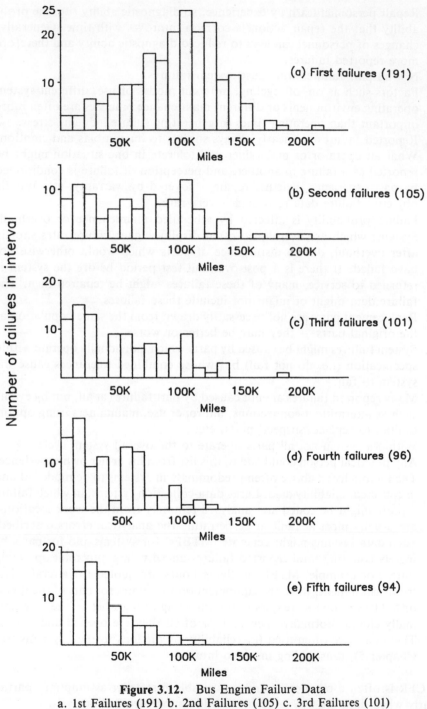

Figure 3.12. Bus Engine Failure Data
a. 1st Failures (191) b. 2nd Failures (105) c. 3rd Failures (101)
d. 4th Failures (96) e. 5th Failures (94)

Example 3.6 (Reprinted from Ref. 4 by courtesy of Marcel Dekker, Inc.)

Data on the miles between major failures (interarrival values) of bus engines are shown plotted in Fig. 3.12. These show the miles between first, second, ..., fifth major failure. Note that the interarrival mileages to the first failures (X_i) are nearly s-normally distributed. Successive interarrival times (second, third, fourth, fifth failures) show a tendency to being exponentially distributed. Nevertheless, the results show clearly that the reliability decreases with successive repairs, since the mean of the interarrival distances is progressively reduced:

Failure No.	$\bar{X}_i$ miles
1	94,000
2	70,000
3	54,000
4	41,000
5	33,000

The importance of this result lies in the evidence that:

1. Repair does not return the engines to an 'as new' condition.
2. Successive X_i's are not IID exponential.
3. The failure rate tends to a constant value only after nearly all engines have been repaired several times. Even after five repairs the steady state has not been reached.
4. Despite the *appearance* of 'exponentiality' after several failures, replacement or more effective overhaul appears to be necessary.

PROBABILITY PLOTTING FOR BINOMIAL DATA

Items such as alarm systems, standby power supply systems and missile components can often be most conveniently considered as either functioning or not when needed, and the reliability data will then be binomially distributed. In these cases we are not concerned with time (or cycles, etc.) to failure, but only with the probability of failure at the instant considered. Examples in reliability using the binomial distribution were covered in Chapter 2.

Binomial probability plotting paper can be used for quickly comparing sets of data or for evaluating data against a standard. Figure 3.13 shows an example of binomial probability paper. It is constructed by having square root scales for the ordinate and the abscissa, so that points lying on a straight line through the origin have the same standard deviation.

To represent a binomial proportion of the paper, the *binomial split* is calculated and plotted, and this is connected to the origin. The binomial split

Figure 3.13. TEAM binomial plotting paper (see also Example 3.6)

is $p/(n-p)$, [or $c/(n-c)$ in the notation used on the TEAM paper], where $p(c)$ is the quantity with one attribute (e.g. defectives) and n is the sample size. For example, a proportion of 10 per cent. would be plotted at 10 on the c scale and 90 on the n scale (not 100). This line then represents all samples with this proportion. s-Confidence limits can be drawn parallel to the datum line through the appropriate points on the s-confidence scale.

Binomial papers with different scale ranges are available. Since the scales are square root, they may both be multiplied by multiples of 100 to provide factors of 10 in the variables.

Example 3.7

An item is required to be 90 per cent reliable. Four samples are tested and give results as shown in Table 3.7. Comment on the results.

Table 3.7. Results of tests on four samples.

Trail	Sample size, n	Failures, c	$n-c$
1	32	5	27
2	20	2	18
3	20	5	15
4	55	10	45

Since not more that 10 per cent. must be defective, the datum point is derived from the binomial split $10/(100-10) = 10/90$. The results are then plotted as shown in Fig. 3.13.

We can now use the s-confidence limit scale to interpret these results, by drawing lines parallel to the requirement line through the s-confidence value required. In this case we can see that trial 2 gives the required proportion. However, the result of trial 1 indicates with higher than 90 per cent. s-confidence that it is not compliant with the requirement and trials 3 and 4 are non-compliant at the 97.5 per cent. confidence level.

Binomial probability paper with an extended n scale can be used as a convenient method of monitoring test data for items with a low proportion failing, such as electronic components, as shown in Example 3.8.

Example 3.8

Two types of custom hybrid integrated circuit are used in an instrument. It is required that not more than 0.02 per cent. of either type fail during production testing. Test data are shown in Table 3.8. 1 is added to the cumulative number of failures, because when plotting binomial data with a small value of p, this gives a better estimate of the population proportion $\hat{p}$. The failure data are shown plotted in Fig. 3.14. The plot shows that, at the 97.5 per cent. confidence level, the type 1 circuit is outside the specification. The trend shows a statistically significant problem in lots 16 to 18. The type 2 circuit also has a higher failure ratio than specified, at the 90 per cent. confidence level.

Binomial probability paper as produced by Chartwell is shown in Fig. 3.15. Sample data are plotted in the same way as on the TEAM paper. The circular scale gives the percentage defective, as $(1 - \sin^2\alpha)$, where α is the value on the degrees scale. This value is read on the vertical scale as shown. For example, Fig. 3.15 shows a plotted binomial split of $10:150$. This intersects at $75.7°$, giving the percentage defective as $(1 - \sin^2 75.7°) = 6.1\%$. The arcsine

Table 3.8. Test data on two types of custom hybrid integrated circuit (see Example 3.8).

Lot no.	No. tested	Cum no. tested	No. failed type 1, c_1	Cum. $c_1 + 1$	No. failed type 2, c_2	Cum. $c_2 + 1$
1	1,184	1,184	1	2	0	1
2	1,152	2,336	0	2	1	2
3	1,328	3,664	1	3	0	2
4	1,287	4,951	0	3	0	2
5	1,463	6,414	0	3	1	3
6	1,510	7,924	1	4	0	3
7	1,291	9,215	1	5	2	5
8	1,384	10,599	0	5	0	5
9	1,570	12,169	0	5	0	5
10	1,648	13,817	0	5	0	5
11	1,586	15,403	2	7	1	6
12	1,353	16,756	1	8	0	6
13	1,422	18,178	0	8	0	6
14	1,338	19,516	0	8	2	8
15	1,546	21,062	1	9	0	8
16	1,284	22,346	2	11	0	8
17	1,172	23,518	2	13	0	8
18	1,390	24,908	3	16	0	8
19	1,521	26,429	0	16	1	9
20	1,441	27,870	0	16	0	9

transform is used to determine s-confidence limits, in conjunction with the standard error scale at the top of the chart. For example, the approximate 2.5 per cent. (single-sided) s-confidence limits are derived by constructing a circle about the binomial split point with a radius of two standard errors. The arcsine transform is then used in the same way to determine the s-confidence limits, in this case 10 per cent. and 3 per cent. s-Confidence limit lines can be drawn as on the TEAM paper by constructing lines parallel to the process line, tangential to the circle.

Bradley binomial paper

Bradley binomial plotting paper allows binomial data to be analysed and s-confidence limits to be derived graphically. Figure 3.16 shows TEAM Bradley binomial plotting paper. The data are plotted by drawing a straight line from the sample size scale to the origin, and the value of the observed probability is marked on this line. Figure 3.16 shows the plotted data for a sample of 55, of which 10 are defective, i.e. 0.18. The s-confidence limits for the population are derived by drawing a horizontal line to intersect the appropriate s-confidence level (or standard deviation) scale. In this case, the 90 per cent. s-confidence limits for the defective proportion of the population would be 0.18 ± 0.086.

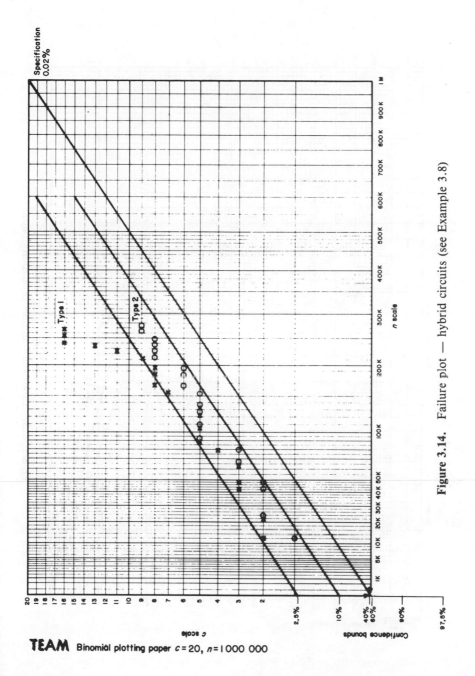

Figure 3.14. Failure plot — hybrid circuits (see Example 3.8)

98

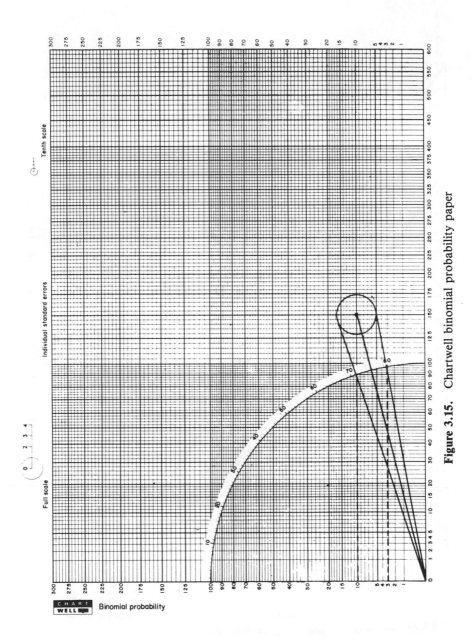

Figure 3.15. Chartwell binomial probability paper

Figure 3.16. TEAM Bradley binomial paper

Bradley binomial paper allows a correction to be made to take account of finite population (lot) sizes, using the sample-to-lot ratio scale at the top of the graph. If in this case the lot size is 165, $n/N = \frac{1}{3} = 0.33$. The correction factor for the s-confidence limits is derived by connecting this point on the n/N scale to the origin and reading the value on the correction factor scale (0.82). The corrected 90 per cent. s-confidence limits for the proportion defective for this lot size are therefore $0.18 \pm (0.086 \times 0.82)$. The np line on the Bradley paper indicates the boundary for points to the left of which the accuracy of s-confidence limit estimation is reduced.

If necessary, the appropriate Bradley plotting paper should be selected in relation to the sample size and the observed probability, to ensure that the plotted data fall to the right of the np line.

COMPUTER PROGRAMS FOR RELIABILITY ANALYSIS

Computer programs are readily available for analysing failure data. Programs providing goodness-of-fit tests and parameter estimation can be written quite easily, and preprogrammed statistical calculators such as the Commodore S61, or programmable calculators such as the Hewlett Packard HP 41 C or Texas Instruments TI 58 or TI 59 can be very useful in this work. Computer programs are available which perform distribution fitting, including data handling (listing, sorting, ranking, censoring) and parameter estimation, and which can provide graphical outputs. Such programs allow analyses to be performed very quickly and accurately. They can be integrated with programs for failure data listing, so that specified data categories can be analysed when required, without the data having to be called out and then reentered into the analysis program. These programs allow the use of extremely powerful techniques which would otherwise be beyond the scope of most engineering departments. However, the ready accessibility of such programs can allow the unaware to make naive and inappropriate analyses. Therefore their use should be accompanied by a good understanding of the underlying techniques, and understandable graphical outputs should always be produced.

BIBLIOGRAPHY

1. J. R. King, *Probability Charts for Decision Making*. Industrial Press, New York (1971).
2. Notes on plotting techniques: TEAM, *Easy Analysis Methods*. Technical and Engineering Aids to Management, Tamworth, New Hampshire.
3. K. A. Dey, *Practical Statistical Analysis for the Reliability Engineer*. Reliability Analysis Center Report SOAR 2 (Rome Air Development Center). Available from National Technical Information Service (NTIS), Springfield, Virginia.
4. H Ascher and H. Feingold, *Repairable Systems Reliability; Inference and Modelling Misconceptions and their Causes*. Marcel Dekker (1984).
5. W Nelson *Applied Life Data Analysis*. John Wiley (1982)

Chapter 4

LOAD–STRENGTH INTERFERENCE

INTRODUCTION

In Chapter 1 we set out the premise that a common cause of failure results from the situation when the applied load exceeds the strength. Load and strength are considered in the widest sense. 'Load' might refer to a mechanical stress, a voltage or internally generated stresses such as temperature. 'Strength' might refer to any resisting physical property, such as hardness, strength, melting point or adhesion. Examples are:

1. A bearing fails when the internally generated loads (due perhaps to roughness, loss of lubricity, etc.) exceed the local strength, causing fracture, overheating or seizure.
2. A transistor gate in an integrated circuit fails when the voltage applied causes a local current density, and hence temperature rise, above the melting point of the conductor or semiconductor material.
3. A hydraulic valve fails when the seal cannot withstand the applied pressure without leaking excessively.
4. A shaft fractures when torque exceeds strength.

Therefore, if we design so that strength exceeds load, we should not have failures. This is the normal approach to design, in which the designer considers the likely extreme values of load and strength, and ensures that an adequate safety factor is provided.

Additional factors of safety may be applied, e.g. as defined in pressure vessel design codes or electronic component derating rules. This approach is usually effective. Nevertheless, some failures do occur which can be represented by the load–strength model. By our definition, either the load was then too high or the strength too low. Since load and strength were considered in the design, what went wrong?

DISTRIBUTED LOAD AND STRENGTH

For most products neither load nor strength are fixed, but are distributed statistically. This is shown in Fig. 4.1(a). Each distribution has a mean value, denoted by $\bar{L}$ or $\bar{S}$, and a standard deviation, denoted by σ_L or σ_s. If an event

102

Figure 4.1 Distributed load and strength: (a) non-overlapping distributions, (b) overlapping distributions

occurs in which the two distributions overlap, i.e. an item at the extreme weak end of the strength distribution is subjected to a load at the extreme high end of the load distribution, such that the 'tails' of the distributions overlap, failure will occur. This situation is shown in Fig. 4.1(b).

For distributed load and strength, we define two factors, the *safety margin* (SM),

$$SM = \frac{\bar{S} - \bar{L}}{(\sigma_S^2 + \sigma_L^2)^{1/2}}$$ (4.1)

and the *loading roughness* (LR),

$$LR = \frac{\sigma_L}{(\sigma_S^2 + \sigma_L^2)^{1/2}}$$ (4.2)

The safety margin is the relative separation of the mean values of load and strength and the loading roughness is the standard deviation of the load; both are relative to the combined standard deviation of the load and strength distributions.

The safety margin and loading roughness allow us, in theory, to analyse the way in which load and strength distributions interfere, and so generate a probability of failure. By contrast, a traditional deterministic safety factor, based upon mean or maximum/minimum values, does not allow a reliability estimate to be made. On the other hand, good data on load and strength properties are very often not available. Other practical difficulties arise in applying the theory, and engineers must always be alert to the fact that people, materials and the environment will not necessarily be constrained to the statistical models being used. The rest of this chapter will describe the theoretical basis of load-strength interference analysis. The theory must be applied with care and with full awareness of the practical limitations. These are discussed later.

Some examples of different safety margin/loading roughness situations are shown in Fig. 4.2. Figure 4.2(a) shows a highly reliable situation: narrow distributions of load and stress, low loading roughness and a large safety margin. If we can control the spread of strength and load, and provide such a high safety margin, the design should be intrinsically failure-free. (Note that we are considering situations where the mean strength remains constant, i.e. there is no strength degradation with time. We will cover strength degradation later.) This is the concept applied in most designs, particularly of critical com-

Figure 4.2 Effect of safety margin and loading roughness. Load L', causes failure of a proportion of items indicated by the shaded area

ponents such as civil engineering structures and pressure vessels. We apply a safety margin which experience shows to be adequate; we control quality, dimensions, etc., to limit the strength variations, and the load variation is either naturally or artificially constrained.

Figure 4.2(b) shows a situation where loading roughness is low, but due to a large standard deviation of the strength distribution the safety margin is low. Extreme load events will cause failure of weak items. However, only a small proportion of items will fail when subjected to extreme loads. This is typical of a situation where quality control methods cannot conveniently reduce the standard deviation of the strength distribution (e.g. in electronic device manufacture, where 100 per cent. visual and mechanical inspection is not always feasible). In this case deliberate overstress can be applied to cause weak items to fail, thus leaving a population with a strength distribution which is truncated to the left (Fig. 4.3). The overlap is thus eliminated and the reliability of the surviving population is increased. This is the justification for high stress burn-in of electronic devices, proof-testing of pressure vessels, etc. Note that the overstress test only destroys weak items. It must not cause

Figure 4.3 Truncation of strength distribution
by screening

weakening (strength degradation) of good items. A decreasing failure rate (DFR) is characteristic of this situation, since as weak items fail ('infant mortality'), the population strength increases and the failure rate decreases.

Figure 4.2(c) shows a low safety margin and high loading roughness due to a wide spread of the load distribution. This is a difficult situation from the reliability point of view, since an extreme stress event will cause a large proportion of the population to fail. Therefore, it is not economical to improve population reliability by screening out items likely to fail at these stresses. The options left are to increase the safety margin by increasing the mean stength, which might be expensive, or to devise means to curtail the load distribution. This is achieved in practice by devices such as current limiters in electronic circuits or pressure relief valves and dampers in pneumatic and hydraulic systems.

ANALYSIS OF LOAD–STRENGTH INTERFERENCE

The reliability of a part, for a discrete load application, is the probability that the strength exceeds the load:

$$R = P(S > L)$$

$$= \int_0^\infty f_L(L) \left[\int_L^\infty f_S(S) \, dS \right] dL$$

$$= \int_0^\infty f_S(S) \left[\int_0^S f_L(L) \, dL \right] dS$$

where $f_S(S)$ is the p.d.f. of strength and $f_L(L)$ is the p.d.f. of load.

Also, if we define y as $S - L$, y is a random variable such that

$$R = P(y > 0)$$

$$= \int_0^\infty \int_0^\infty f_S(y + L) \, f(L) \, dL \, dy \tag{4.3}$$

Normally distributed strength and load

If we consider normally distributed strength and load, so that the c.d.f.s are

$$F_L(L) = \Phi\left(\frac{L - \bar{L}}{\sigma_L}\right)$$

$$F_S(S) = \Phi\left(\frac{S - \bar{S}}{\sigma_S}\right)$$

if $y = S - L$, then $\bar{y} = \bar{S} - \bar{L}$ and $\sigma_y = (\sigma_S^2 + \sigma_L^2)^{1/2}$. So

$$R = P(y > 0)$$

$$= \Phi\left(\frac{\bar{S} - \bar{L}}{\sigma_y}\right) \tag{4.4}$$

Therefore, the reliability can be determined by finding the value of the standard cumulative normal variate from the normal distribution tables. The reliability can be expressed as

$$R = \Phi\left[\frac{\bar{S} - \bar{L}}{(\sigma_S^2 + \sigma_L^2)^{1/2}}\right]$$

$$= \Phi(SM) \text{ (from Eqn 4.1)} \tag{4.5}$$

Example 4.1

A component has a strength which is normally distributed, with a mean value of 5,000 N and a standard deviation of 400 N. The load it has to withstand is also normally distributed, with a mean value of 3,500 N and a standard deviation of 400 N. What is the reliability per load application?

The safety margin is

$$\frac{5,000 - 3,500}{(400^2 + 400^2)^{1/2}} = 2.65$$

From Appendix 1,

$$\Phi(2.65) = 0.996$$

Other distributions of load and strength

The integrals for other distributions of load and strength can be derived in a similar way. For example, we may need to evaluate the reliability of an item whose strength is Weibull-distributed, when subjected to loads that are extreme-value-distributed. These integrals are somewhat complex and the reader is referred to Refs 1 and 2 in Bibliography for details and for tables of the evaluated integrals.

106

EFFECT OF SAFETY MARGIN AND LOADING ROUGHNESS ON RELIABILITY (MULTIPLE LOAD APPLICATIONS)

For multiple load applications:

$$R = \int_0^\infty f_S(S) \left[\int_0^s f_L(L) \, dL \right]^n dS$$

where n is the number of load applications.

Reliability now becomes a function of safety margin and loading roughness, and not just of safety margin. This complex integral cannot be reduced to a formula as Eqn (4.5), but can be evaluated using computerized numerical methods.

Figure 4.4 shows the effects of different values of safety margin and loading roughness on the failure probability per load application for large values of n, when both load and strength are normally distributed. The dotted line shows the single load application case (from Eqn 4.5). Note that the single load case is less reliable per load application than is the multiple load case.

Since the load applications are s-independent, reliability over n load applications is given by

$$R = (1 - p)^n \qquad \text{(from Eqn 2.2)}$$

where p is the probability of failure per load application.

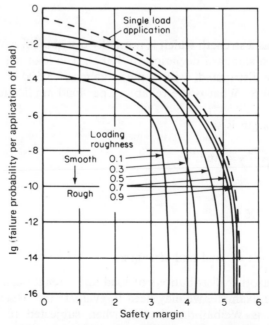

Figure 4.4 Failure probability–safety margin curves when both load and strength are normally distributed (for large n and $n = 1$)

Figure 4.5 Characteristic regions of a typical
failure probability—safety margin curve

For small values of p the binomial approximation allows us to simplify this to

$$R \approx 1 - np \tag{4.6}$$

The reliability for multiple load applications can then be derived, if we know the number of applications, having used Fig. 4.4 to derive the value of p. Once the safety margin exceeds a value of 3 to 5, depending upon the value of loading roughness, the failure probability becomes infinitesimal. The item can then be said to be *intrinsically reliable*. There is an intermediate region in which failure probability is very sensitive to changes in loading roughness or safety margin, whilst at low safety margins the failure probability is high. Figure 4.5 shows these characteristic regions. Similar curves can be derived for other distributions of load and strength. Figures 4.6 and 4.7 shows the failure probability—safety margin curves for smooth and rough loading situations for Weibull-distributed load and strength. These show that if the distributions are skewed so that there is considerable interference, high safety margins are necessary for high reliability. For example, Fig. 4.6 shows that, even for a low loading roughness of 0.3, a safety margin of at least 5.5 is required to ensure intrinsic reliability, when we have a right-skewed load distribution and a left-skewed strength distribution. If the loading roughness is high (Fig. 4.7), the safety margin required is 8. These curves illustrate the sensitivity of reliability to safety margin, loading roughness and the load and strength distributions.

Approximate equivalent safety margin and loading roughness values for Weibull-distributed load and strength situations can be derived by using the following approximations for the standard deviation and the mean of the 'equivalent' normal distribution (i.e. a normal distribution, the cumulative

108

Figure 4.6 Failure probability–safety margin curves for asymmetric distributions (loading roughness = 0.3)

Figure 4.7 Failure probability–safety margin curves for asymmetric distributions (loading roughness = 0.9)

values of which coincide with the cumulative asymmetric Weibull distribution at two points in the tails: 0.9 and 0.99 in the load distribution and 0.01 and 0.10 in the strength distribution), related to the characteristic life of the Weibull distributions. The approximations are

$$\sigma' = 0.3\,\eta \qquad\qquad (4.7)$$
$$\mu' = 0.9\,\eta \qquad\qquad (4.8)$$

(See No. 3 in the Bibliography).

When there is a finite minimum strength γ_S, then the mean strength

$$\bar{S} = \gamma_S + \mu'$$

Two examples of this approach to design are given to illustrate the application to electronic and mechanical engineering.

Example 4.2 (electronic)

A design of a power amplifier uses a single transistor in the output. It is required to provide an intrinsically reliable design, but in order to reduce the number of component types in the system the choice of transistor types is limited. The amplifier must operate reliably at 50 °C.

An analysis of the load demand on the amplifier gives the results in Fig. 4.8. The mean ranking of the load test data is given in Table 4.1. A type 2N2904 transistor is selected. For this device the maximum rated power dissipation is 0.6 W at 25 °C.

The load test data are shown plotted on normal probability paper (Fig. 4.9). A very good fit to the normal distribution is apparent, within the range of the

Figure 4.8 Load data (sampled at 10 s intervals)

Table 4.1 Mean ranking of load test data.

Power (W)	Cumulative percentage time (c.d.f.)
0.1	5
0.2	25
0.3	80
0.4	98.5
0.5	99.95

data. The parameters for the load distribution are

$$\bar{L} = 0.25 \text{ W}$$

$$\sigma_L = 0.065 \text{ W}$$

However, in this case we must consider the combined effects of power dissipation and elevated temperature. The temperature derating guidelines for the 2N2904 transistor advise 3.43 mW K^{-1} linear derating. Since we require the amplifier to operate at $50°$C, the equivalent combined load distribution is s-normal, with the same SD, but with a mean which is $(25 \times 3.43) \text{ mW} = 0.086$ W higher (line 1a in Fig. 4.9). The mean load $\bar{L}$ is now 0.33 W, with an unchanged standard deviation of 0.065 W.

To derive the strength distribution, 100 transistors were tested at 25 °C ambient, for 10 s at each power level (step stress), giving failure data as shown in Table 4.2. These data plot as a straight line (line 2 in Fig. 4.9), indicating a normal distribution with mean power at failure (strength) of 0.97 W and SD of 0.14 W. The data are plotted against the reciprocal (top) probability scale, as we are concerned with the tail of the distribution which intersects the load distribution.

Combining the load–strength data gives

$$\text{LR} = \frac{\sigma_L}{(\sigma_S^2 + \sigma_L^2)^{1/2}} = 0.42$$

$$\text{SM} = \frac{\bar{S} - \bar{L}}{(\sigma_S^2 + \sigma_L^2)^{1/2}} = 4.15$$

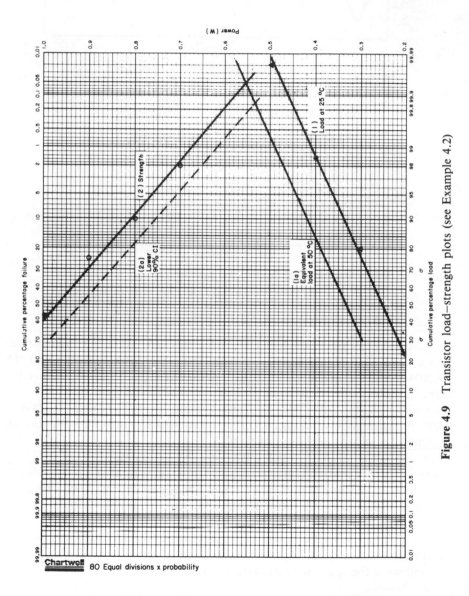

Figure 4.9 Transistor load–strength plots (see Example 4.2)

Table 4.2 Failure data for 100 transistors.

Power (W)	Number failed	Cumulative percentage failure (c.d.f.) (mean ranking)
0.1	0	0
0.2	0	0
0.3	0	0
0.4	0	0
0.5	0	0
0.6	0	0
0.7	2	2
0.8	8	10
0.9	17	25
1.0	35	59
1.2	30	89
1.3	10	99

Therefore

$$R = \Phi(SM)$$
$$= 0.999\ 983$$

This is the reliability per application of load for a single load application. Figure 4.4 shows that for multiple load applications (large n), the failure probability per load application (p) is about 10^{-11} (zone 2 of Fig. 4.5). Over 10^6 load applications the reliability would be about 0.999 99 (from Eqn 4.6).

On the data given, therefore, this would be an unreliable design. (The analysis has been performed on the basis of best estimates. It could have been based upon a lower s-confidence limit for the strength distribution — line 2a in Fig. 4.9.)

In practice, in a case of this type the temperature and load derating guidelines described in Chapter 7 would normally be used. In fact, to use a transistor at very nearly its maximum temperature and load rating (in this case the measured highest load is 0.5 W at 25 °C, equivalent to nearly 0.6 W at 50 °C) is not good design practice, and derating factors of 0.5 to 0.8 are typical for transistor applications. The example illustrates the importance of adequate derating for a typical electronic component. The approach to this problem can also be criticized on the grounds that:

1. The failure (strength) data are sparse at the 'weak' end of the distribution. It is likely that batch-to-batch differences would be more important than the test data shown, and screening could be applied to eliminate weak devices from the population.
2. The extrapolation of the load distribution beyond the 0.5 W recorded peak level is dangerous, and this extrapolation would need to be tempered by engineering judgement and knowledge of the application.

Example 4.3 (mechanical)

A connecting rod must transmit a tension load which trials show to be Weibull-distributed, with the following parameters: $\beta_L = 2.0$, characteristic load, $\eta_L = 4,000$ N, minimum load, $\gamma_L = 0$. Tests on the material to be used show a Weibull strength distribution as follows: $\beta_S = 5.0$, characteristic strength, $\eta_S = 6.8 \times 10^8$ N m^{-2}, minimum strength, $\gamma_S = 1.9 \times 10^8$ N m^{-2}. A large number of components are to be manufactured, so it is not feasible to test each one. What cross-section would ensure intrinsic reliability, assuming that manufacturing quality control will be able to ensure that the material strength is maintained and that the connecting rods are required to operate only for a period well within the fatigue life?

The loading roughness can be derived by using Eqns (4.7) and (4.8). That is:

$$\sigma' = 0.3\eta$$

$$\mu' = 0.9\eta$$

Mean strength $= \gamma_S + \mu_S$

Therefore,

$$\sigma_S = 0.3 \times 6.8 \times 10^8 = 2.04 \times 10^8 \ \text{N m}^{-2}$$

$$\mu_S = 0.9 \times 6.8 \times 10^8 = 6.12 \times 10^8 \ \text{N m}^{-2}$$

Mean material strength $= (1.9 + 6.12) \times 10^8 = 8.02 \times 10^8 \ \text{N m}^{-2}$

$$\sigma_L = 0.3 \times 4,000 = 1,200 \ \text{N}$$

$$\mu_L = \bar{L} = 0.9 \times 4,000 = 3,600 \ \text{N}$$

The mean value of the strength of the rod

$$\bar{S} = \frac{\pi d^2}{4} \times \text{mean material strength}$$

where d is the diameter. The standard deviation of the strength distribution is

$$\frac{\pi d^2}{4} \sigma_S$$

Therefore, the loading roughness (LR) is given by

$$\frac{\sigma_L}{(\sigma_S^2 + \sigma_L^2)^{1/2}} = \frac{1200}{\{[2.04 \times (\pi d^2/4) \times 10^8]^2 + 1,200^2\}^{1/2}}$$

and the safety margin (SM) by

$$\frac{\bar{S} - \bar{L}}{(\sigma_S^2 + \sigma_L^2)^{1/2}} = \frac{[8.02 \times (\pi d^2/4) \times 10^8] - 3,600}{\{[2.04 \times (\pi d^2/4) \times 10^8]^2 + 1,200^2\}^{1/2}}$$

Figures 4.6 and 4.7 show that for $\beta_L = 2.0$ and $\beta_S = 5.0$, we need a safety margin of at least 5, depending upon the loading roughness, for intrinsic

Table 4.3 Results of substituting values of d in the equations for SM and LR.

$d(\times 10^{-3}\text{m})$	5	6	7	8
SM	2.90	3.24	3.43	3.56
LR	0.29	0.20	0.15	0.12

reliability. We can derive a suitable strength value, and hence the correct diameter, by iteration.

First, select a mean strength, $\bar{S} = 6 \times \bar{L}$, say. Then

$$\bar{S} = 6 \times 3{,}600 = 21{,}600 \text{ N}$$

Therefore,

$$21{,}600 = \frac{\pi d^2}{4} \times 8.02 \times 10^8$$

Thus, $d = 5.35 \times 10^{-3}$ m

Substituting values for d in the equations for SM and LR, we can produce Table 4.3.

Since these results are derived by considering the equivalent normal distributions of load and strength, we can use Fig. 4.4 to ascertain the appropriate safety margin/loading roughness combination for intrinsic reliability. An LR of 0.3 requires an SM $>$ 4.5 for intrinsic reliability, and therefore the 5 mm diameter would be too small. On the other hand, the 7 mm diameter gives an SM of 3.43 at an LR of 0.15, which is just about within the intrinsically reliable zone. Therefore a diameter of at least 7 mm should be chosen.

The load and strength distributions are shown plotted in Fig. 4.10, with the strength calculated for a 7 mm diameter. The plot shows the margin available between the cumulative 99.9 per cent. load probability (10,000 N) and the 0.1 per cent. cumulative strength probability (15,000 N). Also, the cumulative probability of the load exceeding the minimum strength of 7,300 N is about 96.5 per cent. which is the reliability per load application for minimum strength items.

PRACTICAL ASPECTS

The examples illustrate some of the limitations of the statistical engineering approach to design. The main difficulty is that, in attempting to take account of variability, we are introducing assumptions that might not be tenable, e.g. by extrapolating the load and strength data to the very low probability tails of the assumed population distributions. We must therefore use engineering knowledge to support the analysis, and use the statistical approach to cater for engineering uncertainty, or when we have good statistical data. For example, in many mechanical engineering applications good data exist or can be obtained on load distributions, such as wind loads on structures, gust loads on

114

Figure 4.10 Connecting rod load–strength plots

aircraft or the loads on automotive suspension components. We will call such loading situations 'predictable'.

On the other hand, some loading situations are much more uncertain, particularly when they can vary markedly between applications. Electronic circuits subject to transient overload due to the use of faulty procedures or because of the failure of a protective system, or a motor bearing used in a hand power drill, represent cases in which the high extremes of the load distribution can be very uncertain. The distribution may be multimodal, with high loads showing peaks, for instance when there is resonance. We will call this loading situation 'unpredictable'.

Obviously it will not always be easy to make a definite classification; for example, we can make an unpredictable load distribution predictable if we can collect a sufficient number of data. The methods described above are meaningful if applied in predictable loading situations. (Strength distributions are more often predictable, unless there is progressive strength reduction, which we will cover later.) However, if the loading is very unpredictable the probability estimates will be very uncertain. When loading is unpredictable we must revert to traditional methods. This does not mean that we cannot achieve high reliability in this way. However, evolving a reliable design is likely to be more expensive, since it is necessary either to deliberately overdesign or to improve the design in the light of experience. The traditional safety factors derived as a result of this experience ensure that a new design will be reliable, provided that the new application does not represent too far an extrapolation.

Instead of considering the distributions of load and strength, we can use discrete maximum/minimum values in appropriate cases. For example, we can use a simple lowest strength value if this can be assured by quality control. In the case of Example 4.3, we could have decided that in practice the strength would not be less than 20,000 N, which would have given a safety factor of 2 above the 99.9 per cent load c.d.f. In other cases we might also assume that for practical purposes the load is curtailed, as in situations where the load is applied by a system with an upper limit of power, such as a hydraulic ram or a human operator. If the load and strength distributions are both curtailed, the traditional safety factor approach is adequate, provided that other constraints such as weight or cost do not make a higher risk design necessary.

The statistical engineering approach can lead to overdesign if it is applied without regard to real curtailment of the distributions. Conversely, traditional deterministic safety factor approaches can result in overdesign when weight or cost reduction must take priority.

In many cases, other design requirements (such as for stiffness) provide intrinsic reliability. The techniques described above should therefore be used when it is necessary to assess the risk of failure in marginal or critical applications.

So far we have taken no account of the possibility of strength reduction with time or with cyclic loading. Of course, fatigue is an important factor in the reliability of many designs in which load and strength must be balanced. The methods described above are only relevant when we can ignore strength reduction, for instance if the item is to be operated well within the safe fatigue life or if no weakening is expected to occur.

TIME-DEPENDENT LOAD–STRENGTH INTERFERENCE

An item may be designed to take account of strength reduction due to factors such as cumulative fatigue damage or corrosion. Additional safety factors are then applied, either to ensure that the cyclic loading does not cause cumulative damage or to provide a suitable safe life. Since fatigue is the most common cause of strength reduction in cyclically stressed components, it will be used as the basis for the rest of this section.

Figure 4.11 is a typical $S–N$ curve, showing cycles to failure at different stress levels. (In keeping with the notation used in this book, we will use the term load in place of stress.) The distributions in Fig. 4.11 indicate sample-to-sample variation, while the curve shows the median values. Below the critical cyclic load L' the life is indefinite, but higher load levels induce cumulative damage, leading ultimately to failure. L is the load which will result in failure in one application. Lower cyclic loads cause cumulative fatigue damage, so that the item is progressively weakened, and the curve indicates the cycles to failure at any cyclic load value between L and L'. The curve indicates the mean value of cyclic load for a given number of cycles to failure (or vice versa). The population would in fact be distributed, as shown. The basic $S–N$ curve shows the simplest situation, in which a uniform cyclic load is applied. In the more general practical case, with randomly distributed loads as shown in Fig. 4.12, the population distribution of cycles to failure will have an additional variance, and we will not know how much damage has been inflicted.

The fatigue life and therefore the reliability can be difficult to predict under these circumstances. Instead of analysing a load–strength situation in which the strength distribution remains fixed, we are now faced with a strength distribution which varies with time, and with the previous load history.

Figure 4.13 shows an $S–N$ diagram, drawn on log–log paper, with the applied load distribution also shown. The load distribution tail extends beyond

Figure 4.11 $S–N$ curve

Figure 4.12 Random overload

L', thus generating fatigue damage, and the mean of the strength distribution is therefore reduced. The strength distribution variance increases as items incur different amounts of fatigue damage. At N', the tails of the load and strength distributions interfere, and we enter the increasing failure rate period.

Design for reliability under potential fatigue conditions therefore means either ensuring that the distributed load does not exceed the critical load or designing for a limited 'safe life', beyond which the item is not likely to be used or will be replaced in accordance with a maintenance schedule. If we can ensure these conditions, then failure should not occur. However, we may need to predict the reliability of an item subject to fatigue after a stated period, in order, for example, to specify a safe life.

The fatigue life of an item can be estimated using *Miner's rule*. This is expressed as

$$\frac{n_1}{N_1} + \frac{n_2}{N_2} + \frac{n_3}{N_3} + \cdots + \frac{n_k}{N_k} = 1$$

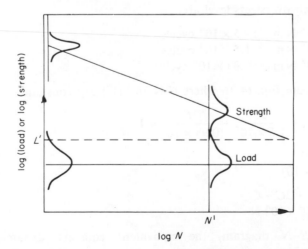

Figure 4.13 Strength deterioration with cyclic overload

that is,

$$\sum_{i=1}^{k} \frac{n_i}{N_i} = 1 \qquad (4.9)$$

where n_i is the number of cycles at a specific load level, above the fatigue limit, and N_i is the median number of cycles to failure at that level, as shown on the $S-N$ curve.

The fatigue life of an item subject to an alternating load with a mean value of zero is

$$N_e = \sum_{i=1}^{k} n_i \qquad (4.10)$$

N_e is called the *equivalent life,* and when used with the $S-N$ diagram gives an equivalent steadily alternating load, at which damage will occur at the same rate as under the varying load conditions.

Example 4.4

Load data on a part indicate that there are three values which exceed the fatigue limit stress of $4.5 \times 10^8 \, \text{N m}^{-2}$. These values occur during operation in the following proportions:

$5.5 \times 10^8 \, \text{N m}^{-2} : 3$
$6.5 \times 10^8 \, \text{N m}^{-2} : 2$
$7.0 \times 10^8 \, \text{N m}^{-2} : 1$

Evaluate the equivalent constant dynamic stress.

The $S-N$ diagram for the material is shown in Fig. 4.14. The cycles to failure at each overstress level are:

$5.5 \times 10^8 \, \text{N m}^{-2} : 9.5 \times 10^4$ cycles
$6.5 \times 10^8 \, \text{N m}^{-2} : 1.5 \times 10^4$ cycles
$7.0 \times 10^8 \, \text{N m}^{-2} : 0.98 \times 10^4$ cycles

Therefore, from Eqn (4.9). where C is an arbitrary constant,

$$\frac{3C}{9.5 \times 10^4} + \frac{2C}{1.5 \times 10^4} + \frac{1C}{0.98 \times 10^4} = 1$$

$$C = 3,746$$

From Eqn (4.10),

$$N_e = 3C + 2C + C$$
$$= 2.25 \times 10^4 \, \text{N}$$

From the $S-N$ diagram, the equivalent constant dynamic stress is $6.3 \times 10^8 \, \text{N m}^{-2}$.

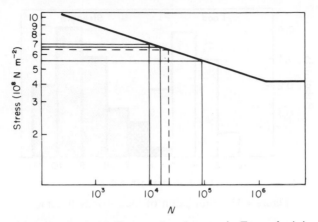

Figure 4.14 $S-N$ diagram for the part in Example 4.4

The instantaneous load–strength interference, and hence the reliabiity, can be evaluated for a particular life value, if we know the load and strength distribution parameters. The equivalent stress evaluated using Miner's rule represents the mean stress. The standard deviation of the stress distribution can be determined by analysis of the load spectrum. The strength distribution parameters can be obtained from fatigue data tables, for different numbers of cycles at given stress levels. However, reliability values calculated in this way are subject to considerable uncertainty. The usual practice in designing for a safe life is to estimate the equivalent cycles to failure and to assign a safe life based upon the expected variation in N_e. However, the predicted safe life should always be confirmed by carrying out life tests, using simulated or actual environments, and actual production items.

The foregoing was a very basic introduction to how fatigue theory relates to reliability. The appropriate material specifications and specialist texts on fatigue should be consulted when necessary. Other weakening processes, such as corrosion, are generally even less predictable, and the reliability effects must therefore be quantified by tests in each case.

USE OF SIMULATION

Distributed load and strength situations can be analysed by using the Monte Carlo simulation technique. In Monte Carlo simulation, the value of a distributed parameter is selected by the generation of a random number, with the probability of a given value being determined by the pre-allocation of random numbers to that variable. As a simple example, if load and strength distributions are as shown in Fig. 4.15, with the interference shown as the cross-hatched area, the system can be modelled by allocating random numbers 0–99 to each distribution. The occurrence of a load value between 4 and 5 units would be signalled by the generation of any of 40 random numbers, say

Figure 4.15 Simulation of load–strength data

36–75. The second random number selection would relate to the strength distribution, so that numbers from 3 to 10 would indicate a strength between 6 and 7 units. The generation of many such paired random numbers would then simulate the interference and the proportion of cases when interference occurred would indicate the reliability. Figure 4.16 shows this as a flow diagram.

The Monte Carlo simulation method is ideally suited for computer application, since a very large number of iterations can be made in a short time. It is also useful when the distributions are such that the analysis of the interference is not tractable. This might be the case if the data on load or strength do not fit a distribution which can be handled analytically, or if assumptions which do not correspond to common distribution functions are made about the expectation of extreme values. Typical of this situation is curtailment due to load protection or quality control. Simulation can also be used to analyse strength deterioration situations such as fatigue.

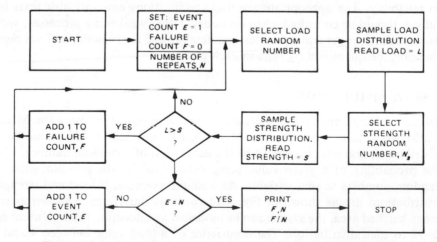

Figure 4.16 Simulation flowchart for load–strength interference

BIBLIOGRAPHY

1. K. C. Kapur and L. R. Lamberson, *Reliability in Engineering Design*. Wiley, New York (1977).
2. C. Lipson and N. J. Sheth, *Statistical Design and Analysis of Engineering Experiments*. McGraw-Hill Kogakusha, Tokyo (1973).
3. A. D. S. Carter, 'Reliability reviewed', *Proc. Inst. Mech. Eng.*, **193** (1979).
4. W. Weibull, *Fatigue Testing and Analysis of Results*. Macmillan, London (1961).

Chapter 5

RELIABILITY PREDICTION AND MODELLING

INTRODUCTION

An accurate prediction of the reliability of a new product, before it is manufactured or marketed, is obviously highly desirable. Depending upon the product and its market, advance knowledge of reliability would allow accurate forecasts to be made of support costs, spares requirements, warranty costs, marketability, etc. It is arguable, on the other hand, that an accurate prediction of reliability implies such knowledge of the causes of failure that they could be eliminated. In fact, a reliability prediction can rarely be made with high accuracy or confidence. Nevertheless, it can often provide an adequate basis for forecasting of dependent factors such as life cycle costs. Reliability prediction can also be valuable as part of the study and design processes, for comparing options and for highlighting critical reliability features of designs.

We can make reliability predictions either by making comparisons with past experience or, in some cases, by analysing the statistics of strength and load. The latter method was covered in Chapter 4. In this chapter we will consider methods of predicting the reliability of systems.

If a type of fighter aircraft has on average 2.5 flying hours between failures, and a new aircraft of similar complexity and performance is being planned, 2.5 flying hours between failures might be a reasonable basis for forecasting. If the existing aircraft is twin-engined, with electronic systems based mainly on discrete component technology, whereas the new aircraft is to have one engine and electronics using integrated circuitry, judgements have to be made on how these changes might affect the failure rate. Instead of comparing only the complete aircraft, we must compare components. What difference will one engine make compared with two? (In this case, trade-off decisions will have been made. For example, two engines might be safer, but a single engine might lead to lower support costs). What will be the difference in failure rate between the old electronic system and the new one? How much more reliable is one integrated circuit compared with ten transistors? The comparisons begin at the level of the overall system and as the system becomes more closely defined extend to more detailed levels. Eventually, for a product which is defined in detail, it becomes necessary to consider the reliability of individual parts.

In order to make comparisons at the part level we need data on the reliability of the parts. We can use these data to predict the likely reliability of the overall system, taking into account functional relationships, utilization, environment and other factors. However, we must first consider some fundamental limitations inherent in the prediction of reliability.

FUNDAMENTAL LIMITATIONS OF RELIABILITY PREDICTION

In engineering and science we use mathematical models for prediction. For example, the power consumption of a new electronic system can be predicted using Ohm's law and the model power = current × voltage. Likewise, we can predict future planetary positions using Newton's laws and our knowledge of the present positions, velocities and masses. These laws are valid within the appropriate domain (e.g. Ohm's law does not hold at temperatures near absolute zero; Newton's laws are not valid at the subatomic level). However, for practical, everyday purposes such deterministic laws serve our purposes well, and we use them to make predictions, taking due account of such practical aspects as measurement errors in initial conditions.

Whilst most laws in physics, for practical predictive purposes, can be considered to be deterministic, the underlying mechanisms can be stochastic. For example, the pressure exerted by a gas in an enclosure is a function of the random motions of very large numbers of molecules. The statistical central limiting theorem, applied to such a vast number of separate random events and interactions, enables us to use the average effect of the molecular kinetic energy to predict the value we call pressure. Thus Boyle's law is really empirical, as are other 'deterministic' physical laws such as Ohm's law. It is only at the level of individual or very few actions and interactions, such as in nuclear physics experiments, that physicists find it necessary to take account of uncertainty due to the stochastic nature of the underlying processes. However, for practical purposes we ignore the infinitesimal variations, particularly as they are often not even measurable, in the same way as we accept the Newtonian view.

Of course some physical systems are comprised of very few components, actions and interactions, with no underlying stochastic mechanism. Thus laws relating to planetary motion or momentum exchange between snooker balls can be considered deterministic because of the small numbers involved. If, however, we attempt to use these deterministic laws to predict the behaviour of a planetary system comprised of 100 planets, or the instantaneous positions of a large number of billiard balls at some time after the first impact, the computational problems become significant and begin to degrade the credibility of the predictions. Physical laws are, therefore, useful predictors of system behaviour either when small numbers of actions and interactions are involved or when very large numbers are involved. For systems involving moderately large numbers the predictive power of empirical and deterministic laws diminishes. This effect is observable in fields such as aerodynamics, meteorology and the study of explosions, for example. Whilst the individual physical and chemical processes are understood and predictable, the large numbers of simultaneous interacting processes swamp our capability to compute predicted values using these simple, known relationships. Very fast computing is applied, and new empirical relationships derived, in our attempts to predict the behaviour of these systems, with results that often fall short of our requirements. Prediction power is increased if we can simplify the problem to a few variables or if we complicate it to a very large number. Thus, we can apply the laws of heat transfer to predict the local atmospheric effect of a rise in sea temperature, and we can use past data to predict the average rainfall on a continent. However, short term local rainfall cannot be predicted accurately.

Prediction power is greater in time-invariant or cyclical systems. The systems mentioned in the last paragraph (weather, aerodynamics, explosives) are time-variant, and this leads to divergent behaviour. Thus predictions of the instantaneous state become progressively less credible as time proceeds (unless a steady state is ultimately achieved, e.g. with an explosion).

It follows from the arguments presented that one or a few measurements of a physical quantity can usually provide us with sufficient information on which to make a prediction of the future state of a physical time-invariant or cyclical system. In other words, very few data are required in order to make confident predictions. We measure the flow of electricity in wire and we assume that the current will remain at that value under the same general conditions. When the quantity is time-variant more measurements might need to be taken before we can predict confidently from the data. It is only when the system is moderately complex that one or a few measurements do not enable us to predict the future state.

The limits of validity of a reliability model

The limitations of expressing laws in physics as mathematical relations are just as applicable to reliability models. We can regard a reliability model as a mathematical relation, that on the basis of deductive or inductive reasoning can be expected to exist between some measure of failure events, and one (usually time) or more explanatory factors. We can see that unlike most laws in physics, which can generally be considered to be concerned with deterministic quantities, we are now dealing with probabilistic quantities. This involves further limitations, principally as a result of how we interpret data. These limitations are imposed because, unlike the situation with a deterministic physical law, in reliability engineering we must deal with systems consisting of moderately large number of parts, actions and interactions. Therefore, we cannot safely imply the lack of variability inherent in systems made up of very large quantities of interacting parts, as typified by the pressure of an enclosed gas or in systems made up of very few parts, such as two snooker balls. Furthermore, failure mechanisms are nearly always time-variant.

Therefore we cannot rely on one or a few data points to enable us to predict future states of a system in terms of whether or not it has failed, or of how many times it might have failed. Also, since reliability data are inherently expensive, we cannot usually collect such large amounts of data to enable us to derive credible empirical relationships. We are forced to make use of relatively limited data, using methods of statistics and probability to derive relationships.

In dealing with probabilistic quantities we no longer consider their true value as being predictable, but rather as being distributed around some expected value. In this case we can no longer regard the relationships observed in data as necessarily reflective of the relationships between quantities. Data, however plentiful, can only be regarded as sample information and thus subject to differing interpretations. There will always be limitations in understanding the connections between data and the quantities they represent.

Therefore, to derive functional relationships between probabilistic quantities, we require considerably more data than are required for deterministic quantities, and then there will always be uncertainties due to sampling variability.

In the absence of an ability to derive reliability models that can be validated by the sheer force of data, an alternative approach is required. One that is often used is to postulate a model and hope that in some way we can confirm this postulate. In logic, such an approach is invalid, in that a postulate is either taken on the basis of common sense to be self-evidently true or in probability terms almost certain. A more logical approach would be to consider a relation found to exist in data as forming the basis of a conjectured model (e.g. regression analysis). However, even this approach is philosophically uncertain in that the basis of the argument would be inductive, and inductive reasoning can be regarded as an invalid basis for proof. The fact remains that without considerable experimentation there will always be uncertainties about the logical connection between probabilistic quantities.

We have also seen (Chapter 3) how the failure probability of an item which has a probabilistic distribution of strength and which is subjected to a distributed load is to a large extent indeterminate, even when we make simplifying, and therefore probably unrealistic, assumptions.

Thus any model used for reliability prediction is subject to severe credibility constraints, related to:

1. Precision.
2. Domain and conditions of applicability.
3. Dim perception of logical connections.
4. Sensitivity to variation.

Another serious criticism can be made of the concept of reliability modelling. Even if sufficient failure data were available to overcome the deficiencies of statistical reliability models (an impossible situation in the practical reliability engineering context) the human element can still invalidate the predictions made. For example, a reliability prediction based on failure data during trials might indicate a particular MTBF in use. However, the model cannot predict that the equipment might be operated by different users or in a different environment, or that different quality standards might be applied to future production. Reliability achievement might depend upon the equipment being improved in some way, typically by the incorporation of reliability improvement modifications. Whether or not such improvements are made (and their effectiveness) is under human control, and therefore it is not feasible even to postulate a plausible reliability model.

Failure data, and therefore any reliability model based upon such data, are seriously affected by human intervention and interpretation. For example, in life testing of electronic components, failure might be defined by a parameter being outside a specified value. However, it is by no means certain that such a parameter excursion will be perceived as a failure in an operating system, since the component might be applied in such a way that the parameter excursion does not affect system performance. Likewise perceived failures of systems are often subject to human interpretation, so that what is considered a failure by a development test engineer might not be reported as such by an operator, or vice versa.

Therefore, any reliability prediction based upon the use of mathematical models, such as in MIL-HDBK-217 for electronics (Chapter 7), or in reliability growth models (Chapter 12) must be treated with some scepticism. See also the comments on reliability analysis methods for repairable systems on pages 89 and 91.

A designer can design an electronic system to a power budget. The power dissipation of the new system can be predicted with credible precision, so that all systems made to the design will dissipate, within reasonable tolerance, the predicted power. However, a designer cannot design for an MTBF, unless he places as much faith in the reliability mathematical model as he does in, say, Ohm's law. The MTBF of each unit cannot be measured as can its power consumption, and there is no reason in logic to believe that they will all show the same MTBFs or patterns of failure over a period of operation, or even that the overall average MTBF or failure pattern will conform to that predicted.

Unfortunately the mathematical modelling approach to reliability prediction (including extrapolation from test data) has been given undue and insufficiently critical attention in the literature and in standards such as MIL-HDBK-217 and MIL-STD-781. (Later chapters deal with these standards in detail.) It is not uncommon to see MTBF predictions such as 5,761 hours. No power dissipation prediction would attempt such precision; 6 kW is an equivalent figure more likely to be stated. The naive presentation of reliability predictions has done much to undermine the credibility of reliability engineering, particularly since so often systems do not achieve the predicted values.

The way ahead

Having identified the fundamental limitations of reliability prediction models, we are still left with the problem that reliability prediction is necessary. We also know that it is possible to make reasonably credible reliability predictions for systems under certain circumstances. These are:

1. The system is similar to systems developed, built and used previously, so that we can apply our experience of what happened before.
2. The new system does not involve significant technological risk (this follows from 1).
3. The system will be manufactured in large quantities, or is very complex (i.e. contains many parts), or will be used over a long time, or a combination of these conditions applies, i.e. there is an asymptotic property.
4. There is a strong commitment to the achievement of the reliability predicted, as an overriding priority.

Thus, we can make credible reliability predictions for a new TV receiver or automobile engine. No great changes from past practice are involved, technological risks are low, they will be built in large quantities and they are quite complex, and in present markets the system must compete with established, reliable products. However, for a new, high technology defence equipment these conditions may not apply, though nowadays the commitment to reliability is generally stronger than in the past.

Note that reliability predictions (in the sense of a reasonable expectation) for the TV receiver and the automobile engine could be made without recourse to statistical or empirical mathematical models. Rather, they could be based upon knowledge of past performance and present conditions, and on management targets and priorities. The reliability prediction does not ensure that the reliability values will be achieved; it is not a demonstration in the way that a mass or power consumption prediction, being based on physical laws, would be. Rather, it should be used as the basis for setting the objective — one which is likely to be attained only if there is a human commitment to it. This is well illustrated by the notable reliability achievement of the US space programme and the Japanese consumer products industry.

Reliability prediction for new, high technology products must, therefore, be based upon identification of objectives and assessment of risks, in that order. This must be an iterative procedure, since objectives and risks must be balanced; the reliability engineer plays an important part in this process, since he must assess whether objectives are realistic in relation to the risks. This assessment can be aided by the educated use of appropriate models and data, which help to quantify the risks. For example, it would not usually be wise to predict a reliability figure greatly in excess of previous experience, and the models and data can be used to quantify this. Once the risks are assessed and the objective quantified, development must be continuously monitored in relation to the reduction of risks through analysis and tests, and to the measured reliability during tests. This is necessary to provide assurance that the objective will be met, if need be by additional management action such as provision of extra resources to solve particular problems. It might even be necessary to change the objective. Once again, the intelligent use of reliability prediction models can help the reliability engineer to predict the future performance. Intelligent use implies that the reliability engineer understands the actual problems involved in the development programme and knows what action will be taken and its likely effectiveness.

RELIABILITY DATA BASES

As mentioned earlier, reliability predictions are often based upon analysis of the individual components of a system, and, for example, summating their individual failure rates to derive a system failure rate prediction. Subject to the limitations described above, this method can be used for predicting likely achievable reliability.

Having determined which parts can contribute to the unreliability of a product, we need data on their reliability in order to evaluate the overall figure. Without reasonably accurate data on the parts it is not practicable to attempt a parts count prediction; the level to which the analysis is taken must be dependent upon the confidence we have in the parts data. In many practical cases very few data are available on the reliability of the components of a system, and the reliability prediction can then be made only on the basis of an overall comparison. For many other products, particularly more complex ones, data on part reliability are available either from the manufacturer's own sources, such as warranty or spares consumption data, or from other reliability data sources. It is necessary, however, to distinguish between genuine failure data and data which may relate to part removal or consumption for reasons other than failure.

The parts reliability data which are available are very variable in quality. A well-known source for electronic device reliability is US MIL-HDBK-217, which is fully described in Chapter 7. As mentioned above, results do not always match the prediction, but since electronic devices are comprised of relatively few basic types and as their application is fairly standard in terms of stress level and environment, it is possible to obtain fairly accurate (within a factor of 2) predictions for systems made up of 500 or more parts, if the prediction is used as the basis of the reliability programme goal. Non-electronic parts reliability data should always be treated with caution, unless they are derived from an identical or very similar application. Very few non-electronic part types have achieved the degree of standardization seen in electronics. Bearings are one of these exceptions, for which reliability data in terms of a *B*-life (see Chapter 3) related to application conditions can be obtained from suppliers. Non-electronic parts reliability data lists are available which purport to show the failure rates of generic parts types such as hydraulic valves, electric motors, gaskets and springs. The variety of types and applications of such parts is so diverse that such data cannot be relied upon as being generally applicable. There is no general unit of operating time for a gasket or a spring, as there might be for a transistor. Also, for most non-electronic parts, the assumption of a constant hazard rate can be grossly misleading.

Non-electronic parts reliability data can be useful in specific prediction applications, such as aircraft, petrochemical plant, photocopying machines or automobiles, when the data are derived from the area of application. However, such data should not be transferred from one application area to another without careful assessment. Even within the application area they should be used with care, since even then conditions can vary widely. An electric motor used to perform the same function, under the same loading conditions as previously in a new design of photocopier, might be expected to show the same failure pattern. However, if the motor is to be used for a different function, with different operating cycles, or even if it is bought from a different supplier, the old failure data will not be appropriate.

We saw in Chapter 4 how reliability can vary by orders of magnitude with small changes in load and strength distributions, and the large amount of uncertainty inherent in estimating reliability from the load–strength model. These real uncertainties must be borne in mind when synthesizing the reliability of a system by considering the likely failure rates of its parts.

A reliability prediction for a system containing many parts is likely to be more accurate than for small system. However, it is common for the erroneous implication to be drawn that, because the individual parts failure data can be combined to provide a reasonably accurate prediction for the reliability of the system, the parts failure data can be used with as much confidence for spare parts provisioning or decisions related to the safety of individual parts. It is important to remember that variances in reliability at the part level will be much greater than the variance at the system level.

SYSTEMS RELIABILITY MODELS

The basic series reliability model

Consider a system composed of two independent components, each exhibiting a constant hazard rate. If the failure of either component will result in failure of the system, the system can be represented by a *reliability block diagram* (Fig. 5.1). (A reliability block diagram does not necessarily represent the system's operational logic or functional partitioning.)

If λ_1 and λ_2 are the hazard rates of the two components, the system hazard rate will be $\lambda_1 + \lambda_2$. Because the hazard rates are constant, the component reliabilities R_1 and R_2, over a time of operation t, are $\exp(-\lambda_1 t)$ and $\exp(-\lambda_2 t)$. The reliability of the system is the combined probability of no

Figure 5.1 Series system

failure of either component, i.e. $R_1R_2 = \exp[-(\lambda_1 + \lambda_2)t]$. In general, for a series of n, s-independent components:

$$R = \prod_{i=1}^{n} R_i \qquad (5.1)$$

where R_i is the reliability of the ith component. This is known as the product rule or series rule (see Eqn 2.2).

$$\lambda = \sum_{i=1}^{n} \lambda_i \quad \text{and} \quad R = \exp(-\lambda t) \qquad (5.2)$$

This is the simplest basic model on which *parts count* reliability prediction is based.

The failure logic model of the overall system may be more complex if there are redundant subsystems or components, but as one considers the components within subsystems the product rule describes the bulk of the reliability relationships.

Active redundancy

The reliability block diagram for the simplest redundant system is shown in Fig. 5.2. In this system, composed of two s-independent parts with reliabilities R_1 and R_2, satisfactory operation occurs if *either* one or *both* parts function. Therefore, the reliability of the system, R, is equal to the probability of part 1 *or* part 2 surviving.

From Eqn (2.6), the probability

$$(R_1 + R_2) = R_1 + R_2 - R_1R_2$$

This is often written

$$1 - (1 - R_1)(1 - R_2)$$

Figure 5.2 Dual redundant system

For the constant hazard rate case,

$$R = \exp(-\lambda_1 t) + \exp(-\lambda_2 t) - \exp[-(\lambda_1 + \lambda_2)t] \tag{5.3}$$

The general expression for active parallel redundancy is

$$R = 1 - \prod_{i=1}^{n} (1 - R_i) \tag{5.4}$$

where R_i is the reliability of the ith unit and n the number of units in parallel. If in the two-unit active redundant system $\lambda_1 = \lambda_2 = 0.1$ failures per 1,000 h, the system reliability over 1,000 h is 0.9909. This is a significant increase over the reliability of a simple non-redundant unit, which is 0.9048. Such a large reliability gain often justifies the extra expense of designing redundancy into systems. The gain usually exceeds the range of prediction uncertainty. The example quoted is for a *non-maintained* system, i.e. the system is not repaired when one equipment fails. In practice, most active redundant systems include an indication of failure of one equipment, which can then be repaired. A maintained active redundant system is, of course, theoretically more reliable than a non-maintained one. Examples of non-maintained active redundancy can be found in spacecraft systems (e.g. dual thrust motors for orbital station-keeping) and maintained active redundancy is a feature of systems such as power generating systems and railway signals.

m-out-of-*n* Redundancy

In some active parallel redundant configurations, m out of the n units may be required to be working for the system to function. This is called m-out-of-n (or m/n) parallel redundancy. The reliability of an m/n system, with n, s-indepedent components in which all the unit reliabilities are equal, is the binomial reliability function (Eqn 2.23):

$$R = 1 - \sum_{i=0}^{m-1} \binom{n}{i} R^i (1 - R)^{n-i} \tag{5.5}$$

or, for the constant hazard rate case (when $\lambda t \ll 1$):

$$R \approx 1 - \frac{1}{(\lambda t + 1)^n} \sum_{i=0}^{m-1} \binom{n}{i} (\lambda t)^{n-i} \tag{5.6}$$

Standby redundancy

Standby redundancy is achieved when one unit does not operate continuously but is only switched on when the primary unit fails. A standby electrical generating system is an example. The block diagram in Figure 5.3 shows another. The standby unit and the sensing and switching system may be considered to have a 'one-shot' reliability R_s of starting and maintaining system function until the primary equipment is repaired, or R_s may be

Figure 5.3 Reliability block diagram for a missile system

time-dependent. The switch and the redundant unit may have dormant hazard rates, particularly if they are not maintained or monitored.

Taking the case where the system is non-maintained, the units have equal constant operating hazard rates λ, there are no dormant failures and $R_s = 1$, then

$$R = \exp(-\lambda t) + \lambda t \exp(-\lambda t) \tag{5.7}$$

The general reliability formula for n equal units in a standby redundant configuration (perfect switching) is

$$R = \sum_{i=0}^{n-1} \frac{(\lambda t)^i}{i!} \exp(-\lambda t) \tag{5.8}$$

If in a standby redundant system $\lambda_1 = \lambda_2 = 0.1$ failure per 1,000 h, then the system reliability is 0.9953. This is higher than for the active redundant system $[R(1,000) = 0.9909]$, since the standby system is at risk for a shorter time. If we take into account less than perfect reliability for the sensing and switching system and possibly a dormant hazard rate for the standby equipment, then standby system reliability would be reduced.

Further redundancy considerations

The redundant systems described represent the tip of an iceberg as far as the variety and complexity of system reliability models are concerned. For systems where very high safety or reliability is required, more complex redundancy is frequently applied. Some examples of these are:

1. In aircraft, dual or triple active redundant hydraulic power systems are used, with a further emergency (standby) back-up system in case of a failure of all the primary circuits.
2. Aircraft electrical flying controls typically utilize triple voting active redundancy. A sensing system automatically switches off one system if it transmits signals which do not match those transmitted by the other two, and there is a manual back-up system. The reliability evaluation must

include the reliability of all three primary systems, the sensing system and the manual system.

3. Fire detection and suppression systems consist of detectors, which may be in parallel active redundant configuration, and a suppression system which is triggered by the detectors.

We must be careful to ensure that single-point failures which can partly eliminate the effect of redundancy are considered in assessing redundant systems. For example, if redundant electronic circuits are included within one integrated circuit package, a single failure such as a leaking hermetic seal could cause both circuits to fail. Such s-dependent failures are sometimes referred to as *common-cause* failures, particularly in relation to systems.

AVAILABILITY OF REPAIRABLE SYSTEMS

Availability is defined as the probability that an item will be available when required, or as the proportion of total time that the item is available for use. Therefore the availability of a repairable item is a function of its failure rate, λ, and of its repair or replacement rate. The proportion of total time that the item is available is the *steady-state availability*. For a simple unit, with a constant failure rate λ and a constant mean repair rate μ, where $\mu = (\text{MTTR})^{-1}$, the steady-state availability is

$$A = \frac{\mu}{\lambda + \mu} = \frac{\text{MTBF}}{\text{MTBF} + \text{MTTR}} \tag{5.9}$$

The instantaneous availability, or probability that the item will be available at time t is equal to

$$A = \frac{\mu}{\lambda + \mu} + \frac{\lambda}{\lambda + \mu} \exp[-(\lambda + \mu)t] \tag{5.10}$$

It is often more revealing, particularly when comparing design options, to consider system *unavailability*:

Steady-state unavailability $= 1 - A$

$$= \frac{\lambda}{\lambda + \mu} \tag{5.11}$$

and

Instantaneous unavailability $= \frac{\lambda}{\lambda + \mu} - \frac{\lambda}{\lambda + \mu} \exp[-(\lambda + \mu)t] \tag{5.12}$

As t becomes large, this approaches the steady-state availability, $\mu/(\lambda + \mu)$. If scheduled maintenance is necessary and involves taking the system out of action, this must be included in the availability formula. The availability of spare units for repair by replacement is often a further consideration, dependent upon the previous spares usage and the repair rate of replacement units.

Table 5.1 Reliability and availability for some systems configurations. (R.H. Myers, K. L. Wong, and H. M. Gordy, *Reliability Engineering for Electronic Systems*,

Reliability configuration	Reliability (no repair) for CFR λ	General system reliability for n blocks
R (single block)	$\exp(-\lambda t)$	
R_1 — R_2 (series)	$\exp[-(\lambda_1 + \lambda_2)t]$	$\displaystyle\prod_{i=1}^{n} R_i$
R_1, R_2 (parallel) — Active	$\exp(-\lambda_1 t) + \exp(-\lambda_2 t) - \exp[-(\lambda_1 + \lambda_2)t]$	$\displaystyle 1 - \prod_{i=1}^{n}(1 - R_i)$
Standby	$\dfrac{\lambda_2 \exp(-\lambda_1 t) - \lambda_1 \exp(-\lambda_2 t)}{\lambda_2 - \lambda_1}$ [b]	$\displaystyle \exp(-\lambda t) \sum_{i=0}^{n-1} \frac{(\lambda t)^i}{i!}$
Active 1/3	$3\exp(-\lambda t) - 3\exp(-2\lambda t) + \exp(-3\lambda t)$	As above
Active 2/3	$3\exp(-2\lambda t) - 2\exp(-3\lambda t)$	$\displaystyle 1 - \sum_{i=0}^{m-1} \binom{n}{i} R^i (1-R)^{n-i}$ [c]
Standby 1/3	$\exp(-\lambda t) + \lambda t \exp(-\lambda t) + \frac{1}{2}\lambda^2 t^2 \exp(-\lambda t)$ [d]	$\displaystyle R \sum_{i=0}^{n-1} \frac{(\lambda t)^i}{i!}$

[a] $\lambda_1 = \lambda_2 = \lambda$. Assumes series repair, i.e. single repair team.
[b] When $\lambda_1 = \lambda_2$ the reliability formula becomes indeterminate. When $\lambda_1 = \lambda_2$ use $R(t) = \exp(-\lambda t) + \lambda t \exp(-\lambda t)$. If $\lambda_1 \approx \lambda_2$ use $\lambda = (\lambda_1 + \lambda_2)/2$. Assumes perfect switching.

Availability is an important consideration in relatively complex systems, such as power stations, satellites, chemical plant and radar stations. In such systems, high reliability by itself is not sufficient to ensure that the system will be available when needed. It is also necessary to ensure that it can be repaired quickly and that essential scheduled maintenance tasks can be performed quickly, if possible without shutting down the system. Therefore maintainability is an important aspect of design for maximum availability, and trade-offs are often necessary between reliability and maintainability features. For example, built-in test equipment (BITE) is incorporated into many electronic systems. This added complexity can degrade reliability and can also result in spurious failure indications. However, BITE can greatly reduce maintenance times, by providing an instantaneous indication of fault location, and therefore availability can be increased. (This is not the only reason for the use of BITE. It can also reduce the need for external test equipment and for training requirements for trouble-shooting, etc.).

R for $\lambda_1 = \lambda_2 = 0.01$, $t = 100$	Steady-state availability, A, repair rate, μ, CFR, λ	General steady-state availability, A, for n blocks	A for $\lambda = 0.01$, $\mu = 0.2$
0.37	$\dfrac{\mu}{\lambda + \mu}$	—	0.950
0.14	$\dfrac{\mu_1\mu_2}{\mu_1\mu_2 + \mu_1\lambda_2 + \mu_2\lambda_1 + \lambda_1\lambda_2}$	$\displaystyle\prod_{i=1}^{n} \dfrac{\mu_i}{\lambda_i + \mu_i}$	0.907
0.60	$\dfrac{\mu^2 + 2\mu\lambda}{\mu^2 + 2\mu\lambda + 2\lambda^2}$ [a]	$1 - \displaystyle\prod_{i=1}^{n} \dfrac{\lambda_i}{\lambda_i + \mu_i}$	0.996
0.74	$\dfrac{\mu^2 + \mu\lambda}{\mu^2 + \mu\lambda + \lambda^2}$ [a]	—	0.998
0.75	$\dfrac{\mu^3 + 3\mu^2\lambda + 6\mu\lambda^2}{\mu^3 + 3\mu\lambda^2 + 6\lambda\mu^2 + 6\lambda^3}$ [a]	As above (active)	0.890
0.31	$1 - \dfrac{1}{(\lambda + \mu)^3}(\lambda^3 + 3\mu\lambda^2 + 3\mu^2\lambda)$ [a]	$1 - \dfrac{1}{(\lambda + \mu)^n} \displaystyle\sum_{i=0}^{m-1} \binom{n}{i}\mu\lambda^{n-i}$ [a,c]	0.300
0.92	$\dfrac{\mu^3 + \mu^2\lambda + \lambda^2\mu}{\lambda^3 + \mu^2\lambda + \lambda^2\mu + \lambda^3}$ [a]	—	0.9999

[c] For m-out-of-n redundancy.
[d] Assumes perfect switching.

Availability is also affected by redundancy. If standby systems can be repaired or overhauled while the primary system provides the required service, overall availability can be greatly increased.

Table 5.1 shows the reliability and steady-state availability functions for some system configurations. It shows clearly the large gains in reliability and steady-state availability which can be provided by redundancy. However, these are relatively simple situations, particularly as a constant failure rate is assumed. Also, for the standby redundant case, it is assumed that:

1. The reliability of the changeover system is unity.
2. No common-cause failures occur.
3. Failures are detected and repaired as soon as they occur.

Of course, these conditions do not necessarily apply, particularly in the case of standby equipment, which must be tested at intervals to determine whether it is serviceable. The availability then depends upon the test interval. Monitoring systems are sometimes employed, e.g. built-in test equipment (BITE) for electronic equipment, but this does not necessarily have a 100 per cent. chance of detecting all failures. In real-life situations it is necessary to consider these aspects, and the analysis can become very complex. Methods for dealing with more complex systems are given at the end of this chapter, and maintenance and maintained ability are covered in more detail in Chapter 9.

Example 5.1

A shipboard missile system is composed of two warning radars, a control system, a launch and guidance system, and the missiles. The radars are arranged so that either can give warning if the other fails, in a standby redundant configuration. Four missiles are available for firing and the system is considered to be reliable if three out of four missiles can be fired and guided. Figure 5.3 shows the system in reliability block diagram form, with the MTBFs of the subsystems. The reliability of each missile is 0.9. Assuming that: (1) the launch and guidance system is constantly activated, (2) the missile flight time is negligible and (3) all elements are s-independent, evaluate: (a) the reliability of the system over 24 h, (b) the steady-state availability of the system, excluding the missiles, if the mean repair time for all units is 2 h and the changeover switch reliability is 0.95.

The reliabilities of the units over a 24 h period are:

Primary radar	0.9762	(failure rate $\lambda_P = 0.001$)
Standby radar	0.9762	(failure rate $\lambda_S = 0.001$)
Launch and guidance	0.9685	(failure rate $\lambda_{LG} = 0.0013$)

(a) The overall radar reliability is given by (from Table 5.1)

$$R_{\text{radar}} = \exp(-\lambda t) + \lambda t \exp(-\lambda t)$$
$$= 0.9762 + (0.001 \times 24 \times 0.9762) = 0.9996$$

The probability of the primary radar failing is $(1 - R_P)$. The probability of this radar failing *and* the switch failing is the product of the two failure probabilities:

$$(1 - 0.9762)(1 - 0.95) = 0.0012$$

Therefore the switch reliability effect can be considered equivalent to a series unit with reliability

$$R_{SW} = (1 - 0.0012) = 0.9988$$

The system reliability, up to the point of missile launch, is therefore

$$R_S = R_{\text{radar}} \times R_{SW} \times R_{LG}$$
$$= 0.9996 \times 0.9988 \times 0.9685 = 0.9670$$

The reliability of any three out of four missiles is given by the cumulative binomial distribution (Eqn 5.5):

$$R_M = 1 - \left[\binom{4}{0} 0.9^0 \times 0.1^4 + \binom{4}{1} 0.9^1 \times 0.1^3 + \binom{4}{2} 0.9^2 \times 0.1^2 \right]$$

$$= 0.9477$$

The total system reliability is therefore

$$R_S' = R_S \times R_M$$
$$= 0.9670 \times 0.9477 = 0.9165$$

(b) The availability of the redundant radar configuration is (see Table 5.1)

$$A_{radar} = \frac{\mu^2 + \mu\lambda}{\mu^2 + \mu\lambda + \lambda^2}$$

$$= \frac{0.5^2 + (0.5 \times 0.001)}{(0.5)^2 + (0.5 \times 0.001) + (0.001)^2}$$

$$= 0.999\,997 \qquad (\text{unavailability} = 3 \times 10^{-6})$$

The availability of the launch and guidance system

$$A_{LG} = \frac{\mu}{\mu + \lambda}$$

$$= \frac{0.5}{0.5 + 0.0013} = 0.9974 \qquad (\text{unavailability} = 2.6 \times 10^{-3})$$

The system availability is therefore

$$A_{radar} \times A_{LG} = 0.9974 \qquad (\text{unavailability} = 2.6 \times 10^{-3})$$

The previous example can be used to illustrate how such an analysis can be used for performing sensitivity studies to compare system design options. For example, a 20 per cent. reduction in the MTBF of the launch and guidance system would have a far greater impact on system reliability than would a similar reduction in the MTBF of the two radars.

In maintained systems which utilize redundancy for reliability or safety reasons, separate analyses should be performed to assess system reliability in terms of the required output and failure rate in terms of maintenance arisings. In the latter case all elements can be considered as being in series, since all failures, whether of primary or standby elements, lead to repair action.

MODULAR DESIGN

Availability and the cost of maintaining a system can also be influenced by the way in which the design is partitioned. 'Modular' design is used in many complex products, such as electronic systems and aeroengines, to ensure that a failure can be corrected by a relatively easy replacement of the defective module, rather than by replacement of the complete unit.

Example 5.2

An aircraft gas turbine engine has a mean time between replacements (MTBR) — scheduled and unscheduled — of 1,000 flight hours. With a total annual flying rate of 30,000 h and an average cost of replacement of $10,000, the annual repair bill amounted to $300,000. The manufacturer redesigned the engine so that it could be separated into four modules, with MTBR and replacement costs as shown in Table 5.2. What would be the new annual cost?

Table 5.2 MTBR and replacement costs for the four modules.

	MTBR (h)	Replacement cost ($)
Module 1	2,500	3,000
Module 2	4,000	2,000
Module 3	4,000	2,500
Module 4	10,000	1,000

With the same total number of replacements, the annual repair cost is greatly reduced, from $300,000 to $72,750 (see Table 5.3).

Table 5.3 Cost per year of replacing the modules.

	Replacements per year	Cost per year ($)
Module 1	12	36,000
Module 2	7.5	15,000
Module 3	7.5	18,750
Module 4	3	3,000
Total	30	$72,750

Note that Example 5.2 does not take into account the different spares holding that would be required for the modular design (i.e. the operator would keep spare modules, instead of spare engines, thus making a further saving). In fact other factors would complicate such an analysis in practice. For example the different scheduled overhaul periods of the modules compared with the whole engine, the effect of wearout failure modes giving non-constatnt replacement rates with time since overhaul, etc. Monte Carlo simulation is often used for planning and decision-making in this sort of situation (see page 119).

BLOCK DIAGRAM ANALYSIS

The failure logic of a system can be shown as a reliability block diagram (RBD), which shows the logical connections between components of the system. The RBD is not necessarily the same as a block schematic diagram of the system's functional layout. We have already shown examples of RBDs for simple series and parallel systems. For systems involving complex interactions construction of the RBD can be quite difficult, and a different RBD will be necessary for different definitions of what constitutes a system failure. A failure mode and effect analysis (FMEA) (see Chapter 6) can be a useful starting point for preparing the RBD, since each failure mode is considered in relation to its effect on the system's output.

Block diagram analysis consists of reducing the overall RBD to a simple system which can then be analysed using the formulae for series and parallel arrangements. It is necessary to assume s-independence of block reliabilities.

The technique is also called block diagram *decomposition*. It is illustrated in Example 5.3.

Example 5.3

The system shown in Fig. 5.4 can be reduced as follows (assuming s-independent reliabilities):

$$R_s = R_1 \times R_2 \times R_B \times R_{10} \times R_C \qquad \text{(from Eqn 5.1)}$$

$$R_B = 1 - [1 - (R_3 \times R_4 \times R_5)][1 - (R_6 \times R_7 \times R_8)](1 - R_9) \qquad \text{(from Eqn 5.4)}$$

$$R_C = 1 - \frac{3 \times 2}{3 \times 2} R_{11}^0 (1 - R_{11})^3 + \frac{3 \times 2}{2} R_{11}(1 - R_{11})^2 \qquad \text{(from Eqn 5.5)}$$

$$= 1 - (1 - R_{11})^3 + 3R_{11}(1 - R_{11})^2$$

Cut and tie sets

Complex RBDs can be analysed using cut set or tie set methods. A cut set is produced by drawing a line through blocks in the system to show the minimum number of failed blocks which would lead to system failure. Tie sets (or path sets) are produced by drawing lines through blocks which, if all were working, would allow the system to work. Figure 5.5 illustrates the way that cut and tie sets are produced. In this system there are three cut sets and two tie sets.

Figure 5.4 Block diagram decomposition

140

Figure 5.5 (a) Cut sets and (b) tie sets

Approximate bounds on system reliability as derived from cut sets and tie sets, respectively, are given by

$$R_s > 1 - \sum_{i=1}^{N} \prod_{j=1}^{n} (1 - R_i) \tag{5.12}$$

$$R_s < \sum_{i=1}^{T} \prod_{j=1}^{n} R_i \tag{5.13}$$

where N is the number of cut sets, T is the number of tie sets and n_j is the number of blocks in the jth cut set or tie set.

Example 5.4

Determine the reliability bounds of the system in Fig. 5.5 for

$$R_1 = R_2 = R_3 = R_4 = 0.9.$$

Cut set:

$$R_s > 1 - [(1 - R_1)(1 - R_3) + (1 - R_2)(1 - R_3) + (1 - R_4)]$$
$$> 1 - [3 - R_1 - R_2 - 2R_3 - R_4 + R_1R_3 + R_2R_3]$$
$$> 1 - 0.12 = 0.88$$

Tie set:

$$R_s < R_1R_2R_4 + R_3R_4$$
$$< 1.54 \text{ (i.e. } < 1.0)$$

For comparison, the exact reliability is

$$R_s = [1 - (1 - R_1R_2)(1 - R_3)]R_4$$
$$= R_3R_4 + R_1R_2R_4 - R_1R_2R_3R_4$$
$$= 0.883$$

The cut and tie set approaches are not used for systems as simple as in Example 5.4, since the decomposition approach is easy and gives an exact result. However, since the derivation of exact reliability using the decomposition approach can become an intractable problem for complex systems, the cut and tie set approach has its uses in such applications. The approximations converge to the exact system reliability as the system complexity increases, and the convergence is more rapid when the block reliabilities are high. Tie sets are not usually identified or evaluated in system analysis, however.

Cut and tie set methods are suitable for computer application. Their use is appropriate for the analysis of large systems in which various configurations are possible, such as aircraft controls, power generation, or control and instrumentation systems for large plant installations. The technique is subject to the constraint (as is the decomposition method) that all block reliabilities must be s-independent.

STATE-SPACE ANALYSIS (MARKOV ANALYSIS)

If a system or component can be in one of two states (e.g. failed, non-failed), and if we can define the probabilities associated with these states on a discrete or continuous basis, the probability of being in one or other at a future time can be evaluated using state-space (or state-time) analysis. In reliability and availability analysis, failure probability and the probability of being returned to an available state, or failure rate and repair rate, are the variables of interest.

The best-known state-space analysis technique is Markov analysis. The Markov method can be applied under the following major constraints;

1. The probabilities of changing from one state to another must remain constant, i.e. the process must be homogeneous. Thus the method can only be used when a constant hazard rate assumption can be justified.
2. Future states of the system are independent of all past states except the immediately preceding one. This is an important constraint in the analysis of repairable systems, since it implies that repair returns the system to an 'as new' condition.

Nevertheless, Markov analysis can be usefully applied to system reliability, safety and availability studies, particularly to maintained systems for which BDA is not directly applicable, provided that the constraints described above are not too severe. The method is used for analysing complex systems such as power generation and communications. Computer programs can conveniently be written for Markov analysis.

The Markov method can be illustrated by considering a single component, which can be in one of two states: failed (F) and available (A). The probability of transition from A to F is $P_{A \rightarrow F}$ and from F to A is $P_{F \rightarrow A}$. Figure 5.6 shows the situation diagrammatically. This is called a *state transition* or a *state-space*

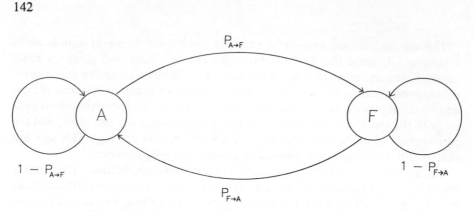

Figure 5.6 Two-state Markov state transition diagram

diagram. All states, all transition probabilities and probabilities of remaining in the existing state ($= 1 -$ transition probability) are shown. This is a *discrete* Markov chain, since we can use it to describe the situation from increment to increment of time. Example 5.5 illustrates this.

Example 5.5

The component in Fig. 5.6 has transitional probabilities in equal time intervals as follows:

$$P_{A \to F} = 0.1$$
$$P_{F \to A} = 0.6$$

What is the probability of being available after four time intervals, assuming that the system is initially available?

   This problem can be solved by using a *tree diagram* (Fig. 5.7).

   The availability of the system in Example 5.5 is shown plotted by time interval in Fig. 5.8. Note how availability approaches a steady state after a number of time intervals. This is a necessary conclusion of the underlying assumptions of constant failure and repair rates and of *s*-independence of events.

   Whilst the transient states will be dependent upon the initial conditions (available or failed), the steady state condition is independent of the initial condition. However, the rate at which the steady state is approached is dependent upon the initial condition and on the transition probabilities.

Complex systems

The tree diagram approach used above obviously becomes quickly intractable if the system is much more complex than the one-component system described, and analysed over just a few increments. For more complex systems, matrix methods can be used, particularly as these can be readily solved by computer programs. For example, for a single repairable component the probability of

143

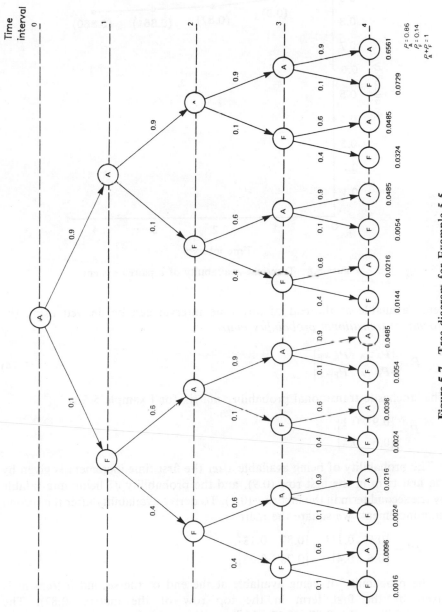

Figure 5.7 Tree diagram for Example 5.5

Time
interval

0   1   2   3   4

$P_A = 0.86$
$P_F = 0.14$
$P_A + P_F = 1$

Figure 5.8 Transient availability of repaired system

being available at the end of any time interval can be derived using the *stochastic transitional probability matrix*:

$$P = \begin{vmatrix} P_{A \to A} & P_{A \to F} \\ P_{F \to A} & P_{F \to F} \end{vmatrix} \tag{5.14}$$

The stochastic transitional probability matrix for Example 5.5 is

$$P = \begin{vmatrix} 0.9 & 0.1 \\ 0.6 & 0.4 \end{vmatrix}$$

The probability of being available after the first time increment is given by the first term in the first row (0.9), and the probability of being unavailable by the second term in the first row (0.1). To derive availability after the second time increment, we square the matrix:

$$P^2 = \begin{vmatrix} 0.9 & 0.1 \\ 0.6 & 0.4 \end{vmatrix}^2 = \begin{vmatrix} 0.87 & 0.13 \\ 0.78 & 0.22 \end{vmatrix}$$

The probability of being available at the end of the second increment is given by the first term in the top row of the matrix (0.87). The unavailability $= 1 - 0.87 = 0.13$ (the second term in the top row).

For the third time increment, we evaluate the third power of the probability matrix, and so on.

Note that the bottom row of the probability matrix raised to the power 1, 2, 3, etc., gives the probability of being available (first term) or failed (second term) if the system started from the failed state. The reader is invited to repeat the tree diagram (Fig. 5.7). starting from the failed state, to corroborate this. Note also that the rows always summate to 1; i.e. the total probability of all states. (Revision notes on simple matrix algebra are given in Appendix 7.)

If a system has more than two states (multicomponent or redundant systems), then the stochastic transitional probability matrix will have more than 2×2 elements. For example, for a two-component system, the states could be:

State	Component 1	Component 2	
1	A	A	
2	A	$\bar{A}$	
3	$\bar{A}$	A	A: available
4	$\bar{A}$	$\bar{A}$	$\bar{A}$: unavailable

The probabilities of moving from any one state to any other can be shown on a 4×4 matrix. If the transition probabilities are the same as for the previous example, then:

$P_{1 \to 1} = 0.9 \times 0.9 = 0.81$

$P_{1 \to 2} = 0.9 \times 0.1 = 0.09$

$P_{1 \to 3} = 0.1 \times 0.9 = 0.09$

$P_{1 \to 4} = 0.1 \times 0.1 = 0.01$

$P_{2 \to 1} = 0.9 \times 0.6 = 0.54$

$P_{2 \to 2} = 0.9 \times 0.4 = 0.36$

$P_{2 \to 3} = 0.1 \times 0.6 = 0.06$

$P_{2 \to 4} = 0.1 \times 0.4 = 0.04$

$P_{3 \to 1} = 0.6 \times 0.9 = 0.54$

$P_{3 \to 2} = 0.6 \times 0.1 = 0.06$

$P_{3 \to 3} = 0.4 \times 0.9 = 0.36$

$P_{3 \to 4} = 0.4 \times 0.1 = 0.04$

$P_{4 \to 1} = 0.6 \times 0.6 = 0.36$

$P_{4 \to 2} = 0.6 \times 0.4 = 0.24$

$P_{4 \to 3} = 0.4 \times 0.6 = 0.24$

$P_{4 \to 4} = 0.4 \times 0.4 = 0.16$

and the probability matrix is

$$P = \begin{vmatrix} P_{1\to1} & P_{1\to2} & P_{1\to3} & P_{1\to4} \\ P_{2\to1} & P_{2\to2} & \cdots & \\ P_{3\to1} & \vdots & \vdots & \vdots \\ P_{4\to1} & \cdots & \cdots & P_{4\to4} \end{vmatrix}$$

$$= \begin{vmatrix} 0.81 & 0.09 & 0.09 & 0.01 \\ 0.54 & 0.36 & 0.06 & 0.04 \\ 0.54 & 0.06 & 0.36 & 0.04 \\ 0.36 & 0.24 & 0.24 & 0.16 \end{vmatrix}$$

The first two terms in the first row give the probability of being available and unavailable after the first time increment, given that the system was available at the start. The availability after 2, 3, etc., intervals can be derived from P^2, P^3, ..., as above.

It is easy to see how, even for quite simple systems, the matrix algebra quickly diverges in complexity. However, computer programs can easily handle the evaluation of large matrices, so this type of analysis is feasible in the appropriate circumstances.

Continuous Markov processes

So far we have considered discrete Markov processes. We can also use the Markov method to evaluate the availability of systems in which the failure rate and the repair rate (λ, μ) are assumed to be constant in a time continuum. The state transition diagram for a single repairable item is shown in Fig. 5.9.

In the steady state, the stochastic transitional probability matrix is:

$$P = \begin{vmatrix} 1 - \lambda & \lambda \\ \mu & 1 - \mu \end{vmatrix} \tag{5.15}$$

The instantaneous availability, before the steady state has been reached, can be derived using Eqn (5.11).

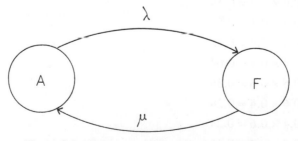

Figure 5.9 State–space diagram for a single-component repairable system

The methods described in the previous section can be applied for evaluating more complex, continuous Markov chains. Markov analysis can also be used for availability analysis, taking account of the holdings and repair rate of spares. The reader should refer to Ref. 11 in the Bibliography for details of the Markov method as applied to more complex systems.

Limitations, advantages and applications of Markov analysis

The Markov analysis method suffers one major disadvantage. As mentioned earlier, it is necessary to assume constant probabilities or rates for all occurrences (failures and repairs). It is also necessary to assume that events are s-independent. These assumptions are hardly ever valid in real life, as explained in Chapter 3 and earlier in this chapter. The extent to which they might affect the situation should be carefully considered when evaluating the results of a Markov analysis of a system.

Markov analysis requires knowledge of matrix operations. This can result in difficulties in communicating the methods and the results to people other than reliability specialists. The severe simplifying assumptions can also affect the credibility of the results.

Markov analysis is fast when run on computers and is therefore economical once the inputs have been prepared. The method is used for analysing systems such as power distribution networks and logistic systems.

MONTE CARLO SIMULATION

Monte Carlo simulation (see page 119) can also be used for system reliability and availability modelling, using suitable computer programs. Since Monte Carlo simulation involves no complex mathematical analysis, it is an attractive alternative approach. It is a relatively easy way to model complex systems, and the input algorithms are easy to understand. There are no constraints regarding the nature of input assumptions on parameters such as failure and repair rates, so non-constant values can be used. It is also easy to model aspects such as queueing rules for repairs, repair priorities and 'cannibalization' (the use of serviceable spare parts from unserviceable systems).

One problem with Monte Carlo analysis is its expensive use of computer time. Since every event (failure, repair, movement, etc.) must be sampled for every unit of time, using the input distributions, a simulation of a moderately large system over a reasonable period of time can require hours of computer run-time. Also, since the simulation of probabilistic events generates variable results, in effect simulating the variability of real life, it is usually necessary to perform a number of runs in order to obtain estimates of means and variances of the output parameters of interest, such as availability, number of repairs arising and repair facility utilization. On the other hand, the effects of variations can be assessed.

Figure 5.10 Monte Carlo model flowchart outline

A flow diagram showing the failure and repair logic for a typical system is shown in Fig. 5.10, and the results of simulation runs are shown in Fig. 5.11.

RELIABILITY APPORTIONMENT

Sometimes it is necessary to break an overall system reliability requirement down to individual subsystem reliabilities. This is common in large systems such as avionics, particularly when different design teams or subcontractors are involved. The main contractor or system design team leader requires early assurance that subsystems will have reliabilities which will match the system

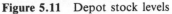

Figure 5.11 Depot stock levels

requirement, and therefore the appropriate values have to be included in the subsystem specifications.

The starting point for reliability apportionment is a reliability block diagram for the system drawn to show the appropriate system structure. The system requirement is then broken down in proportions which take account of the complexity, risk and existing experience related to each block. It is important to take account of the uncertainty inherent in such an early prediction, and therefore the block reliabilities need not aggregate to the system requirement, but to some higher reliability. The apportionment and specifications derived from it should take account of different operating conditions of subsystems. For example, a radar system might have a subsystem which operates for only half of the total system operating time, and therefore this should be shown on the RBD, and the failure rate apportioned to it should be related clearly either to the operating time of the system or of the subsystem.

STANDARD METHODS FOR RELIABILITY PREDICTION

The most commonly used standard reference for reliability prediction is US MIL-STD-756 (Ref. 1 in the Bibliography). It is cited in other standards covering reliability programme management in defence and non-defence work in the United States and in Europe. MIL-STD-756 covers the following aspects of reliability prediction:

1. The prediction plan as part of the reliability programme plan, related to the various phases of the development programme. It requires that an initial reliability prediction is performed as early as possible as part of the feasibility study, to compare options and validate the concept. Subsequent

predictions are performed to update the initial work, using more detailed design data as they become available, leading to a final prediction involving full stress analysis where applicable.

2. Procedures for performing the reliability prediction, including modelling methods and data sources. MIL-HDBK-217 is the referenced source for electronic equipment parts count and stress analysis reliability prediction (see Chapter 7).

3. Procedures for documenting and reporting the predictions.

A further useful reference is NASA-CR-1129 (Ref. 3 in the Bibliography), which describes the approaches to reliability prediction for NASA space programmes. Here the stress is laid on reliability prediction as a method of comparing designs, and there is less emphasis on quantitative methods. A British Standard, BS 5760, (Ref. 4) also describes reliability prediction methods and relates them to the reliability programme.

CONCLUSIONS

System reliability (and availability) prediction and modelling can be a frustrating exercise, since even quite simple systems can lead to complex reliability logic when redundancy, repair times, testing and monitoring are taken into account. On the other hand, the parameters used, particularly reliability values, are usually very uncertain, and can be highly variable between similar systems. This chapter has described some approaches, but it must be realized that the results can be very sensitive to parameter changes. For example, a single common-cause failure, overlooked in the analysis, might have a probability of occurrence which would completely invalidate the reliability calculations for a high reliability redundant system. In real life, availability is often determined more by spares holdings, administrative times (transport, documentation, delays, etc.) than by 'predictable' factors such as mean repair time. Therefore predictions and models of system reliability and availability should be used as a form of design review, to provide a disciplined framework for considering factors which will affect reliability and availability, and the sensitivity to changes in assumptions. Critical aspects can then be highlighted for further attention, and alternative system approaches can be compared. Also, the predictions should be used as the basis for setting the reliability and availability goals for the project.

BIBLIOGRAPHY

1. US MIL-STD-756: *Reliability Prediction*, Issue B. Available from the National Technical Information Service, Springfield, Virginia.
2. US MIL-HBK-217: *Reliability Prediction for Electronic Systems*. Available from the National Technical Information Service, Springfield, Virginia (1980).
3. NASA-CR-1129: *Reliability Prediction*. Available from the National Technical Information Service, Springfield, Virginia.

4. British Standard, BS 5760: *Reliability of Systems, Equipments and Components*. British Standards Institution, London.
5. K. C. Kapur and L. R. Lamberson, *Reliability in Engineering Design*. Wiley, New York (1977).
6. C. Singh and R. Billington, *System Reliability Modelling and Evaluation*. Hutchinson, London (1977).
7. R. H. Myers, K. L. Wong and H. M. Gordy (Eds), *Reliability Engineering for Electronic Systems*. McGraw-Hill, New York (1964).
8. F. F. Martin, *Computer Modeling and Simulation*, Wiley, New York (1968).
9. A. K. S. Jardine, *Maintenance, Replacement and Reliability*. Pitman, London (1973).
10. E. J. Henley and H. Kumanoto, *Reliability Engineering and Risk Assessment*. Prentice-Hall, Englewood Cliffs (1981).
11. R. Billinton and R. N. Allen, *Reliability Evaluation of Engineering Systems*. Plenum Press, New York (1982).

Chapter 6

RELIABILITY IN DESIGN

INTRODUCTION

The reliability of a product is strongly influenced by decisions made during the design process. Deficiencies in design affect all items produced and are progressively more expensive to correct as development proceeds. It is often not practicable or economic to change a design once production has started. It is therefore essential that design disciplines are used which minimize the possibility of failure and which allow design deficiencies to be detected and corrected as early as possible. In Chapter 4 the basic requirements for failure-free design were laid down, i.e. adequate safety margins, protection against extreme load events and protection against strength degradation. The design must also take account of all other factors that can affect reliability, such as production methods, use and maintenance.

The design process must therefore be organized to ensure that failure-free design principles are used and that any deviations from the principles are detected and corrected. Failure-free design is the only acceptable principle for any reliability-conscious project team. Anything less will be reflected in the acceptance of failures throughout the development and production cycle, and a low rate of improvement. The designer must produce designs which will not fail if manufactured and used as specified. In order to be able to do this test data may be needed to reduce uncertainties. Any subsequent failures can then be firmly classified as design deficiencies which escaped the review or test system, or as being due to manufacturing failures or overload. Failure-free design therefore involves prevention, check and cure.

This chapter (and Chapter 7 on the reliability of electronics) does not cover the range of knowledge necessary for good design. The designer must be aware of the materials, processes, components, production methods, design rules and guidelines, costs, and much else in order to create a good design. These chapters cover only fairly general topics, which should be known by reliability engineers working on most types of project. Design and development engineers should also be aware of some general principles which can assist in the creation of failure-free designs.

FAILURE PREVENTION

Load–strength analysis and safety margins

Chapter 4 covered the theoretical basis of load–strength analysis for distributed loads and strengths. All components in a design subject to load should be analysed with respect to the load and strength distributions. The consideration of load must take account of combined loads, e.g. temperature and vibration and their possible interactions.

The effect of combined loads may be very difficult to analyse and combined environment step-stress testing as described in Chapter 11 may be the most practicable way to obtain confidence in the design. Whenever the loads on a design are analysed, it is important that possible interactions are considered, since design to protect against only one load when interactions exist can be very optimistic. A high frequency cyclic stress may be superimposed upon a low frequency stress, or a loading stress may be combined with a high temperature which affects material strength, lubricity or some other property. Combined stresses can have effects out of proportion to their separate contributions, both in terms of instantaneous effects and strength degradation effects. Chapter 13 describes analytical techniques for investigating the effects of combined stress.

Strength distributions must be estimated in relation to material and process specifications and tolerances, quality control procedures and possible strength degradation. Whenever an adequate safety margin cannot be derived with reasonably high assurance from the analysis, tests should be designed to provide the necessary assurance.

An essential element of design to ensure adequate safety margins is correct tolerancing. Tolerancing, with appropriate quality control of manufacture, provides control over the resulting strength distributions. Analysis may then be based on worst case strength, rather than on an anticipated strength distribution. This will give a conservative estimate of reliability, which may be desirable, and will also simplify the analysis.

Load protection

Protection against extreme loads is not always possible, but should be considered whenever practicable. In many cases the maximum load can be pre-determined, and therefore no special protection is called for. However, in many other loading situations extreme external loads can occur and can be protected against. Standard products are available to provide protection against, for example, overpressure in hydraulic or pneumatic systems, impact loads or electrical overload. When overload protection is provided, the reliability analysis is perfomed on the basis of the maximum load which can be anticipated, bearing in mind the tolerances of the protection system. In appropriate cases, loads which can occur when the protection system fails must also be considered.

However, in most practical cases it will be sufficient to design to withstand a predetermined load and to accept the fact that loads above this will cause failure. The probability of such loads occurring must be determined for a full reliability analysis to be performed. It may not always be practicable to determine the distribution of such extreme events, but data may be available either from failure records of similar items, or from test or other records.

Where credible data are not available, the worst design load case must be estimated. The important point is that the worst design case is estimated and specified. One of the commonest causes of failure is the use of safety factors related to average load conditions, without adequate consideration having been given to the extreme conditions which can occur during use of the product.

Protection against strength degradation

Strength degradation, in its many forms, can be one of the most difficult aspects to quantify in design reliability analysis. Strength degradation due to fatigue in metals is well understood and documented, and therefore reliability analysis involving metal fatigue, including the effects of stress raisers such as notches, corners, holes and surface finish, can be performed satisfactorily, and parts can be designed to operate below the fatigue limit, or for a defined safe life.

However, other weakening mechanisms are often more complex. Combined stresses may accelerate damage or reduce the fatigue limit. Corrosion and wear are dependent upon environments and lubrication, the effects of which are therefore often difficult to forecast. If complete protection is not possible, the designer must specify maintenance procedures for inspection, lubrication or scheduled replacement.

Reliability analysis of designs with complex weakening processes is often impracticable. Tests should then be designed to provide the required data by generating failures under known loading conditions.

DESIGN ANALYSIS METHODS

Despite discipline, training and care, it is inevitable that occasional oversights or errors will occur in new designs. Design analysis methods have been developed to highlight critical aspects and to focus attention on possible shortfalls.

Design analyses are sometimes considered tedious and expensive. In most cases the analysis will show that nearly all aspects of the design are satisfactory, and much more effort will have been expended in showing this than in highlighting a few deficiencies. However, the discovery of a very few deficiencies at an appropriately early stage can save far more than the costs that might be incurred by having to modify the design at a later stage, or by having to live with the consequences of the defect. Therefore, well-managed design analyses are extremely cost-effective. The tedium and expense can be greatly reduced by good planning and preparation. In this section, we will describe the main design analysis techniques available. Their place in the overall design review process and the way they should be managed are also covered. The main reliability design analysis techniques are:

1. Reliability prediction (covered in Chapter 5).
2. Load–strength review.
3. Failure mode, effects and criticality analysis.
4. Fault tree analysis.
5. Parameter variation analysis.

LOAD–STRENGTH ANALYSIS

Load and strength will have been considered as part of the reliability prediction. The load–strength analysis is a procedure to ensure that all load and strength aspects have been considered in deriving the design, and if necessary in planning of tests. Quantitative analysis of the interactions of distributed load and strength, as described in Chapter 4, might be part of this procedure.

Table 6.1 is an example of a hypothetical load–strength analysis for a mechanical and electrical assembly. For electronic systems a different procedure is used, combined with the reliability prediction; this is covered in Chapter 7. The example shows approaches that can be used for different aspects of the analysis. Event probabilities can be expressed as full distributions, or as the likelihood of a particular limiting case being exceeded. The former is more appropriate when the load(s) can cause degradation, or if a more detailed reliability assessment is required. Both examples show typical, though rather simple, cases where the effects of combined loads might have been overlooked but for the analysis. For example, the solenoid might be supplied with a manufacturer's rating of 28V operating, ± 2 V, and a maximum ambient temperature of 45°C. A room temperature test of the solenoid might have confirmed its ability to function with a 32 V supply without overheating. However, the combined environment of + 45 °C and 32 V supply, albeit an infrequent occurrence, could lead to failure.

The mechanical example is less easy to analyse and testing is likely to be the best way of providing assurance, if the assembly is critical enought to warrant it.

FAILURE MODE, EFFECTS AND CRITICALITY ANALYSIS (FMECA)

Failure mode, effects and criticality analysis (FMECA) is probably the most widely used and most effective design reliability analysis method. The usual reference is US MIL-STD-1629 (Procedures for Performing a Failure Mode, Effects and Criticality Analysis). The principle of FMECA is to consider each mode of failure of every component of a system and to ascertain the effects

Table 6.1 Load–strength analysis example

Item (Matl, function)	Worst case load/ combined load	Frequency/ probability of occurrence	Data source	Combined effect	Strength	Remarks
Rivet (× 4) (aluminium, fixing bracket to plastic frame)	1. 50 N total, axial 2. 40 N, lateral impact 3. Temperature 0–35 °C	Continuous See load distribution annex 1		— Combine with 1. *Degradation* NIL	Plastic frame is weak link NIL effect	Life test to confirm Thickness of plastic frame at bracket attachment may be critical feature
Solenoid coil	1. 32 V (at 27 °C) 2. 45 °C ambient	$1/10^4$ h $1/10^2$ h	Operating data Data on power supply variation	72 °C	Insulation limited to 70 °C	Overvoltage protection or improved cooling needed

on system operation of each failure mode in turn. Failure effects may be considered at more than one level, e.g. at subsystem and at overall system level. Failure modes are classified in relation to the severity of their effects.

An FMECA may be based on a hardware or a functional approach. In the hardware approach actual hardware failure modes are considered (e.g. resistor open circuit, bearing seizure). The functional approach is used when hardware items cannot be uniquely identified or in early design stages when hardware is not fully defined. In this approach function failures are considered (e.g. no feedback, memory lost). Note that a functional failure mode can become a hardware failure effect in a hardware-approach FMECA. An FMECA can also be performed using a combination of hardware and functional approaches.

MIL-STD-1629 provides two basic methods (101,102) for performing an FMECA. In both an FMECA worksheet is generated to document the analysis (Figs 6.1 and 6.2). Method 101 is a non-quantitative method, which serves to highlight failure modes whose effects would be considered important in relation to severity, detectability, maintainability or safety.

Method 102 (criticality analysis) includes consideration of failure rate or probability, failure mode ratio and a quantitative assessment of criticality, in order to provide a quantitative criticality rating for the component or function. The *failure mode criticality number* is

$$C_m = \beta\alpha\lambda_p t \tag{6.1}$$

where
β = conditional probability of loss of function or mission
α = failure mode ratio (for an item, $\sum_\alpha = 1$)
λ_p = part failure rate (for repairable systems)
t = operating or at-risk time of item

$\lambda_p t$ can be replaced by failure probability, $1-\exp(-\alpha\lambda_p t)$, when λ_p might be 1/MTTF for a non-repairable system.

The *item criticality number* is the sum of the failure mode criticality numbers for the item.

Steps in performing an FMECA

An effective FMECA can be performed only by an engineer or team of engineers having thorough knowledge of the system's design and application. The first step therefore is to obtain all the information available on the design. This includes specifications, drawings, computer-aided design (CAD) data, stress analysis, test results, etc., to the extent they are available at the time. For a criticality analysis, the reliability prediction data must also be available or they might be generated simultaneously.

A system functional block diagram and reliability block diagram (page 129) should be prepared, if not already available, as these form the basis for preparing the FMECA and for understanding the completed analysis.

System _____
Indenture level _____
Reference drawing _____
Mission _____

Date _____
Sheet _____ of _____
Compiled by _____
Approved by _____

Identification number	Item/functional identification (nomenclature)	Function	Failure modes and causes	Mission phase/operational mode	Failure effects			Failure detection method	Compensating provisions	Severity class	Remarks
					Local effects	Next higher level	End effects				

Figure 6.1 MIL-STD-1629 worksheet for method 101

System _____
Indenture level _____
Reference drawing _____
Mission _____

Date _____
Sheet _____ of _____
Compiled by _____
Approved by _____

Identification number	Item/functional identification (nomenclature)	Function	Failure modes and causes	Mission phase/ operational mode	Severity class	Failure probability / Failure rate data source	Failure effect probability	Failure mode ratio (a)	Failure rate (λ_p)	Operating time (t) (β)	Failure mode crit $C_m = \beta a \lambda_p'$	Item crit $C_r = \Sigma(C_m)$	Remarks

Figure 6.2 MIL-STD-1629 worksheet for method 102

If the system operates in more than one phase in which different functional relationships or item operating modes exist, these must be considered in the analysis. The effects of redundancy must also be considered by evaluating the effects of failure modes assuming that the redundant subsystem is or is not available.

An FMECA can be performed from different viewpoints, such as safety, mission success, availability, repair cost, failure mode or effect detectability, etc. It is necessary to decide, and to state, the viewpoint or viewpoints being considered in the analysis. For example, a safety-related FMECA might give a low criticality number to an item whose reliability seriously affects availability, but which is not safety critical.

The FMECA is then prepared, using the appropriate worksheet, and working to the item or subassembly level considered appropriate, bearing in mind the design data available and the objectives of the analysis. For a new design, particularly when the effects of failure are serious (high warranty costs, reliability reputation, safety, etc.) the analysis should take account of all failure modes of all components. However, it might be appropriate to consider functional failure modes of subassemblies when these are based upon existing designs, e.g. modular power supplies in electronic systems, particularly if the design details are not known.

When CAD is used for design the facility should be used to aid the assessment of failure effects. For example, electronic circuit analysis programs such as SPICE, HILO and LASAR can be used to determine the effects on circuit performance of failure modes simulated by the program. In mechanism design the effects of distortion, fracture, etc., can be accurately assessed. Use of CAD for this work can provide more accurate and objective assessment than can be obtained by study of circuit diagrams and drawings.

The FMECA should be started as soon as initial design information is available. It should be performed iteratively as the design evolves, so that the analysis can be used to influence the design and to provide documentation of the eventually completed design. Design options should be separately analysed, so that reliability implications can be considered in deciding on which option to choose. Test results should be used to update the analysis.

Uses for FMECA

FMECAs can be used very effectively for several purposes, in addition to the prime one of identifying safety or reliability critical failure modes and effects. The main uses are:

1. Preparation of diagnostic routines such as flowcharts or fault-finding tables. The FMECA provides a convenient listing of the failure modes which produce particular failure effects or symptoms, and their relative likelihoods of occurrence.
2. Preparation of preventive maintenance requirements. The effects and

likelihood of failures can be considered in relation to the need for scheduled inspection, servicing or replacement. For example, if a failure mode has an insignificant effect on safety or operating success, the item could be replaced only on failure, rather than at scheduled intervals, to reduce the probability of failure.

3. Design of built-in test (BIT), failure indications and redundancy. The failure detectability, including BIT, viewpoint is an important one in FMECA of systems which include these features.
4. For analysis of testability, particularly for electronic subassemblies and systems, to ensure that hardware can be economically tested and failures diagnosed, using automatic or manual test equipment.
5. For development of software for automatic test and BIT.
6. For retention as formal records of the safety and reliability analysis, to be used as evidence if required in reports to customers or in product safety litigation.
7. A FMECA can be performed specifically to consider the possibility of production-induced failures, e.g. wrong diode orientation. Such a production FMECA can be very useful in test planning and in design for ease of production.

It is important to coordinate these activities, so that the most effective use can be made of the FMECAs in all of them, and to ensure that FMECAs are available at the right time and to the right people.

Computerized FMECA

Computer programs have been developed for performance of FMECA. Using a computer program instead of FMECA worksheets allows FMECAs to be produced more quickly and accurately, and greatly increases the ease of editing and updating to take account of design changes, design options, different viewpoints, and different input assumptions. Like any other computer-aided design technique, computerized FMECA frees engineers to concentrate on engineering, rather than on tedious compilation, so that for the same total effort designs can be more thoroughly investigated, or less effort can be expended for the same depth of analysis. Also, by eliminating the more tedious and time-consuming aspects of the work, engineers' motivation and effectiveness is increased.

Computerized FMECA enables more perceptive analysis to be performed. Failure effects can be ranked in criticality order, at different system levels, in different phases of system operation and from different viewpoints. Report preparation can be partly automated and sensitivity analyses quickly performed.

Figure 6.3 shows part of a computerized FMECA, performed using the PREDICTOR® FMECA program, on a hypothetical anti-aircraft missile control system. PREDICTOR is available from Management Sciences Inc., Alberquerque, NM.

Mission failure effects (MFE)

MFE	Effect	Loss probability
1	No effect	0
2	True no control	1
3	False no control	1
4	Intermittent control	0.4
5	Unstable control	0.7
6	Control flutter	0.5

SFE no.	System effect	MFE no.	System effect failure probability	Percent of all groups failure probability	Mission loss prob.	System criticality	Percent contribution to all groups criticality
2	No drive to control fins	2	0.3713E-05	19.69	1.0	0.3713E-05	24.75
7	T no output to solenoid	2	0.2495E-05	13.23	1.0	0.2495E-05	16.63
5	Drive to one extremity of travel	2	0.1500E-05	7.95	1.0	0.1500E-06	9.99
8	F no output to solenoid	3	0.1500E-05	7.95	1.0	0.1500E-05	9.99
10	Unstable output to solenoid	5	0.1976E-05	10.48	0.7	0.1383E-05	9.22
11	Low output to solenoid	4	0.2686E-05	14.24	0.4	0.1074E-05	7.16
13	No fin movement	2	0.1000E-05	5.30	1.0	0.1000E-05	6.66
3	Intermittent drive	5	0.1383E-05	7.33	0.7	0.9661E-06	6.45
12	Jitter on output to solenoid	6	0.9000E-06	4.77	0.5	0.4500E-06	2.99
4	Slow drive	4	0.1080E-05	5.72	0.4	0.4320E-06	2.87
9	Incorrect sensing of input demand	2	0.3952E-06	2.09	1.0	0.3952E-06	2.63
6	Drive ceases prematurely	4	0.2240E-06	1.18	0.4	0.8962E-07	0.59

Figure 6.3 Part of output listing from computerized FMECA (PREDICTOR®)

FAULT TREE ANALYSIS (FTA)

Fault tree analysis (FTA) is a reliability/safety design analysis technique which starts from consideration of system failure effects, referred to as 'top events'. The analysis proceeds by determining how these can be caused by individual or combined lower level failures or events. It differs from FMECA in being strictly a top-down approach and in considering multiple failures as a matter of course.

Standard symbols are used in constructing an FTA to describe events and logical connections. These are shown in Fig. 6.4. Figure 6.5 shows a simple FTA for an aircraft internal combustion engine, in which the top event is failure to start. There are two ignition systems in an active parallel redundant configuration. The FTA shows that failure to start can be caused by either fuel flow failure, carburettor failure or ignition failure (three-input OR gate). At a lower level total ignition failure is caused by failure of ignition systems 1 and 2 (two-input AND gate). The reliability block diagram of the system is shown at Fig. 6.6.

In addition to showing the logical connections between failure events in relation to defined top events, FTA can be used to quantify the top event probabilities, in the same way as in block diagram analysis (page 138). Failure probability derived from the reliability prediction values can be assigned to the failure events, and cut set and tie set methods can be applied to evaluate system failure probability.

Note that a different FTA will have to be constructed for each defined top event which can be caused by different failure modes or different logical connections between failure events. In the engine example, if the top event is 'unsafe for flight' then it would be necessary for both ignition systems to be available before take-off, and gate Q1 would have to be changed to an OR gate.

The FTA shown is very simple; a representative FTA for a system such as this, showing all component failure modes, or for a large system, such as a flight control system or a chemical process plant, can be very complex and impracticable to draw out and evaluate manually. Computer programs are used for generating and evaluating large FTAs. The use of computer programs for FTA provides the same advantages of effectiveness, economy and ease of iterative analysis as described for FMECA. Many FTA programs have been written, and several are generally available. Examples are FTAP (University of California, Berkeley) and RESULTS (Management Sciences Inc., Albuquerque, NM).

Common mode failures

A common mode (or common cause) failure is one which can lead to the failure of all paths in a redundant configuration. Identification and evaluation of common mode failures is very important, since they might have a higher probability of occurrence than the failure probability of the redundant system when only individual path failures are considered. In the design of redundant

164

EVENTS

BASIC EVENT

A basic fault event that requires no further development. Is s-independent of other events.

BASIC EVENT

Is s-dependent on lower events developed as a separate fault tree.

BASIC EVENT

Is s-dependent upon lower events, but not developed downwards.

COMBINATION EVENT

An event that results from the combination of basic events through the input logic gates.

TRANSFERRED EVENT

OUT

A line from the apex of the triangle indicates a transfer in; a line from the side denotes a transfer out.

IN

LOGIC

SWITCH

Used to include or exclude parts of the fault tree, which may or may not apply to certain situations.

AND GATE

Failure (next higher event) will occur if all inputs fail (parallel redundancy).

OR GATE

Failure (next higher event) will occur if any input fails (series reliability).

INHIBIT GATE

INHIBIT gates describe a causal relationship between one fault and another. The input event directly produces the output event if the indicated condition is satisfied.

Figure 6.4 Standard symbols used in fault tree analysis

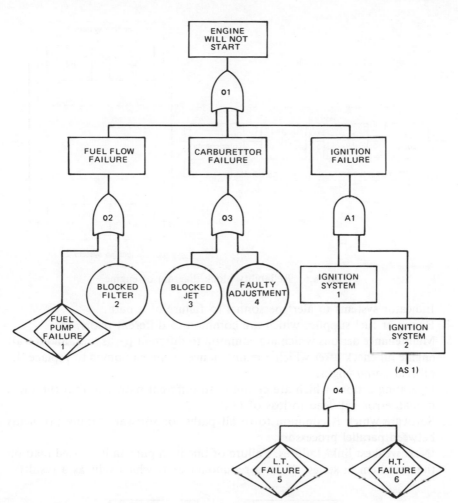

Figure 6.5 FTA for engine (incomplete)

systems it is very important to identify and eliminate sources of common mode failures, or to reduce their probability of occurrence to levels an order or more below that of other failure modes.

For example, consider a system in which each path has a reliability $R = 0.99$ and a common mode failure which has a probabiliy of non-occurrence $R_{CM} = 0.98$. The system can be designed either with a single unit or in a dual redundant configuration (Fig. 6.7 a and b). Ignoring the common mode failure, the reliability of the dual redundant system would be 0.9999. However, the common mode failure practically eliminates the advantage of the redundant configuration.

Examples of sources of common mode failures are:

1. Changeover systems to activate standby redundant units.
2. Sensor systems to detect failure of a path.

Figure 6.6 Reliability block diagram of engine

3. Indicator systems to alert personnel to failure of a path.
4. Power or fuel supplies which are common to different paths.
5. Maintenance actions which are common to different paths, e.g. an aircraft engine oil check after which a maintenance technician omits to replace the oil seal on *all* engines.
6. Operating actions which are common to different paths, so that the same human error will lead to loss of both.
7. Software which is common to to all paths, or software timing problems between parallel processors.
8. 'Next weakest link' failures. Failure of one item puts an increased load on the next item in series, or on a redundant unit, which fails as a result.

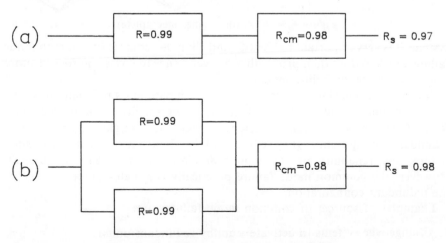

Figure 6.7 Effect of common mode failure

Common mode failures can be very difficult to foresee, and great care must be taken when analysing safety aspects of systems to ensure that possible sources are identified.

Enabling events

An enabling event is one which, whilst not necessarily a failure or a direct cause of failure, will cause a higher level failure event when accompanied by a failure. Like common mode failures, enabling events can be important and difficult to anticipate, but they must be considered in FTA. Examples of enabling events are:

1. Warning systems disabled for maintenance, or because they create spurious warnings.
2. Controls incorrectly set.
3. Operating or maintenance personnel following procedures incorrectly, or not following procedures.
4. Redundant elements being out of action due to maintenance.

RELIABILITY PREDICTIONS FOR FMECA AND FTA

Since FMECA and FTA are performed primarily to identify critical failure modes and to evaluate design options, failure rate or reliability values which could be considered as realistic worst cases should be used. Standard methods, such as MIL-STD-1629 for FMECA, stipulate the reliability prediction methods to be used, e.g. MIL-HDBK-217 for electronics, so these must obviously be used when such standards are called for in a development contract. However, it is very important to appreciate the large amount of uncertainty inherent in reliability prediction, particularly at the level of individual failure events (see Chapter 5). Therefore, worst-case or pessimistic reliability values should always be used as input assumptions for failure modes which are identified as critical, or which might be critical if the pessimistic assumption proved to be realistic. Generally, the more critical the failure mode the more pessimistic should be the worst-case reliability assumptions. The use of computer programs for FMECA and FTA greatly facilitates this type of sensitivity analysis.

PARAMETER VARIATION AND TOLERANCE ANALYSIS

It is often necessary to consider the effects of changes or uncertainties in the values of parameters upon the performance of the system. For example, in production, part tolerances can affect performance. Tolerances and parameter drift can combine to cause out-of-specification performance and thus failure.

There are two possible approaches to designing for parameter variation and tolerances. The traditional approach is to consider the worst case. For example,

168

if a shaft must fit into a bore, the shaft and bore diameters and tolerances might be specified as: shaft, 20 ± 0.1 mm; bore, 20.2 ± 0.1 mm, in order to ensure that all shafts fit all bores. However, if we use statistical tolerancing, much closer combined tolerances can be achieved. In this case, if the tolerance limits are based upon machining processes which produce parts with *s*-normally distributed diameters, of which 2.5 per cent. are oversize and 2.5 per cent. are undersize (2σ limits), the probability of a shaft and bore having an interference would be

$$0.025 \times 0.025 = 0.000625$$

On the other hand, most combinations will result in a fairly loose fit. Figure 6.8 shows the situation graphically. Twenty-five per cent. of combinations will have fits greater than 0.2 mm.

If, however, statistical tolerancing were used, we could design for much closer nominal diameters and still have an acceptably low probability of interference. If the shaft nominal diameter was set at 20.1 mm (dotted line on Fig. 6.8), the interference probability P_I can be calculated as (see Eqn 4.5):

$$P_I = 1 - \Phi\left[\frac{D_1 - D_2}{(\sigma_1^2 + \sigma_2^2)^{1/2}}\right]$$

In this case, if 2σ is 0.1 mm then σ is 0.05 mm. Therefore,

$$P_I = 1 - \Phi\left[\frac{20.2 - 20.1}{(0.05^2 + 0.05^2)^{1/2}}\right]$$

$$= 0.08$$

This is a very simple example to illustrate the principle. Reference 10 in the Chapter 2 Bibliography covers statistical tolerancing in more detail. For systems such as electronic circuits, where tolerances of many components must be considered, statistical analysis of parameter tolerances or drift can provide more economic designs, since the probability of several parts being near their specification limits is much lower than for one or two parts. Statistical tolerancing can also result in lower production costs, since part and subassembly

Figure 6.8 Shaft–bore interference

test specifications do not need to be as tight and thus there will be fewer test rejects. However, statistical tolerancing must be based on the correct models, and it is not always safe to assume that variables are *s*-normally distributed. For example, many electronic parts are sorted, with parts whose values lie close to the nominal being sold as low tolerance parts. Thus the distribution of values of a lot from which such parts have been removed would be bimodal, with no parts being within the sorted range of the nominal value.

PARTS, MATERIALS AND PROCESSES (PMP) REVIEW

All new parts, materials and processes called up in the design should be identified. 'New' in this context means new to the particular design and production organization. The designer is likely to assume that a part or material will perform as specified in the brochure and that processes can be controlled to comply with the design. The reliability and quality assurance (QA) staff must ensure that this faith is well founded. New parts, materials and processes must therefore be assessed or tested before being applied, so that adequate training for production people can be planned, quality control safeguards set up and alternative second sources located. New parts, materials and processes must be formally approved for production and added to the approved lists.

Materials and processes must be assessed in relation to reliability. The main reliability considerations are:

1. *Cyclical loading*. Whenever loading is cyclical, including frequent impact loads, fatigue must be considered.
2. *External environment*. The environmental conditions of storage and operation must be considered in relation to factors such as corrosion and extreme temperature effects.
3. *Wear*. The wear properties of materials must be considered for all moving parts in contact.

There is such a wide variation of material properties, even among categories such as steels, aluminium alloys, plastics or rubbers, that it is not practicable to generalize about how these should be considered in relation to reliability. Material selection will be based upon several factors; the design review procedure should ensure that the reliability implications receive the attention appropriate to the application.

NON-MATERIAL FAILURE MODES

Most reliability engineering is concerned with material failure, such as caused by load—strength interference and strength degradation, and previous chapters have dealt with the quantitative analysis of these situations. However, there is a large class of failure modes which are not related to this type of material failure, but which can have consequences which are just as serious. Examples of these are:

1. Fasteners which secure essential panels and which can be insecurely fastened due to wear or left unfastened without being detected.
2. Wear in seals, causing low pressure leaks in hydraulic systems.
3. Resistance increase of electrical contacts due to arcing and accretion of oxides.
4. Failure of protective surfaces, such as paints, metal plating or anodized surfaces.
5. Distortion of pins, or intermittent contact, on multipin electrical connectors.
6. Drift in electronic component parameter values.
7. Noisy electronic components.
8. Other personnel-induced failures such as faulty maintenance, handling or storage, e.g. omitting to charge electrolytic capacitors kept in long term storage, which can result in reduced charge capacity in use.

All of these modes can lead to perceived failures. Failure reporting systems always include a proportion of such failures. However, there is usually more scope for subjective interpretation and for variability due to factors such as skill levels, personal attitudes and maintenance procedures, especially for complex equipment. Statistical analysis of such failure data can be performed in the same way as for material failures. For example, electronic component parameter drift can be analysed to produce data for use in design, in the same way as shown for life data in Chapter 3, and variance analysis can show up features such as s-significant differences between the reliabilities by different users of the same equipment. However, non-material failures can be harder to assess at the design stage, and often do not show up during a test programme. Design reliability assessments should address these types of failure, even though it may be impracticable to attempt to predict the frequency of occurrence in some cases, particularly for personnel-induced failures.

Human reliability

The term 'human reliability' is sometimes used to describe the probability of human fallibility resulting in system failure. When people are involved in the operation or maintenance of systems, the designer must take account of the interfaces between human and machine performance. This is particularly important when safety is involved. Safety assessments must extend beyond consideration of single failures to take account of multiple failures, some of which may be undetectable or undetected until too late. Human reliability is often a feature of these situations, when a failure occurs. The designer might correctly identify the need for checks for dormant failures, or action to be taken in the case of failure, but the assessment should address the questions:

1. What will be the effect if the human does fail to perform as required?
2. Are the actions required clearly specified (in the maintenance schedule), easy to perform and — in the case of emergency actions — unambiguous?
3. During manufacture, test, operation and repair, is it easy to do the right thing and difficult (impossible) to do the wrong thing?

Modern materials and techniques have reduced the need for scheduled maintenance intervention in many areas, such as lubrication of bearings and calibration of electronic equipment. On the other hand, the wider consumer, industrial and military use of complex equipment and the consequent demand for safe, easy, low maintenance operation make the assessment of human factors in reliability and maintainability increasingly important in new designs. In performing safety assessments, e.g. fault tree analysis, human failure possibilities should be included.

CRITICAL ITEMS LIST

The critical items list is a summary of the items shown by the other analyses to be likely either to have an appreciable effect on the product's reliability or to involve uncertainty. Its purpose is to highlight these items and summarize the action being taken to reduce the risks. The initial list will be based upon the design analyses, but updates will take account of test results, design changes and service data as the project develops. The critical items list is a top document for management reporting and action as it is based upon the 'management by exception' principle and summarizes the important reliability problems. Therefore, it should not usually include more than ten items and these should be ranked in order of criticality, so that management attention can be focused upon the few most important problems. It could be supported by a Pareto chart (Chapter 13) to show the relative importance, when there are sufficient data. The critical items list should provide only identification of the problem and a very brief description and status report, with references to other relevant reports.

MANAGEMENT OF DESIGN REVIEW

The review techniques described must be made part of a disciplined design sequence, or they will merely generate work and not advance the objective of more reliable design. To be effective, they must be performed by people who understand the design. This does not necessarily mean the designer, for two reasons. First, the analyses are an audit of his work and therefore an independent assessment is generally more likely to highlight aspects requiring further work than would be the case if the designer were reviewing his own work. Second, the analyses are not original work in the same sense as is the design.

The designer is paid to be creative and time spent on reassessing this effort is non-productive in this sense. The designer may, however, be the best qualified to perform much of the analysis, since he knows the problem, assessed the options, carried out all the design calculations and created the solution. On the other hand, the creative talent may not be the best at patiently performing the rather tedious review methods.

The best solution to this situation is for the person performing the reviews to work closely with the designer and to act as his 'Devil's advocate' during the creative process. In this way, the designer and the reviewer work as a team, and problem areas are highlighted as early as possible. The organization of reliability engineering staff to provide this service is covered in Chapter 14. The reviewer should ideally be a reliability engineer who can be respected by the designer as a competent member of a team whose joint objective is the excellence of the design. Since the reliability engineer is unlikely to spend as much time on one design as the designer, one reliability engineer can usually cover the work of several designers. The ratio obviously depends upon the reliability effort considered necessary on the project and on the design disciplines involved.

By working as a team, the design and reliability staff can resolve many problems before the formal analysis reports are produced, and agreement can be reached on recommendations, such as the tests to be performed. Since the reliability engineer should plan and supervise the tests, the link is maintained. Also, the team approach makes it possible for designs to be adequately reviewed and analysed before drawings are signed off, beyond which stage it is always more difficult and more expensive to incorporate changes.

Unfortunately, this team approach is frequently not applied, and design and reliability staff work separately in preparing analyses and criticizing one another's work at a distance, either by memo or over the conference table. Design review techniques then lose credibility, as do reliability staff. The main victim is the design itself, since the protagonists usually prosper within their separate organizations. If the organization is such that it is not possible to set up a team approach to design reliability, it is usually better not to try to apply the design review methods at all, since the effect will almost certainly be too little, too late.

To be of continuing value, the design analyses must be updated continually as design and development proceed. Each formal review must be based upon analyses of the design as it stands and supported by test data, parts assessments, etc. The analyses should be scheduled as part of the design programme, with design reviews scheduled at suitable intervals. The reviews should be planned well in advance, and the designer must be fully aware of the procedure. All people attending must also be briefed in advance, so they do not waste review time by trying to understand basic features. To this end, all attendees must be provided with a copy of all formal analysis reports (reliability prediction, load–strength analysis, PMP review, maintainability analysis, critical items list, FMECA, FTA) and a description of the item, with appropriate design data such as drawings. The designer should give a short presentation of his or her solution and clear up any general queries. Each analysis report should then form a separate agendum, with the queries and recommendations as the subjects for discussion and decision. If experience has generated a checklist appropriate to the design, this could also be run through, but see the comments below.

With this procedure, nearly all aspects requiring further study or decision will have been discussed before, during the continuous, informal process of the team approach to preparing the analyses. The formal review then becomes a decision-making forum, and it is not bogged down with discussion of trivial points. This contrasts markedly with the type of design review meeting which is based largely upon the use of checklists, with little preparatory work. Such reviews become a stolid march through the checklist, many of whose questions might be irrelevant to the design. They can become a substitute for thinking.

Three golden rules for the use of checklists should be:

1. Use them in the design office, not during the formal design review meetings.
2. Ensure that they are relevant and up to date.
3. Avoid vague questions such as 'Has maintenance been considered?', or even 'Are the grease points accessible?' 'What access is provided for lubrication?' would be a better question, since it calls for detailed response, not a simple affirmative.

The design review team should consist of staff from sales, production, QA, and specialists in key design areas. The people on the spot are the designer and his reliability engineer team member (who may belong to the QA department). The chairman should be the project manager or another person who can make decisions affecting the design, e.g. the chief designer. Sometimes design reviews are chaired by the procuring agency, which may require the option of attending. A design review which is advisory and has no authority is unlikely to be effective, and therefore all those attending must be concerned with the project (apart from specialists called in as advisers).

Formal design review meetings should be scheduled to take place when sufficient information is available to make the meeting worth while and in time to influence future work with the minimum of interference with project schedules and budgets. Three reviews are typical, based upon initial designs, completion of development testing and production standard drawings. Each review authorizes transition to the next phase, with such provisos deemed to be necessary, e.g. design changes, additional tests. The design reviews should be major milestones in a project's evolution. They are not concerned solely with reliability, of course, but reliability engineers have considerably influenced the ways that modern design reviews are conducted, and design reviews are key events in reliability programmes.

CONFIGURATION CONTROL

Configuration control is the process whereby the exact design standard of a system is known. Configuration control applies to hardware and to software. Effective configuration control ensures that, for example, the specifications and sources of components, and the issue numbers of drawings, can be readily identified for a particular system. Configuration control is very important in the development and production of systems, and it is mandatory for projects such as in aerospace (civil and military) and defence. Formal control should start after the first design review.

Configuration control is important to reliability, since it allows failures to be traced back to the appropriate design standard. For example, failures might occur in a component machined to a particular tolerance; the configuration control system should enable this cause to be identified.

BIBLIOGRAPHY

1. C. O. Smith, *Introduction to Reliability in Design*. McGraw-Hill, New York (1976).
2. US MIL-STD-1629: *Failure Mode and Effects Analysis*. Available from the National Technical Information Service, Springfield, Virginia.
3. British Standards, BS 5760: *Reliability of Systems, Equipment and Components*. British Standards Institution, London.
4. R. W. Hertzberg, *Deformation and Fracture Mechanics of Engineering Materials*. (2nd Ed) Wiley, New York (1983).
5. J. A. Collins, *Failure of Materials in Mechanical Design*. Wiley, New York (1981).

Chapter 7

ELECTRONIC SYSTEMS RELIABILITY

INTRODUCTION

The reliability of electronic systems requires to be covered as a topic on its own in reliability engineering, for several reasons. Reliability engineering and management grew up largely in response to the problems of electronic equipment reliability and many of the techniques have been developed from electronics applications. The design and construction of an electronic system, more than any other branch of engineering, involves the utilization of very large numbers of components which are similar, but over which the designer and production engineer have relatively little control. For example, for a given logic function a particular integrated circuit device might be selected. Apart from choosing a functionally identical device from a second source, the designer usually has no option but to use the catalogued item. The reliability of the device used can be controlled to a large extent in the procurement and manufacturing phases but, as will be explained, mainly by quality control methods. The circuit designer generally has little control over the design reliability of the device. This trend has become steadily more pronounced from the time that complex electronic systems started to be produced. As the transistor gave way to the integrated circuit (IC) and progressively with the advent of large scale integration (LSI) and very large scale integration (VLSI), the electronic system designer's control over some of the major factors influencing reliability has decreased. However, this is changing in some respects as system designs are increasingly implemented on custom-designed or semi-custom integrated circuits. This aspect is covered in more detail on page 180. This is not to say that the designer's role is diminished in relation to reliability. Rather, the designer of an electronic system must be, more than in most other branches of engineering, a member of a reliability team, involving people from production, quality control, test planning, reliability engineering and others. Without such a team approach, his or her functionally correct design could be highly unreliable. It is less likely that the designer of a functionally correct hydraulic or mechanical system could be as badly let down. It is important to understand the reasons for this difference.

For the great majority of electronic components and systems the major determinant of reliability is quality control of all the production processes.

This is also true of non-electronics (subject in both cases to the items being used within specification). However, most non-electronic equipments can be inspected and tested to an extent sufficient to assure that they will operate reliably. Electronic components cannot be easily inspected, since they are nearly always encapsulated. In fact, apart from X-ray inspection of parts for ultra-high reliability products, internal inspection is not generally possible. Since electronic components, once encapsulated, cannot be inspected, and since the size and quantity of modern components dictates that very precise dimensions be held at very high production rates, it is inevitable that production variations will be built into any component population. Gross defects, i.e. failure to function within specification, can easily be detected by automatic or manual testing. However, it is the defects that do not immediately affect performance that are the major causes of electronic component unreliability.

Consider a typical failure mechanism in an electronic component: a weak mechanical bond of a lead-in conductor to the conducting material of the device. This may be the case in a resistor, a capacitor, a transistor or an IC. Such a device might function satisfactorily on test after manufacture and in all functional tests when built into a system. No practical inspection method will detect the flaw. However, because the bond is defective it may fail at some later time, due to mechanical stress or overheating due to a high current density at the bond. Several other failure mechanisms lead to this sort of effect, e.g. flaws in semiconductor material and defective hermetic sealing. Similar types of failure occur in non-electronic systems, but generally they do not predominate.

The typical 'electronic' failure mechanism is a wearout or overstress failure of a defective item. In this context, 'good' components do not fail, since the application of specified loads during the anticipated life will not lead to failure. While every defective item will have a unique life characteristic, depending upon the nature of the defect and the load(s) applied, it is possible to generalize about the nature of the failure distributions of electronic components. Taking the case of the defective bond, its time to failure is likely to be affected by the voltage applied across it, the ambient temperature around the device and mechanical loading, e.g. vibration. Other failure mechanisms, say a flaw in a silicon crystal, may be accelerated mainly by temperature variations. In semiconductor devices increased temperature results in higher currents being passed, and positive feedback effect can lead to 'thermal runaway'. Defects in devices can result in high localized current densities, leading to failure when the critical value for a defective device is exceeded. In general, experience shows that the predominant failure-accelerating stresses for most electronic components are temperature and voltage. Also, electronic components characteristically have a decreasing hazard rate, due to the fact that as weaker components fail and are rejected the proportion of defectives in the population is reduced. Similarly, repairable electronics systems usually show a decreasing failure rate trend as defective components fail and are replaced or system manufacturing defects such as solder defects show up and are repaired. Wearout is seldom a characteristic of electronic components or systems.

Of course, by no means all electronic system unreliability is due to defective components. The designer still has the task of ensuring that the load applied to components in the system will not exceed rated (or derated values), under steady-state, transient, test or operating conditions. Since electronic component reliability depends upon temperature, design to minimize temperatures, particularly localized 'hot spots', is necessary. Thus the designer is still subject to the reliability disciplines covered in Chapter 6.

Electronic system failures can be caused by mechanisms other than load exceeding strength. For example, parameter drifts in components, short circuits due to solder defects or inclusions in components, high resistance relay or connector contacts and dielectric breakdown in capacitors are examples of failures which may not be caused by load. We will consider these failure modes later, appropriate to the various components and processes which make up electronic systems.

RELIABILITY OF ELECTRONIC COMPONENTS

Microelectronic devices

Microelectronic devices vary enormously in complexity, from relatively simple small-scale integrated (SSI) circuits containing fewer than ten logic gates to very large scale integration (VLSI) microprocessors and large memory devices, containing thousands of transistor logic gates or memory bits.

Medium scale integration (MSI) and large scale integration (LSI) refer to devices with 10 to 100 gates, repectively. VLSI devices can have over 10,000 gates on a single 4×4 mm square silicon chip. VLSI devices have line widths of the conducting tracks and semiconductor gate geometries as narrow as $2 \mu m$; future devices will utilize line widths of less than $1 \mu m$.

The prime reasons for the increasing level of integration, apart from the obvious one of reducing the volume requirements, have been to decrease power consumption and to increase operating speed. Increasing levels of integration have also led to great reductions in the cost per function and improvements in reliability per function. Figure 7.1 shows the reliability trend for digital ICs.

The reduced power dissipation has been just about matched by the increases in packing density, so that power dissipation per unit volume has remained roughly constant. Therefore the problem of keeping junction temperatures low in densely packed systems can still be a difficult one.

The quality control of IC manufacture has improved markedly, largely due to competition, and this has contributed to the improvements in reliability. This aspect is covered in more detail in Chapter 10.

The main competing directions of the basic silicon technology for digital ICs have been between variations of bipolar or transistor–transistor logic (TTL),

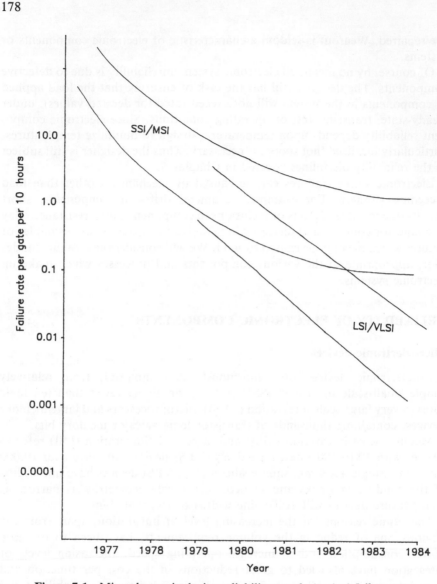

Figure 7.1 Microelectronic device reliability trends (typical failure rate contribution per gate)

the basis of the 5400 and 7400 series of devices, and variations of metal oxide semiconductor (MOS) fabrication, e.g. complementary MOS (CMOS), the basis of the 4000 series. TTL transistor action takes place within the silicon material, whereas MOS transistor action occurs mainly on the surface of the chip. TTL has always been best for speed whilst MOS, particularly CMOS, has had lower power dissipation.

Reliability aspects of TTL and MOS are shown in Table 7.1. The relative sensitivity of MOS to damage from surface corrosion, from ionizing radiation

Table 7.1 TTL and MOS digital IC technologies.

TTL	MOS
Speed High speed, particularly 5400S/7400S (Schottky) and HS (high-speed Schottky) version	Lower speed, but new devices nearly as fast as TTL HS
Power Higher power requirements, but LS (low-power Schottky) and ELS (extra-low-power Schottky) nearly as good as CMOS	Lower power, particularly CMOS
Electrostatic damage (ESD) sensitivity Not as sensitive as MOS	Very sensitive
Radiation hardness (gamma rays) High	Low (but silicon on sapphire (SOS) CMOS has high radiation hardness)
Temperature sensitivity Relatively low (high activation energy)	Relatively high
Surface corrosion sensitivity Relatively low	Relatively high

and from electrostatic discharge is due to the fact that MOS devices, though very reliable, have a higher temperature sensitivity of time to failure, due to the lower activation energies of the main failure mechanisms. However, their lower power dissipation offsets this, since MOS junction temperatures can usually be kept lower than for TTL.

Generally speaking, if there are no overriding performance reasons for selecting either TTL or MOS (such as speed, power or radiation hardness), CMOS should be selected if heat dissipation is likely to be a problem.

The connections between transistors on the chip and to the input pins are made via a network of aluminium conductor tracks, by condensing a thin layer of metal (*metallization*) on to the surface.

Different layers of the diffusion process are electrically isolated by layers of silicon dioxide (SiO_2). This is called *passivation*. Finally, the entire die surface, apart from connector pads for the wire bonds, is protected with a further SiO_2 layer. This process is called *glassivation*.

Linear ICs (e.g. amplifiers, converters) are made using the same basic methods, and the same reliability considerations generally apply.

New IC technologies are being developed which will affect reliability. Very high speed or very high performance ICs (VHSIC, VHPIC) are being developed in the United States and the United Kingdom, with the objectives of increasing speed, packaging density, radiation hardness and reliability, primarily for military applications. Gallium arsenide ICs are being developed, as GaAs provides higher speed and lower power dissipation than silicon-based ICs, due to higher electron mobility in GaAs.

Custom and Semi-custom IC design

There is an increasing trend for ICs to be designed for specific applications. Standard ICs such as microprocessors and memories will always be used, but many circuits can be more economically implemented by using custom or semi-custom designs for many of the functions. Common examples of these applications are in electronic cameras, calculators and TV teletext decoders.

In a semi-custom IC, all fabrication processes on the chip are previously completed, leaving arrays of transistors to be interconnected by a conductor pattern designed for the particular application. In a full custom design, however, the chip is designed and manufactured entirely for the specific application.

Semi-custom ICs are more economical than full custom ICs in relatively low quantities (less than 10,000), but there is less flexibility of design and the utilization of chip area is less economical. Full custom ICs are usually economical when large quantities are to be used, since the design and development costs are high. Both design approaches rely heavily on CAD though the semi-custom method is easier to implement.

Semi-custom and custom ICs introduce important reliability aspects. The electronic system designer is no longer selecting 'black boxes', i.e. standard ICs, from a catalogue, but is designing the system at the component (or functional group) level. Reliability (and maintainability/testability) analysis must be performed to this level, not just to input and output pins. Since design changes are very expensive, it is necessary to ensure that the circuit is reliable and testable the first time. Particular aspects that need to be considered are:

1. Satisfactory operation under the range of operating inputs and ouputs. It is not usually practicable to test a LSI or VLSI design exhuastively, due to the very large number of different operating states (analogous to the problem of testing software, see Chapter 8), but the design must be tested under the widest practicable range of conditions, particularly for critical functions.
2. The effects of failures on system functions. Different failure modes will have different effects, with different levels of criticality. For example, some failure modes might have no effect until certain specific conditions are encountered, whilst others might cause total and obvious loss of all functions. FMECAs and, where appropriate, stress analyses should therefore be performed on new IC designs, and the designs should be formally reviewed.
3. The effects of system software on the total system operation. For example, the extent to which the software is designed to compensate for specified hardware failures, by providing failure indications, selecting alternate operating paths, etc.

4. The need for and methods for providing built-in redundancy, both to increase production test yield and to improve reliability, particularly for critical applications.
5. Testability of the design. The ease with which circuits can be tested can greatly influence production costs and reliability, since untested functions present particular reliability hazards.

The CAD methods used for IC design include facilities for assessing reliability and testability. For example, failure modes can be simulated at the design stage and the effects evaluated, so stress analysis and FMECA can be integrated with the design process. Circuit design CAD programs also include facilities for testability analysis.

Microelectronics packaging

There are two main methods of packaging IC chips. In hermetic packaging the die is attached to the package base, usually by soldering, wire bonds are connected between the die connector pads and the leadout conductors, and the package is then sealed with a lid. Packages are either ceramic or metal. Plastic encapsulated ICs (PEICs or PEDs) use an epoxy or silicone encapsulant.

The most common package form is the dual-in-line package (DIL or DIP), with pin spacing of 0.1 inch (2.5 mm). The pins are inserted into holes in the printed circuit board (PCB) and soldered, or into a DIL socket which allows easy removal and reinsertion.

PEICs are cheaper than hermetics ICs, and therefore tend to be used in domestic and much commercial and industrial equipment. However, PEICs are not suitable for high temperature operation (above 50 °C ambient). They are therefore not normally approved for use in military equipment and they cannot be high temperature screened. They also tend to suffer a life dependent (wearout) failure mode due to moisture ingress, either by absorption through the encapsulation material or along the plastic/metal boundary of the leads. The moisture provides a medium for electrolytic corrosion at the interfaces of conductor tracks and wire bonds, or of the conductor tracks themselves through any gaps or holes in the glassivation layer. No plastic encapsulant is totally impervious to moisture ingress, though modern materials and process controls have reduced the problem. Therefore PEICs are not recommended for use in high moisture environments or where long life is important, e.g. in military or aerospace applications.

The packaging techniques used for the first 20 years or so of IC manufacture are giving way to new methods, primarily in order to enable more circuitry to be packaged in less volume. The conventional DIL package requires a PCB area many times larger that the chip surface area. The leadless chip carrier (LCC) package is an example of a *surface-mounted* package, which has leadouts around the periphery. These are reflow soldered to the PCB conductor tracks (or to a ceramic substrate which is in turn soldered to the PCB

182

conductor tracks) rather than being inserted through PCB holes as with DIP. (In reflow soldering the components are placed on the PCB or substrate, and the assembly is heated to melt the solder.) The leadouts are on a 0.05 in (1.25 mm) spacing. Use of LCCs reduces PCB surface area by about four times.

A more recent development is the pin grid array (PGA) package. Leadouts are taken to an array of solder bumps on the underside of the package, on a 0.05 in (1.25 mm) grid. As with the LCC package, connection is made to the PCB or substrate by reflow soldering. Figure 7.2 shows examples of the packaging techniques described above. Other packaging methods are being developed for special applications, such as vertical mounting for high density computer circuitry.

The main reliability implication of the new IC packaging technologies is the fact that, as the volume per function is decreased, the power dissipation per

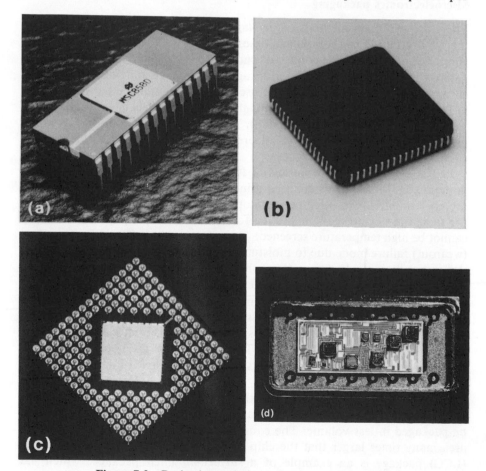

Figure 7.2 Packaging techniques for integrated circuits
(Courtesy National Semiconductor Corporation)

unit volume increases. This leads to difficult thermal management problems in order to prevent junction temperatures attaining levels above which reliability would be seriously affected. Liquid cooling of assemblies is now necessary in some applications such as in some military and high speed computing equipment, and there is likely to be a trend to more powerful cooling systems in future.

Another reliability aspect of the new methods is the use of reflow soldering to the surface of the PCB or substrate. The solder connections cannot be inspected as can DIP solder connections; with a PGA they cannot be seen at all. Also, thermal shock can cause the solder joints to fail in shear since there is no mechanical compliance as with the pins on a DIP. Therefore the solder process must be very carefully controlled, and subsequent burn-in and reliability tests (see Chapters 10 and 11) must be designed to ensure that good joints are not damaged by the test conditions.

Hybrid microelectronic packaging

Hybrid microelectronic packaging is a technique for mounting unencapsulated semiconductor and other devices on a ceramic substrate on which connections, resistors and capacitors are laid down. Hybrids are now also made using LCC or PGA packaged ICs. Resistors are made by screen-printing with conductive ink and laser-trimming to obtain the desired values. Connections from the gold conducting paths to the device pads are made using fine gold or aluminium wire and ultrasonic bonding in the same way as within an encapsulated device. The complete assembly is then encased in an hermetic package. See Figure 7.2.

Hybrid packaging provides certain advantages over conventional printed circuit board construction, for special applications. It is very rugged, since the complete circuit is encapsulated, and it allows higher density packaging than PCB mounting of components. However, it is not practicable to repair hybrid circuits, since clean room conditions and special equipment are necessary. Therefore they are suited for systems where repair is not envisaged except by replacing hybrid modules and discarding defective ones. Hybrid circuits are used in missile electronics, automotive engine controls and severe environment industrial conditions, and in many other applications where compact assembly is required. Hybrid circuits can be custom designed and manufactured or standard catalogue hybrids can be used.

According to the MIL-HDBK-217 model, a hybrid circuit will have a similar failure rate to an equivalent conventional circuit. However, hybrid circuit manufacture suffers from the disadvantage that the production processes involve handling of bare IC chips and working around the bare chips during the wire bonding and test processes to a greater extent, and at much lower production rates, than for discretely packaged ICs. Testing and screening is therefore more expensive and often not as effective. However, the use of LCC and PGA packaged ICs within the hybrid package removes these drawbacks, since the LCC/PGA packages can be tested and screened before mounting to the substrate, and they are also much less easy to damage.

Due to their relatively large size, the number of internal bonds and the long package perimeter, hybrids tend to suffer from inclusion of contamination and conducting particles, bond failure and sealing problems more than do discrete devices on equivalent PCB circuits. Therefore, very stringent production and quality control are required if the potential reliability of hybrid circuits is to be realized. MIL-STD-883 includes the same screening techniques for hybrid microelectronics as for discrete ICs, and the European and UK specifications (CECC, BS 9450) include similar requirements.

Microelectronic component attachment

Microelectronic components in dual-in-line packages can either be soldered on the PCBs or plugged into IC sockets which are soldered in place. Plugging ICs into sockets provides three major advantages from the test and maintainance points of view:

1. Failed components can easily be replaced, with less danger of damaging the PCB or other components.
2. Testing and diagnosis is usually made much easier and more effective if complex devices such as microprocessors are not in place.
3. It is much easier to change components which are subject to modifications, such as PROMS and custom or semi-custom components.

On the other hand, there are some drawbacks which can override these advantages in certain circumstances. These are:

1. Heat transfer is degraded, so it might not be possible to derate junction temperatures adequately.
2. There might be contact problems in high vibration, shock or contamination environments.

IC sockets are therefore used on most repairable systems, particularly mass-produced systems for which test costs must be kept low. However, they are not used to the same extent in high reliability, few-of-a-kind systems, such as spacecraft, or in systems such as avionics and mobile military equipment.

Microelectronic device failure modes

For hermetically sealed semiconductor devices, there are no inherent wearout failure modes. That is, there are no failure mechanisms which depend upon operating or non-operating time, within a correctly manufactured device. The only ways in which a hermetically sealed semiconductor device can fail is if

Table 7.2 Microcircuit device failure modes.

	Failure mode	Caused by	Activation energy range (eV)	Protection
Die-related	Internal open circuits:			
	1. Electromigration,	Ionic diffusion of conductor material, leading to 'opens' in tracks; increasingly severe problems with VLSI due to low cross-sectional area of tracks	0.5–1.2	Minimize current density in conductors; minimise operating temperature; quality control of metallization process
	2. Other causes	Voltage overstress, e.g. electrostatic discharge.	—	Circuit protection, quality control.
	Internal short-circuits	Voids in passivation layers; 'punch through' of dielectric layers	—	Quality control of passivation processes; voltage derating;
	Incorrect transistor action	Bulk silicon or oxide defects	0.3–0.5	Transient protection Burn-in
	Surface corrosion	Moisture ingress; internal contamination	0.3–0.6	Quality control of packaging and of die fabrication processes
	Radiation damage	Ionizing radiation, e.g. gamma rays	—	Selection of radiation hard devices
Bond. failure	Broken or lifted bond wire	Poor bonding; corrosion	—	Quality control
Package-related	Non-hermetic package, leading to corrosion; internal contamination or particle inclusion	Production processes	—	Quality control

it is overloaded beyond its design rating (temperature, voltage or current) or if there is a defect which causes immediate or progressive weakening.

Therefore IC reliability is very dependent upon quality control of the manufacturing processes and the effectiveness of the screening techniques used to remove defective devices. Operating and environmental conditions, particularly high temperature operation, affect reliability by accelerating the onset of failure due to defects or by leading to overstress (e.g. electrostatic discharge damage or other voltage transients).

The main causes of IC failure are shown in Table 7.2. The acceleration of temperature-dependent failure mechanisms with increased temperature is related to the process activation energy, measured in electron volts (eV) in accordance with the Arrhenius model (p.188). A high activation energy indicates high temperature acceleration and vice versa.

Microelectronic device screening

Since quality is such an important factor in device reliability, standard production quality screening methods have been established to eliminate weak devices. Different screening levels have been standardized to enable a cost-effective choice to be made between cost and reliability. (Remember that device reliability affects production costs as well as in-use costs.)

Screening methods for ICs are covered by US MIL-STD-883: *Test Methods and Procedures for Microelectronics* and a British Standard, BS 9400. MIL-STD-883 lays down three basic screen levels, each of which includes required inspections and tests. Table 7.3 summarizes the screening requirements for each test level.

Device quality depends upon the control exercised over the production processes. Screening cannot be 100 per cent. effective in eliminating defective devices. Therefore standards also exist to control specifications and manufacturing processes and manufacturers must obtain and retain approval of their production lines in order to conform to these standards. The qualification standards for microelectronics are:

USA. Standard military microelectronics specifications are given in MIL-M-38510, with separate sections ('slash-sheets') for each device type.

Europe. The CECC specification system is very similar to MIL-M-38510, and also includes test methods based upon MIL-STD-883. In the United Kingdom a British Standard, BS 9400, covers the same aspects (a BS in the 9000 series is given a suffix 'E' to denote that it has been harmonized with the equivalent European standard). The CECC and BS systems use the same screen levels (A, B, C) as MIL-STD-883.

Table 7.3 Microelectronic device screening requirements[a].

Screen	Defects effective against	Screen level applicability MIL-STD-883/BS 9400		
		A	B	C
Pre-encapsulation visual inspection (30–200 × magnification)	Bonds, contamination, chip surface defects, wire bond positioning	100%	100%	100%
Stabilization bake	Bulk silicon defects, metalization defects, stabilizes electrical parameters	100%	100%	100%
Temperature cycling	Package seal defects, weak bonds, cracked substrate	100%	100%	100%
Constant acceleration (20 000 g)	Chip adhesion, weak bonds, cracked substrate	100%	100%	100%
Leak tests	Package seal	100%	100%	100%
Electrical parameter tests (pre-burn-in)	Surface and metallization defects, bond failure, contamination/particles	100%	100%	100%
Burn-in test (168 h, 125 °C with applied a.c. voltage stress)	Surface and metallization defects Weak bonds	100% 240 h	100% 160 h	—
Electrical test (post-burn-in)	Parameter drift	100%	100%	—
X-ray	Particles, wire bond position	100%	—	—

[a]This table is not comprehensive, and the reader should refer to the appropriate standard to obtain full details of applicable tests.

International. The International Electrotechnical Commission (IEC) is attempting to establish a worldwide approvals system, which will inevitably be based upon the existing US and European systems.

Effect of screening on IC reliability

The effect of screening on reliability is expressed by the π_Q factor. Table 7.4 shows the failure note multiplier (π_Q) values attributed to the various screen levels by MIL-HDBK-217 (see later) and typical relative cost multipliers. Table 7.4 shows that, based upon analysed data from many millions of hours of test and use, device screening is very important in electronic systems reliability. B-level screening is mandatory for military electronic equipment in the United States and Europe, and is often used for high quality commerical and industrial equipment, particularly when operation in severe environments is involved. S-level screening is used where very high reliability is paramount, e.g. spacecraft, submarine cable repeaters and some medical electronic equipment. Some users of very large numbers of devices, e.g. TV set manufacturers, obtain high quality devices not by specifying a screen but by using their buying power to enforce a requirement for a very low proportion defective and by

Table 7.4 Microcircuit device screening effectiveness and cost.

Screen level (MIL-HDBK-217)	Specification	π_Q	Typical relative cost
A	MIL-M-38510, Class S	0.5	8–20
B,B-0	As above, Class B	1.0,2.0	
B-1	MIL-STD-883 method 5004, Class B	3.0	4–6
B-2	Vendor equivalent of B-1	6.5	
C	MIL-M-38510, Class C	8.0	
C-1	Vendor equivalent of C	13.0	2–4
D	Commercial (i.e. no special screening), hermetic package	17.5	1
D-1	Commercial plastic encapsulated	35.0	0.5

relying on the suppliers to derive their own most cost-effective procedures. More information on electronic device and assembly screening and burn-in is given in Chapter 10.

Other electronic components

Other electronic component types are primarily 'active' devices (transistors, diodes) and 'passive' devices such as resistors, capacitors, inductors, PCBs, connectors, etc. In general, these discrete components are very reliable, since they are much simpler than ICs. The factors that most affect reliability are the same as for ICs, namely thermal and electrical stress, and quality control methods. Specific reliability considerations of discrete device types are covered in the next section in this chapter, on reliability prediction for electronic systems using MIL-HDBK-217.

ELECTRONIC SYSTEM RELIABILITY PREDICTION

The physics of many degradation processes affecting the reliability of electronic components is such that the process rate can be described by the Arrhenius model:

$$\lambda_b = K \exp \left(\frac{-E}{kT} \right) \tag{7.1}$$

where λ_b is the process rate (component 'base' hazard rate), E the activation energy (eV) for the process, k Boltzmann's constant (1.38×10^{-23} J K^{-1} or 8.63×10^{-5} eV K^{-1}), T the absolute temperature (K) and K is a constant.

This model is used for failure rate prediction for electronic systems. The standard failure rate model used is that developed in the United States by USAF Rome Air Development Center and published as US MIL-HDBK-217 (*Reliability Prediction for Electronic Systems*) (Ref. 13 in the Bibliography). E/k is then replaced by A. A very large database of electronic equipment and component reliability is used to derive constant hazard rate models for different component types, based upon the Arrhenius equation.

MIL-HDBK-217 in fact quotes 'failure rate' models, not hazard rate models. Since most data used for MIL-HDBK-217 is derived from repairable systems, and since the methods used are mainly applied to such systems, the models really relate to the failure rate contribution, to a system, by that component type, i.e. the failure rate of the 'socket' into which that component type fits (see Chapter 3). MIL-HDBK-217 assumes independent, identically exponentially distributed (IID exponential) times to failure for all components. The use of a constant failure rate (CFR) for reliability prediction of electronic systems is based upon this assumption, which is not always valid in practice, as has been shown in earlier chapters. However, it is acceptable for most work in this area, because:

1. A CFR assumption makes system reliability prediction relatively easy, since an additive (parts count) method can be used.
2. For most maintained equipment, as repairs are carried out and as modules and components are exchanged, after a period the system might consist of parts with different ages. Also, maintenance induces failures, which tend to have a constant rate of occurrence. Therefore there might be an overall tendency to a CFR.
3. It is much easier to calculate an assumed CFR from data on systems than to derive the parameters of a two-parameter (e.g. Weibull) distribution. Also, the data are seldom numerous enough to allow derivation with adequate statistical or engineering confidence.
4. For logistics planning purposes, e.g. avionics spares planning for a fleet of aircraft, the CFR model is adequate both for prediction and for monitoring of performance. MTBF is the function usually used in such cases.
5. Predicted reliabiity is subject to such wide margins of error that the IID/CFR assumptions do not make much difference. In the rest of this discussion, therefore, we will use the term failure rate, as in MIL-HDBK-217.

The general MIL-HDBK-217 failure rate model is of the form:

$$\lambda_p = \lambda_b \pi_Q \pi_E \pi_A \cdots$$

where λ_b is the base failure rate related to temperature and π_Q, π_E, π_A,... are factors which take account of part quality level, equipment environment, application stress, etc.

The values of the base failure rates and the various factors in MIL-HDBK-217 are kept up-to-date by analysis of failure data on components and systems. However, it must be stressed that failure rate prediction by this method only gives an estimate of the perfomance that can be achieved or exceeded if there is an adequate reliability and quality programme. Nominally identical equipments may have widely different failure rates in service, for several reasons, notably differences in manufacturing quality standards and methods of use. Similar equipments (i.e built to similar specifications with similar components) may have very different reliabilities due to one having design shortcomings, not taken into account in the failure rate prediction model. The failure rate model is based upon constant or average stress levels (e.g. temperature and voltage). It does not take into account the possible incidence of transient overstress, stress cycling and other factors that can lead to failure. The model is also based on the assumption that all devices are prone to failure. This is of course not true, since the only devices in a circuit that are likely to fail under the correct operating conditions are those which are defective in some way, as the great majority of electronic device types have no inherent wearout failure modes. The model takes account of device quality. However, the objective in electronic design should be intrinsic reliability (see Chapter 4). The MIL-HDBK-217 model provides a yardstick with which to compare designs and options, based upon practical experience. It is subject to the same qualifications as described for reliability prediction techniques in Chapter 5. However, it is often a requirement in development contracts, particularly for military electronic equipment, and is widely accepted as being the best method generally available.

Thus the MIL-HDBK-217 method of stress analysis failure rate prediction is very useful as a design review procedure to enable the effects of various stresses and other factors to be quantified, albeit more in the sense of a comparison or sensitivity analysis than as an explicit prediction of the failure rate which will be achieved. The failure rate which is actually achieved will depend upon many factors, such as design to protect from overstress, quality of manufacture, screening (burn-in) methods used, in addition to the factors considered in the stress analysis. The reliability figure derived can be used as a target for development testing or for in-service achievement. It is also essential to understand that it does not represent a ceiling. Many systems have been produced which are much more reliable than MIL-HDBK-217 would indicate. Therefore, this analysis is but one element of the reliability programme for an electronic system.

MIL-HDBK-217 is very widely used for electronic system reliability prediction and component stress analysis outside the United States and in non-defence work. For example it is the required method according to UK Defence Standard 00–41.

MIL-HDBK-217 includes (Part 3) a simplified method of failure rate prediction, which omits the stress analysis. This parts count method can be useful in the early design stages of a system, before component stress information is available, to provide a rough early guide to likely achievable reliability.

The subsequent sections include data derived from MIL-HDBK-217, Issue D, Notice 1 (January 1983). The data are subject to change as MIL-HDBK-217 is updated and therefore should not be used for preparing reliability predictions when MIL-HDBK-217 is stipulated as the source. The data given summarize only the most important features of MIL-HDBK-217 and are therefore not complete.

Quality levels, π_q

The quality of components used in an electronic system has a large influence on reliability. The objective of a components quality effort is to eliminate from the population those components which initially do not function correctly and those which have defects which may lead to failure later during use. In MIL-HDBK-217 the quality factor, π_Q, is used to account for the effect of quality on failure rate.

The main QC determinants of component reliability are the degree of screening applied to components, over and above basic functional test, and the control of the production process. MIL-HDBK-217 therefore allocates π_Q values appropriate to these factors.

System environmental effects, π_E

Experience shows that electronic systems have failure rates which depend upon the use environment. For example, a radio set in a fighter aircraft will usually fail more often than a similar set in a fixed ground environment. This is due to the higher stresses (vibration, temperature, shock, switching cycles) experienced in the aircraft installation. MIL-HDBK-217 takes account of these system level stresses by applying an environmental factor, π_E. The environmental factor covers all the failure-inducing stresses not otherwise included in the component failure rate models.

Microelectronics failure rate model

The MIL-HDBK-217 failure rate model for microelectronic devices is

$$\lambda_p = \pi_Q \pi_L [C_1 \pi_T \pi_V + (C_2 + C_3) \pi_E]/10^6 h$$

where π_Q and π_E are as described above; π_V is a voltage derating stress factor, equal to unity for all technologies except 12 V and 18 V CMOS; π_L is a learning factor, equal to unity except for new, relatively unproved devices, when $\pi_L = 10$; C_1 and C_2 are complexity factors based upon the chip gate count, or bit count for memory devices, or transistor count for linear devices; C_3 is a complexity factor based upon packaging aspects (number of pins, package type); and π_T is the temperature factor, a function of the maximum junction temperature within the device in the operating environment. π_T must be estimated from the expected ambient temperature and the load applied to the device output, or derived from measurements of the device case temperature. It is shown plotted against temperature for some typical device types in Fig. 7.3, whilst Fig. 7.4 shows C_1 and C_2 plotted against gate count. Similar models are used for memory devices and microprocessors.

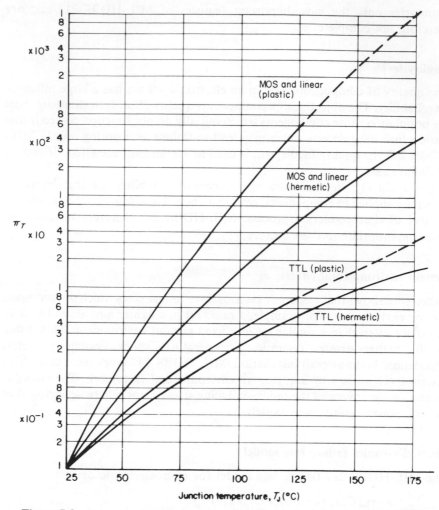

Figure 7.3 π_T versus temperature — microelectronic devices. $T_J \approx$ ambient temperature $+25\,^{\circ}$C (hermetic) and $+60\,^{\circ}$C (plastic). Note that plastic-encapsulated devices are not normally used above $50\,^{\circ}$C ambient

Transistors and diodes

The MIL-HDBK-217 failure rate model for these discrete semiconductors is

$$\lambda_p = \lambda_b \pi_E \pi_A \pi_Q \pi_R \pi_{S2} \pi_C / 10^6 h$$

λ_b, the base failure rate, is shown plotted against temperature in Fig. 7.5 for npn and pnp silicon transistors, against various values of power stress ratio (PSR) (power dissipation/rated power). Figure 7.5 indicates the maximum combined temperature/PSR combinations recommended for reliable operation, e.g. at a PSR of 0.6, an npn transistor should not be operated at more

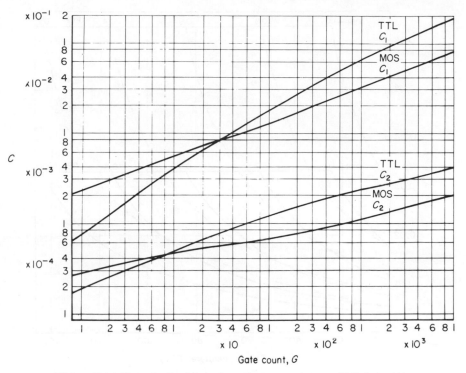

Figure 7.4 Complexity factors versus gate count — digital devices

than 80 °C. Other defintions are: π_R is power rating factor for transistors; π_{S2} is a function of voltage stress ratio (VSR) (applied voltage/rated voltage) shown plotted in Fig. 7.6; π_C is a complexity factor for transistors to cover other than a single transistor in a package; and π_A relates to the various applications, e.g. power, high frequency switching or rectifying.

Transistor and diode screening methods are similar to those for microelectronic devices. Standard specifications and screening methods are covered in MIL-S-19500, MIL-STD-750 and Joint Army–Navy (JAN) specifications (USA) and BS 9300 (UK). π_Q values for the various quality levels are as shown in Table 7.5.

Resistors

The MIL-HDBK-217 failure rate model for fixed value resistors is

$$\lambda_p = \lambda_b \pi_E \pi_R \pi_Q / 10^6 \text{h}$$

λ_b is shown plotted against temperature in Fig. 7.7, for fixed composition and film type resistors, at different power stress ratios. π_R is a resistance factor.

As would be expected, failure rate increases with both increasing temperature and increasing electrical stress, since the predominant failure

Figure 7.5 λ_b versus temperature for silicon transistors (hermetic encapsulated) for various power stress ratios (PSR)

Figure 7.6 π_{S2} versus voltage stress ratio for transistors (for PSR $\geqslant 0.25$). When PSR < 0.25, $\pi_{S2} = 0.3$

Table 7.5 Transistor and diode screening effectiveness.

	π_Q	
Quality level	Transistors	Diodes
JAN TXV (includes 100% pre-cap visual inspection, burn-in)	0.12	0.15
JAN TX (100% burn-in)	0.24	0.3
JAN, BS 9300	1.2	1.5
Commercial, hermetic or glass encapsulated	6.0	7.5
Commercial, plastic encapsulated	12.0	15.0

Figure 7.7 λ_b versus temperature for fixed resistors for various PSR.

mode of most resistors is an open circuit caused by fusing of the conducting material, due to the combined effects of inherent defects and heating. Fixed resistors generally have no wearout failure modes and reliability is very dependent upon quality control. π_Q values vary from 0.03 (established reliability, CECC, BS 9000) to 5 for commercial devices.

Variable resistors are less reliable and should be avoided whenever practicable. Wire-wound variables are more robust and less prone to be electrically noisy than film variables. Therefore they should be used unless the higher discrimination provided by the film type is necessary. On the other hand, the need for very fine adjustment should be avoided whenever practicable.

Capacitors

The MIL-HDBK-217 failure rate formula is

$$\lambda_p = \lambda_b \pi_E \pi_{CV} \pi_{SR} \pi_Q / 10^6 \text{ h}$$

where π_{CV} is a function of capacitance and π_{SR} a series resistance factor used for some electrolytic capacitors. Figure 7.8 shows how failure rate varies with temperature and voltage stress ratio for two typical capacitor types.

π_Q values vary widely, depending on capacitor type. Failure rate variation with temperature also depends on the type, as do the predominant failure modes.

Other electronic components

MIL-HDBK-217 includes failure rate prediction models for other component types, such as inductors, switches, transformers, relays, PCBs and connectors. Refer to MIL-HDBK-217 and the Bibliography for details.

Reliability prediction worksheet

To facilitate and standardize MIL-HDBK-217 stress analysis, a reliability prediction worksheet can be used. Figure 7.9 shows such a form as used for part of a circuit. USAF Form ADTC 610 is similar in layout.

Figure 7.8 λ_b versus temperature for fixed capacitors for various VSR.

				Assembly drg no.			Sub assembly drg no.					Lower assembly drg no.								Page of pages
System							Reliability prediction worksheet													
Circuit ref.	Description part	Part type	Part no	Value or gate/bit/ TX count	Mfr code	Qty	Voltage		Power/current		Stress ratios	T(°C)	λ_b $\times 10^{-6}$	π_Q	π_E	π	π	π	π	$\lambda_p \times$ qty $\times 10^{-6}$
							Rated or:	Applied C	Rated C_2	Applied C_3										
R1–R25	Resistor	R-11804		47 K	L	38			250 mW	25	0.1	50	0.0098	1	2	R 1				0.49
C1–C17	Capacitor	C-11015		0.1 µF	S	17	100	1C			0.1	50	0.00077	0.1	2	CV 1.6				0.0042
C18	Capacitor	C-39003		10 µF	S	2	25	1C			0.4	50	0.011	0.1	2	CV 1.3	SR0.1		C 1	0.0006
D1	Diode, Zener			5.6 V	P	2					0.2	50	0.0042	3	5	A 1				0.126
T1	Transistor	2N3055			P	1					0.6		0.039	1.2	5	A 1.5	A 1.5	S.20.3		0.105
X1	IC TTL DIG	54151		17	P	1		0.0048	0.0006	0.0048		50		3	2.5	T 0.54	R 1			0.0483
X2	IC microproc	1802A		1375	B	1		0.032	0.0013	0.011		50		3	2.5	T 1.7	L 1	L 1		0.2555

Remarks	Failure rate total this page 1.0276
	Failure rate total
	Issue no Date
	Compiled by

Figure 7.9 Reliability prediction worksheet for electronic equipment

Computerized stress analysis

Computer programs have been written to enable MIL-HDBK-217 stress analysis to be performed with less effort than by manual calculation and compilation. Such programs store all the MIL-HDBK-217 formulae and parameter values, so that component data input is greatly eased and all calculations are performed automatically. Apart from the obvious advantages of reduced effort and greater accuracy, simplifying the analysis is a great benefit when performing design tradeoffs. For example, it might be decided to evaluate the likely effect on failure rate of increasing cooling air flow to reduce component temperatures. Likewise, hardware changes can be quickly evaluated and the analysis can be kept up-to-date with changes during the development phase with relatively little effort. The MIL-HDBK-217 PREDICTOR program (Management Sciences Inc., Albuquerque, New Mexico) is a good example. It is available through several computer bureaux. Programs are also available for popular microcomputers, and Fig. 7.10 shows a typical output.

MTBF Calculation-MIL-HDBK-217D

DRG Number: MANPACK
Ambient temperature: 55 °C

C.ref	Spec	Pins	Gates	S	D T	PI Q	FR	Qty	Tot FR
IC 1		16	160		30	6.5	.4510	2	.9020
IC 2		8	34		30	6.5	1.2860	1	1.2860
IC 3		40	1375		25	6.5	1.6663	1	1.6660
IC 4		24	16384		25	6.5	2.8227	2	5.6450
TR 6				0.30	15	1.2	0.0237	2	0.0470
TR 7				0.30	15	1.2	0.0086	10	0.0860
D10				0.30	10	1.5	0.0087	10	0.0860
D12				0.30	10	3.0	0.0508	3	0.1520
C14				0.20	0	3.0	0.0246	20	0.4910
C15				0.20	0	3.0	0.0426	10	0.4250
C16				0.20	0	3.0	0.2662	10	2.6610
C17				0.20	0	3.0	0.9633	4	3.8530
R18				0.30	25	5.0	0.0795	100	7.9510
L20					30	4.0	0.0333	10	0.3330
T21					30	12.0	0.4767	2	0.9530
S23				0.30	0	1.0	0.5743	2	1.1480
RL24				0.30	10	1.0	1.1902	1	1.1900
CON25					5	1.0	0.1392	10	1.3910
CON26					5	1.0	0.0604	4	0.2410
27					0	1.0	0.0125	600	7.5130
28					0	1.0	0.0190	120	2.2770
29					0	1.0	0.0005	480	0.2410
MISC30					0	1.0	0.0200	1	0.0200

Total FR: 40.568 FPMH
MTBF: 24,649 HRS.

Figure 7.10   Typical output from a MIL-HDBK-217 program (RELPRED on HP 75/9816)

Device failure modes

The more important device failure modes have been covered in the preceding sections on component reliability. The main failure modes must be considered during the design so that their effects can be minimized. Circuit FMECAs and system reliability block diagrams should also take account of likely failure modes. Table 7.6 summarizes the main failure modes for the most common device types. The failure modes listed are not exhaustive and the device types

Table 7.6 Device failure modes.

Type	Main failure modes	Typical approximate proportions (per cent)
Microcircuits[a]		
Digital logic	Outputs SA 0 or SA 1	80
	No function	20
Linear	Parameter drift	20
	No output	70
	Hard over output	10
Transistors		
	Low gain	20
	Open-circuit	30
	Short-circuit	20
	High leakage collector-base	30
Diodes		
Rectifier, general purpose	Short-circuit	10
	Open-circuit	20
	High reverse current	70
Resistors		
Film, fixed	Open-circuit	30
	Parameter change	70
Composition, fixed	Open-circuit	10
	Parameter change	90
Variables	Open-circuit	30
	Intermittent	10
	Noisy	10
	Parameter change	50
Relays		
	No transfer	20
	Intermittent	70
	Short-circuit	10
Capacitors		
Fixed	Short-circuit	60
	Open-circuit	20
	Excessive leakage	10
	Parameter change	10

[a]Microprocessor and memory devices are covered in Chapter 8.

listed are only a summary of the range. The failure mode proportions can vary considerably depending on type within the generic headings, application, rating and source. Devices used within a particular design should be individually assessed, e.g. a resistor rated very conservatively is likely to have a reduced relative chance of failing open-circuit.

Circuit design should take account of the likely failure modes whenever practicable. Capacitors in series will provide protection against failure of one causing a short-circuit and resistors in parallel will provide redundancy against one failing open. Blocking diodes are often arranged in series to protect against shorts.

RELIABILITY IN ELECTRONIC SYSTEM DESIGN

Introduction

The designer of an electronic system has three main responsibilities, from the point of view of protecting the system and its components from failure-inducing stresses:

1. To ensure that as far as is practicable, no part in the system can be overstressed, e.g. by voltage switching transients, during operation or testing.
2. To ensure that all parts are electrically derated to the maximum extent compatible with other design constraints to cater for part and parameter variations, and transient conditions.
3. To ensure that device operating temperatures are kept as low as practicable.

In addition to these primary considerations, there are other aspects of circuit and system design which can be applied to improve reliability. By reducing the number of different part types, the parts screening effort can be reduced and designs become easier to check. This also generates cost savings in production and use. Redundancy can also be designed into circuits. Circuits and systems must be designed for ease of production, inspection and test; otherwise defects can be built in and will be difficult to detect and correct. Whenever practicable the need for adjustments or fine tolerances should be avoided.

Not all of the means of achieving reliable electronic design are complementary. For example, derating, redundacy and the inclusion of additional protective devices or circuits are not compatible with reducing complexity and the number of part types. The various design options relevant to reliability must be considered in relation to their effectiveness, cost and the consequences of failure. Stress analysis reliability prediction is a part of the assessment process, but the analyst must beware of the implication inherent in part failure rate modelling that eliminating parts from a circuit will necessarily improve reliability.

The sections which follow outline some of the methods available to ensure high reliability. They are by no means comprehensive: circuit designers should consult their own organizations' design rules, component application notes, the Bibliography to this chapter and other relevant sources. However, what follows is intended as a guide to the points which reliability engineers and circuit designers need to consider.

Circuit protection

Electronic components are often prone to damage by short duration voltage transients, caused by switching of loads, capacitive or inductive effects, static electricity, power supply ripple, testing, etc. Small semiconductor components such as ICs and low power transistors are particularly vulnerable, owing to their very low thermal inertias. MOS devices are very vulnerable to static electricity, and require special protection.

Logic devices which interface with inductive or capacitive loads, or which 'see' test connections, require transient voltage protection. This can be provided by: a *capacitor* between the voltage line to be protected and ground, to absorb high frequency transients, *diode protection,* to prevent voltages from rising beyond a fixed value (clamping), and *series resistances,* to limit current values. Figures 7.11 and 7.12 show typical arrangements for the protection of a TTL logic device and a transistor. IC protection is also provided by transmitting logic signals via a light-emitting diode (LED) and optical transducer combination, called an opto-isolator or opto-coupler.

The transient voltage levels which can cause failure of semiconductor devices are referred to as VZAP. VZAP values depend upon transient duration. Passive devices can also be damaged by transient voltages, but the energy levels required are much higher than for small semiconductor devices. Therefore passive devices do not normally need individual protection.

Derating

The sections on device reliability indicated the stresses which are most likely to accelerate damage, leading to failure, and which therefore require the most

Figure 7.11 TTL device protection. Diode D1 prevents the input voltage from rising above the power supply voltage. Capacitor C1 absorbs high frequency power supply transients

Figure 7.12 Transistor protection. Resistor R1 limits
the base collector current I_{BC} and capacitor C1 absorbs
power supply high frequency transients

attention for derating. Electronic design rules should contain specific derating instructions and any proposed component rating above these should be subject to design review investigation and justification. Typical derating guidelines for maximum expected values are summarized in Table 7.7. However, MIL-HDBK-217 base failure rate (λ_b) tables should be used to provide a more detailed basis for considering the effects on failure rate of those single and combined stresses which are included in the failure rate models. Derating guidelines obviously should depend upon the criticality of the device or circuit. Table 7.7 also shows typical derating for high reliability applications.

Thermal design

Figures 7.3, 7.5, 7.8 and 7.9 show how failure rate increases rapidly as device temperatures are increased. Therefore design to minimize operating temperatures is an essential aspect of the development of electronic systems. Since the maximum temperature within a device depends upon both electrical load and heat flow, the following aspects should be considered:

1. Electrical loads on the outputs of devices should be derated. For logic circuits the number of other device inputs driven by an output ('fan-out') should be derated to less than 80 per cent. of the maximum rated value, and this factor should also be applied to the load on linear device outputs. Power supply and input voltages should be derated to a value which is nevertheless within the rated limits. In every case the likely variations of load and input power supply values must be considered when assessing the rating of electrical parameters.
2. Sufficient cooling air (or other medium) must be provided to allow heat generated within devices to be dissipated at a reasonable temperature, or adequate provision must be made for convection and radiation cooling. Printed circuit board assemblies are normally mounted vertically to allow

Table 7.7 Device derating guidelines.

Device type	Parameter	Max. rating	
		Normal	Hi-rel
Microelectronics			
Digital	Power supply, input voltages	Derated but within	
	Output current (load,	performance spec.	
	fan-out)	0.8	0.8
	Junction temp. (hermetic): TTL	130 °C	100 °C
	(hermetic): CMOS	110 °C	90 °C
	(plastic)	100 °C	70 °C
	Speed	0.8	0.8
Linear	Power supply	Derated but within	
	Voltage	performance spec.	
	Input voltage	0.8	0.7
	Junction temp. (hermetic)	110 °C	90 °C
	(plastic)	100 °C	70 °C
Transistors			
Silicon (general purpose)	Collector current	0.8	0.5
	Voltage V_{cc}	0.8	0.6
	Junction temp. (hermetic)	120 °C	100 °C
	(plastic)	100 °C	80 °C
Silicon (power)	Collector current	0.8	0.6
	Voltage V_{cc}	0.8	0.6
	Voltage (reverse bias)	0.9	0.8
	Junction temp. (hermetic)	130 °C	110 °C
	(plastic)	110 °C	90 °C
Diodes			
Silicon (general purpose)	Forward current, voltages	0.8	0.5
Zener	Junction temp.	120 °C	100 °C
Resistors			
	Power dissipation	0.8	0.5
	Operating temp.	Rated−20 °C	Rated−40 °C
Capacitors			
	Voltage	0.8	0.5
	Operating temp.	Rated−20 °C	Rated−40 °C
Relays and switches			
Resistive or capacitive load	Current	0.8	0.5
Inductive load	Current	0.5	0.3
Motor	Current	0.3	0.2
Filament	Current	0.2	0.1

cooling flow to be assisted by convection. 'Hot spots' must be minimized by ensuring uniform coolant flow and by providing adequate heat sinks and heat flow paths. For example, a copper 'thermal ladder' is sometimes designed into PCBs to provide a good heat flow path from components to the edge of the PCB and thence to the outer wall of the assembly.

3. System layout should be arranged to minimize heat input to microelectronic devices. For example, devices should not be located downstream (from the heat flow point of view) of power supplies or power transistor heat sinks.

Sneak circuit analysis (SCA)

A sneak circuit is an unwanted connection in an electrical or electronic circuit, not caused by component failure, which leads to an undesirable circuit condition or which can inhibit a desired condition. Sneak circuits can be inadvertently designed into systems when interfaces are not fully specified or understood, or when designers make mistakes in the design of complex circuitry.

SCA is based on the identification within the system of circuit 'patterns' which can lead to a sneak circuit or sneak open circuit. The five basic patterns are shown in Fig. 7.13.

Any electrical circuit can be considered as being made of combinations of these patterns. Each pattern is analysed to detect if conditions could arise, either during normal operation or due to a fault in another part of the system, that will cause a sneak circuit or sneak open. For example, in the power dome or the combination dome the power sources could be reversed if S1 and S2 are closed.

Sneak circuits are of five types:

1. *Sneak paths*. Current flows along an unexpected route.
2. *Sneak opens*. Current does not flow along an expected route.
3. *Sneak timing*. Current flows at the incorrect time or does not flow at the correct time.
4. *Sneak indications*. False or ambiguous indications.
5. *Sneak labels*. False, ambiguous or incomplete labels on controls or indicators.

When potential sneak conditions are identified, they must be validated by test or detailed investigation, reported and corrective action must be considered.

SCA is a tedious task when performed on relatively large systems. However, it has proved to be very beneficial in safety analysis and in assessing the integrity of controls in aircraft and industrial systems. Computer programs have been developed for SCA, and a MIL-HDBK is being prepared.

Electromagnetic compatibility (EMC) and Electrostatic damage (ESD)

Digital microelectronic systems can be very prone to external or internal electromagnetic interference and damage. For example, voltage transients can corrupt data or timing on data busses, or even cause permanent damage, such as fusing of conductor tracks. Sources of instantaneous or cyclic transients can be static discharge, lightning, power supply spikes, RF radiation from arcing of motors, switches or relays, or RF radiation from radio, TV or radar transmitters.

Figure 7.13 SCA basic patterns

EMC can present very difficult reliability and safety problems, even in quite simple systems such as vending machines. Protection is provided by careful electromagnetic screening, bonding and earthing, as well as by isolation techniques such as the use of opto-isolators on data lines.

ESD can be caused by personnel handling devices or PCBs during production or maintenance. Devices can be damaged in such a way that they are not made immediately inoperative, but fail in later use. ESD precautions are very important during device manufacture and handling, and during equipment manufacture and maintenance. For example, personnel involved should always be electrically earthed, and devices and PCBs should be kept in anti-static packages during storage and transport.

Testability analysis

The importance of testability in the design of microcircuits was described on page 181. The same basic techniques are used for digital system assemblies as for ICs. However, whilst ICs can only be tested via their external connections, higher assemblies such as PCBs can be tested by probing the PCB surface, either using in-circuit testers or by manual probing. Therefore PCBs and assemblies must be designed for ease, economy and coverage of testing, both functional and diagnostic. This involves including additional circuitry and test points, as well as designing with the capabilities of the test equipment in mind. Testability is a very important feature of modern electronic system design. Lack of attention to this aspect can seriously affect production and maintenance costs. It can also affect apparent reliability, since assemblies which are difficult to diagnose are more likely to be damaged during repair.

Testability analysis involves review of the design by the designer and the test engineer, to ensure that good practices are followed and that test coverage is adequate with the test methods and equipment which will be used.

Testability analysis is performed by circuit simulation programs and test generation programs such as HILO and LASAR. Testability figures of merit are derived to show the extent to which the possible failure modes can be detected and diagnosed.

Parameter variation and tolerances

In any circuit where individual part parameters combine to make up circuit parameters, variations in part parameters will affect circuit function. Wherever practicable, circuit values should have tolerances wide enough to take account of combined part parameter tolerances. Statistical tolerancing is covered in Chapter 6. Electronic CAD programs can be used for performing parameter variation analysis.

Redundancy

Redundancy techniques are covered in Chapter 5. In electronic circuit and system design it is possible to apply redundancy at any level from individual components to subsystems. Decisions on when and how to design-in redundancy depend upon the criticality of the system or function and must always be balanced against the need to minimize complexity and cost. However, it is often possible to provide worthwhile reliability improvements by using redundant circuit elements, at relatively little cost, owing to the low cost of most modern devices.

Design simplification

Like all good engineering design, electronic system designs must be kept as simple as practicable. The motto often quoted is KISS — 'Keep it simple, stupid'. In electronics, design simplification is mainly a matter of minimizing the number of components to perform a required function. Reducing the number of components should improve reliability as well as reduce production costs. However, the need to provide adequate circuit protection and component derating, and where necessary redundancy, should normally take priority over component count reduction.

Minimizing the number of component types is also an important aspect of design simplification. It is inevitable that when a number of designers contribute to a system, different solutions to similar design problems will be used, resulting in a larger number of component types and values being specified than is necessary. This leads to higher production costs, and higher costs of maintenance, since more part types must be bought and stocked. It can also reduce reliability, since quality control of bought-in parts is made more difficult if an unnecessarily large number of part types must be controlled.

Design rules can be written to assist in minimizing component types, by contraining designers to preferred standard approaches. Component type reduction should also be made an objective of design review, particularly of initial designs, before prototypes are made or drawings frozen for production. The US Navy Standard Electronic Module (SEM) programme, as described in MIL-M-28787, provides a range of highly reliable standard modules, designed to reduce logistics costs.

Design for production, test and maintenance

Electronic systems which are easy to manufacture, test and maintain are likely to be more reliable, all other things being equal, than similar systems without these properties. Attention to these aspects also leads, of course, to reduced costs of manufacturing and maintenance. Producability, testability, and maintainability must be made the subject of design rules and design review. Aspects of good electronic design which contribute to these are listed below:

1. Avoid the necessity for adjustments, e.g. potentiometers, whenever possible. Adjustable components are less reliable than fixed-value components and are more subject to drift.
2. Avoid 'select on test' situations, where components must be selected on the basis of measured parameter values. Specify components which can be used anywhere within the tolerance range of the applicable parameters and do not rely on parameters which are listed as typical rather than guaranteed. Where 'typical' tolerances must be used, ensure that the appropriate component screening is performed before assembly.
3. Ensure that adjustments are easily accessible at the appropriate assembly level.
4. Partition circuits so that subassemblies can be tested and diagnosed separately. For example, if several measured values require amplification, analogue-to-digital conversion, logic treatment and drivers for displays, it is better to include all functions for each measured value on one PCB rather than use a PCB for each function, since fault diagnosis is made easier and repair of one channel does not affect the performance or calibration of others.

Production and test aspects are covered in more detail in Chapter 10 and maintenance in Chapter 9.

BIBLIOGRAPHY

1. J. E. Arsenault and J. A. Roberts (Eds), *Reliability and Maintainability of Electronic Systems*. Computer Science Press Inc., Rockville, Maryland (1980).
2. M. J. Howes and D. V. Morgan, *Reliability and Degradation in Semiconductor Devices and Circuits*. Wiley (1981).
3. *Reliability Handbook*. National Semiconductor Corp. (1979).
4. *Proceedings of the Annual International Reliability Physics Symposium*. IEEE.
5. R. I. Anderson, *Reliability Design Handbook*, RDH 376. Reliability Analysis Center, USAF/RADC (1978).
6. R. H. Myers, K. L. Wong and H. M. Gordy, *Reliability Engineering for Electronic Systems*. Wiley, New York (1964).
7. US MIL-HDBK-338: *Electronic Reliability Design Handbook*. Available from the National Technical Information Service, Springfield, Virginia.
8. US MIL-STD-883: *Test Methods and Procedures for Microelectronic Devices*. Available from the National Technical Information Service, Springfield, Virginia.
9. US MIL-STD-750: *Test Methods for Semiconductor Devices*. Available from the National Technical Information Service, Springfield, Virginia.
10. US MIL-STD-202: *Test Methods for Electronic and Electrical Component Parts*. Available from the National Technical Information Service, Springfield, Virginia.
11. US-MIL-M-38510: *General Specification for Microcircuits*. Available from the National Technical Information Service, Springfield, Virgina.

12. US-MIL-S-19500: *General Specification for Semiconductor Devices.* Available from the National Technical Information Service, Springfield, Virginia.
13. British Standard, BS 9000: *Electronic Components of Assessed Quality.* British Standards Institution, London.
    US MIL-HDBK-217: *Reliability Prediction for Electronic Systems.* USAF Rome
14. Air Development Center, available from the National Technical Information Service, Springfield, Virgina.
15. US MIL-HDBK-251: *Electronic System Reliability—Design Thermal Applications.* Available from the National Technical Information Service, Springfield, Virginia.
16. A. D. Kraus and A. Barr-Cohen, *Thermal Analysis and Control of Electronic Equipment.* McGraw-Hill (1983).
17. *Thermal Guide for Reliability Engineers.* Rome Air Development Center Report TR-82-172. (NTIS)
18. *Microelectronics and Reliability.* Pergamon Press (published quarterly).
19. F. Jensen and N. E. Petersen, *Burn-in: An Engineering Approach to the Design and Analysis of Burn-in Procedures.* Wiley (1983).

Chapter 8

SOFTWARE RELIABILITY

INTRODUCTION

Software is now part of the operating system of a very wide range of products and systems, and this trend is accelerating with the opportunities presented by low cost microprocessor devices. In most cases, the fact that computer programs take over functions previously performed by hardware results in enhanced reliability, since software does not fail in the way that hardware does. Performing functions with software leads to less complex and more robust hardware. Each copy of a computer progam is identical to the original, so failures due to variability cannot occur. Also, software does not degrade, except in a few special senses,* and when it does it is easy to restore it to its orginal standard.

Nevertheless, software can fail to perform the function intended, due to undetected errors in the program. Errors can cause system failure if they lead to faults which prevent the system from operating correctly. When a software error does exist, it exists in all copies of the program, and if it is such as to cause failure in certain circumstances, it will always do so when those circumstances occur. Therefore software errors can be extremely serious.

Since most programs consist of very many individual statements and logical paths, the scope for error is large. A software reliability effort is concerned with minimizing the existence of errors, by imposing programming disciplines, by checking, and by testing. It is also important to minimize the effects of errors, by designing the system to be 'fault tolerant'.

The term 'software reliability' is not universally accepted, since it is argued that reliability implies a probability, whereas a program either contains one or more errors in which case the probability of failure in certain circumstances is unity, or it contains no errors, in which case the probability of failure is zero. Similarly it is argued that 'rate of failure' has no meaning in a software sense. A good program will run indefinitely without failure, and so will all copies of it. A program with an error which causes failure will always fail when that part of the program is executed under the failure conditions. However, we will consider the system user's point of view. The user will observe system failures, and will be equally affected whether they are caused by hardware failures or by software

---

*Data or programs stored in some media can degrade. Magnetic media such as discs and tape are susceptible to electromagnetic fields, or even to being closely packed for long periods. Semiconductor memories using very large scale integration (VLSI) can suffer changes in the voltage state of individual memory cells due to naturally occurring α-particle bombardment. In each case a refresh cycle will restore the program.

errors. In some cases it might not even be possible to distinguish between hardware and software causes.

There are several ways by which hardware and software reliability differ. Some have already been mentioned. Table 8.1 lists the differences.

Table 8.1 Hardware and software reliability.

Hardware	Software
1. Failures can be caused by deficiencies in design, production, use and maintenance.	1. Failures are primarily due to design errors, with production (copying), use and maintenance (excluding corrections) having negligible effect.
2. Failures can be due to wear, or other energy-related phenomena. Sometimes warning is available before failure occurs.	2. There is no wearout phenomenon. Software failures occur without warning.
3. Repairs can be made which might make the equipment more reliable.	3. The only repair is by redesign (reprogramming), which, if it removes the errors and introduces no others, will result in higher reliability.
4. Reliability can depend upon burn-in or wearout phenomena, i.e. failure rates can be decreasing, constant or increasing with respect to operating time.	4. Reliability is not so dependent. Reliability improvement over time may be effected, but this is not an operational time relationship. Rather it is a function of the effort put into detecting and correcting errors.
5. Failure can be related to the passage of operating (or storage) time.	5. Reliability is not time-related in this way. Failures occur when a program step or path which is in error is executed.
6. Reliability is related to environmental factors.	6. The external environment does not affect reliability, except insofar as it might affect program inputs.
7. Reliability can be predicted in theory from knowledge of design and usage factors.	7. Reliability cannot be predicted from any physical bases, since it entirely depends upon human factors in design. Some *a priori* approaches have been proposed (see below).
8. Reliability can sometimes be improved by redundancy.	8. Reliability cannot be improved by redundancy if the parallel program paths are identical, since if one path fails the others will have the same error. It is possible to provide redundancy by having parallel paths, each with different programs written and checked by different teams.
9. Failures can occur to components of a system in a pattern which is to some extent predictable from the stresses on the components, and other factors. Reliability critical lists and Pareto analysis of failures are useful techniques.	9. Failures are not usually predictable from analysis of separate statements. Errors are likely to exist randomly throughout the program, and any statement may be in error. Reliability critical lists and Pareto analysis of failures are not appropriate.

SOFTWARE FAILURE MODES

Software errors ('bugs') can arise from the specification, the software system design and from the coding process.

Specification errors

Typically more than half the errors recorded during software development originate in the specification. Since software is not perceivable in a physical sense, there is little scope for common sense interpretation of ambiguities, inconsistencies or incomplete statements. Therefore software specifications must be very carefully developed and reviewed. The software specification must describe fully and accurately the requirements of the program. The program should reflect the requirements exactly. There are no safety margins in software design as in hardware design. For example, if the requirement is to measure 9 ± 0.5 V and to indicate if the voltage is outside these tolerances, the program will do precisely that. If the specification was incorrectly formulated, e.g. if the tolerances were not stated, the out-of-tolerance voltage would be indicated at this point every time the measured voltage varied by a detectable amount from 9 V, whether or not the tolerances were exceeded. Depending upon the circumstances this might be an easily detectable error, or it might lead to unnecessary checks and adjustments because the out-of-tolerance indication is believed. This is a relatively simple example. Much more serious errors, such as a misunderstanding of the logical requirement of the program, can be written into the specification. This type of error can be much harder to correct, involving considerable reprogramming, and is much more serious in effect.

The specification must be logically complete. Consider the statement: 'Sample inputs A, B and C. If any one exceeds by $> \pm 10$ units the average of the other two, feed forward the average of these two. Indicate failure of the out-of-tolerance input. If the out-of-tolerance condition does not exist, feed forward the average of the three inputs.'

This is an example of a two-out-of-three majority voting redundant system. The logic is shown in the flow diagram, Fig. 8.1. Consider the situation when the values of A, B and C are 100, 90 and 120. The values of the derived parameters, and route taken by the program, are shown in Fig. 8.1. In this case, two fault conditions exist, since both B and C exceed the average of the other two inputs. The program will indicate a B failure, as the algorithm, compares B before C. The specification has not stated what should happen in the event of more than one input being out of tolerance. The program will work as shown in the algorithm, in the sense that an input will always be available, but the system may not be safe. The flowchart complies with the specification, but it probably does not reflect the real wishes of the specification writer. A software specification must cover all the possible input conditions and output requirements, and this usually requires much more detailed consideration than for a hardware specification.

Figure 8.1 Voting redundant system

The specification must be consistent. It must not give conflicting information or use different conventions in different sections.

The specification must not include requirements that are not testable, e.g. accuracy or speed requirements that are beyond the capability of the hardware.

The specification should be more than just a description of program requirements. It should describe the structure to be used, the program test requirements and documentation needed during development and test, as well as basic requirements such as the programming language, and inputs and outputs. (Program structure, test and documentation will be covered later.) By adequately specifying these aspects, a framework for program generation will be created which minimizes the possibilities for creating errors, and which ensures that errors will be found and corrected.

Software system design

The software system design follows from the specification. The system design may be a flowchart and would define the program structure, test points, limits, etc. Errors can occur as a result of incorrect interpretation of the specification, or incomplete or incorrect logic.

An important reliability feature of software system design is *robustness*, the term used to describe the capability of a program to withstand error conditions without serious effect, such as becoming locked in a loop or 'crashing'. The robustness of the program will depend upon the design, since it is at this stage that the paths to be taken by the program under error conditions are determined.

Software code generation

Code generation is a prime source of errors, since a typical program involves a large number of code statements.

Typical errors can be:

1. Typographical errrors.
2. Incorrect numerical values, e.g. 0.1 for 0.01.
3. Omission of symbols, e.g. parentheses.
4. Inclusion of expressions which can become indeterminate, such as division by a value which can become zero.

SOFTWARE STRUCTURE AND MODULARITY

Structure

Structured programming is an approach that constrains the programmer to using certain clear, well-defined approaches to program design, rather than allowing total freedom to design 'clever' programs which might be complex, difficult to understand or inspect, and prone to error. A major source of error in programs is the use of the GOTO statement for constructs such as loops and branches (decisions). The structured programming approach therefore discourages the use of GOTOs, requiring the use of control structures which have a single entry and a single exit. For example, the simple branch instruction in Fig. 8.2 can be programmed (using BASIC) in either an unstructured or a structured way as shown. The unstructured approach can lead to faults

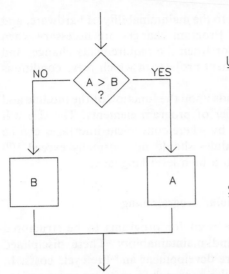

Unstructured:

If A > B goto A [line number]
else goto B [line number]
A [line number]
B [line number]

Structured:

If A > B then A [subroutine]
else B [subroutine]

Figure 8.2 Structured versus unstructured programming

if the wrong line number is given (e.g. if line numbers are changed as a result of program changes), and it is difficult to trace the subroutines (A, B) back to the decision point.

On the other hand, the structured approach eliminates the possibility of line number errors, and is much easier to understand and to inspect.

Structured programming leads to fewer errors, and to clearer, more easily maintained software. On the other hand, structured programs might be less efficient in terms of speed or memory requirements.

Modularity

Modular programming breaks the program requirement down into separate, smaller program requirements, or modules, each of which can be separately specified, written and tested. The overall problem is thus made easier to understand and this is a very important factor in reducing the scope for error and for easing the task of checking. The separate modules can be written and tested in a shorter time, thus reducing the chances of changes of programmer in mid-stream.

Each module specification must state how the module is to interface with other parts of the program. Thus, all the inputs and outputs must be specified. Structured programming might involve more preparatory work in determining the program structure, and in writing module specifications and test requirements. However, like good groundwork in any development programme, this effort is likely to be more than repaid later by the reduced overall time spent on program writing and debugging, and it will also result in a program which is easier to understand and to change. The capability of a program to be modified fairly easily can be compared to the maintainability of hardware, and it is often a very important feature. Program changes are necessary when logical corrections have to be made, or when the requirements change, and there are not many software development projects in which these conditions do not arise.

The optimum size of a module depends upon the function of the module and is not solely determined by the number of program elements. The size will usually be determined to some extent by where convenient interfaces can be introduced. As a rule of thumb, modules should not normally exceed 100 separate statements or lines of code in a high level language.

Requirements for structured and modular programming

Major software customers specify the need for programs to be structured and modular, to ensure reliability and maintainability. These disciplined approaches can greatly reduce software development and life cycle costs. In the United Kingdom, MASCOT (modular approach to software construction, operation and test) and, primarily in the United States, APSE (Ada programming and support environment) are methods imposed to ensure that the great difficulties and costs generated by the unstructured approaches used in the past are obviated. Reference 2 in the Bibliography covers structured and modular programming in more detail.

PROGRAMMING STYLE

Programming style is an expression used to cover the general approach to program design and coding. Structured and modular programming are aspects of style. Other aspects are, for example, the use of REM (remark) statements in the listing to explain the program, 'defensive' programming in which routines are included to check for faults, and the use of simple constructs whenever practicable. Obviously, a disciplined programming style can have a great influence on software reliability and maintainability, and it is therefore important that style is covered in software design guides and design reviews, and in programmer training. Programming style covered in detail in Ref. 7.

FAULT TOLERANCE

Programs can be written so that faults do not cause serious problems or complete failure of the program. We have mentioned 'robustness' in connection with program design, and this is an aspect of fault tolerance. A program should be able to find its way gracefully out of a fault condition and indicate the source. This can be achieved by programming internal tests, or checks of cycle time, with a reset and fault indication if the set conditions are not met. Where safety is a factor, it is important that the program sets up safe conditions when a fault occurs. For example, a process controller could be programmed to set up

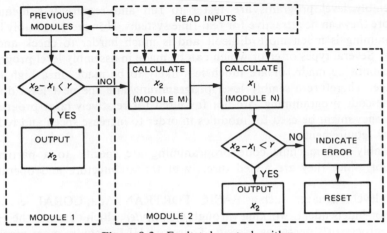

Figure 8.3 Fault tolerant algorithm

known safe conditions and indicate a problem, if no output is generated in two successive program cycle times or if the output value changes by more than a predetermined amount.

Fault tolerance can also be provided by program redundancy. For high integrity systems separately coded programs can be arranged to run simultaneously on separate but connected controllers, or in a time-sharing mode on one controller. A voting or selection routine can be used to select the output to be used. The effectiveness of this approach is based on the premise that two separately coded programs are very unlikely to contain the same coding errors, but of course this would not provide protection against a specification error. Redundancy can also be provided within a program by arranging that critical outputs are checked by one routine, and if the correct conditions are not present then they are checked by a different routine (Fig. 8.3).

These software techniques can also be used to protect against hardware failures, such as failure of a sensor which provides a program input. For example, failure of a thermostat to switch off a heating supply can be protected against by ensuring that the supply will not remain on for more than a set period, regardless of the thermostat output. This type of facility can be provided much more easily with software than with hardware, at no extra material cost, and therefore the possibility of increasing the reliability and safety of software controlled systems should always be analysed in the specification and design stages.

LANGUAGES

The selection of the computer language to be used can affect the reliability of software. There are two main approaches which can be used:

1. Assembly level programming.
2. High level (or high order) language (HLL or HOL) programming.

Assembly level programs are faster to run and require less memory; therefore they can be attractive for real-time systems. However, assembly level programming is much more difficult and is much harder to check and to modify. Several types of error which can be made in assembly level programming cannot be made, or are much less likely to be made, in a high level language. Therefore assembly level programming and the next level down, machine code programming, are not favoured for relatively large programs, though they might be used for modules in order to increase speed and reduce memory requirements.

Assembly and machine code programming are specific to a particular processor, since they are aimed directly at the architecture and operating system.

High level languages, such as BASIC, FORTRAN, Ada, CORAL, etc., are processor-independent, working through a *compiler* which converts the HLL to that processor's operating system. Therefore HLLs require more memory (the compiler itself is a large program) and they run more slowly. However, it is much easier to program in HLLs, and the programs are much easier to inspect and correct.

Selection of an HLL also has reliability and maintainability implications, for two main reasons:

1. In some languages, constructs which are frequent sources of errors are more likely to be used. For example, BASIC, while popular and easy to learn, allows GOTO statements to be used for many control structures. GOTO errors are very common. PASCAL and Ada discourage the use of GOTOs, using less error-prone constructs such as IF...THEN...ELSE or DO...WHILE instead.
2. The older HLLs (FORTRAN, BASIC) do not encourage structured programming. PASCAL, Ada and the UK language for defence systems, CORAL 66, strongly encourage structured programming. Ada and CORAL 66 are the standard HLLs in the United States and the United Kingdom for military systems, their use being mandatory in order to ensure system reliability and maintainability.

Since HLLs must work through a compiler, the reliability of the compiler affects system reliability. Generally speaking, though, compilers are reliable once fully developed, since they are so universally used. Compilers for new HLLs and/or new processors sometimes cause problems for the first few years until all errors are corrected.

REAL-TIME SYSTEMS

A real-time system is one in which the software must operate at the speed demanded by the system inputs and outputs. A chess program or a circuit simulation program, for example, will run when executed, and it is not critical exactly how long it takes to complete the run. However, in an operational system such as a process controller or an autopilot, it is essential that the software is ready to accept inputs and completes tasks at the right times. In real-time systems the processor and input and output functions are synchronized by the system clock. The software must be designed so that functions are correctly timed in relation to the system clock pulses.

Timing errors are a common cause of failure, particularly during development, in real-time systems. Timing errors are often difficult to detect, particularly by inspection of code. Timing errors can be caused by hardware faults or by interface problems. However, logic test instruments (logic analysers) can be used to show exactly when and under what conditions system timing errors occur, so that the causes can be pinpointed.

DATA RELIABILITY

Data reliability (or information integrity) is an important aspect of the reliability of software-based systems. When digitally coded data are transmitted, there are two sources of degradation:

1. The data might not be processed in time, so that processing errors are generated. This can arise, for example, if data arrive at a processing point (a 'server', e.g. a microprocessor or a memory address decoder) at a higher rate than the server can process.
2. The data might be corrupted in transmission or in memory by digital bits being lost or inverted, or by spurious bits being added. This can happen if there is noise in the transmission system, e.g. from electromagnetic interference or defects in memory.

System design to eliminate or reduce the incidence of failures due to processing time errors involves the use of queueing theory, applied to the expected rate and pattern of information input, the number and speed of the 'servers', and the queueing disciplines (e.g. first-in first-out (FIFO), last-in first-out (LIFO), etc.). Also, a form of redundancy is used, in which processed data are accepted as being valid only if they are repeated identically at least twice, say, in three cycles. This might involve some reduction in system processing or operating speed.

Data corruption due to transmission or memory defects is checked for and corrected using error detection and correction codes. The simplest and probably best known is the parity bit. An extra bit is added to each data word, so that there will always be an even (or odd) number of ones (even (or odd) parity). If an odd number of ones occurs in a word, the word will be rejected or ignored. More complex error detection codes, which also provide coverage over a larger proportion of possible errors and which also correct errors, are also used. Examples of these are Hamming codes and BCH codes.

Ensuring reliable data transmission involves trade-offs in memory allocation and operating speed. References 2 and 9 provide further information on this topic.

SOFTWARE CHECKING

Few programs run perfectly the first time they are tested. The scope for error is so large, due to the difficulty that the human mind has in setting up perfectly logical structures, that it is routine for programmers to have to spend a long time debugging a new program until the basic errors are eliminated. Modern high level language compilers contain error detection, so that many logical, syntactical or other errors are displayed to the programmer, allowing them to be corrected before an attempt is made to load the program. Automatic error correction is also possible in some cases, but this is limited to certain specific types of error. When the program is capable of being run, it is then necessary to confirm that it fulfills all requirements of the specification and that it will run under all anticipated input conditions.

To confirm that the specification is satisfied, the program must be checked against each item of the specification. For example, if a test specification calls for an impedance measurement of 15 ± 1 Ω, only a line-by-check of the program listing is likely to discover an error that calls for a measurement tolerance of $+ 1$ Ω, $- 0$ Ω. Program checking can be a tedious process, but it is made much easier if the program is structured into well-specified and understandable modules, so that an independent check can be performed quickly and comprehensively. Like hardware design review procedures, the cost of program checking is usually amply repaid by savings in development time at later stages. The program should be checked in accordance with a prepared plan, which stipulates the tests required to demonstrate specification compliance.

Formal program checking, involving the design team and independent people, is called a *structured walkthrough,* or a *code review.*

FMECAs of software-based systems

It is not practicable to perform an FMECA on software, since software 'components' do not fail. The nearest equivalent to an FMECA is a code review, but whenever an error is detected it is corrected so the error source is eliminated. With hardware, however, we cannot eliminate the possibility of, say, a transistor failure.

In performing an FMECA of a software-based hardware system it is necessary to consider the failure effects in the context of the operating software, since system behaviour in the event of a hardware failure might be affected by the software. This is particularly the case in systems utilizing built-in-test software, or when the software is involved in functions such as switching redundancy, displays, warnings and shut-down.

Software sneak analysis

The sneak circuit analysis (SCA) method described in Chapter 7 for evaluating circuit conditions that can lead to system failure is also applicable to software. Since a section of code does not fail but performs the programmed functions whether or not they are the intended ones, there is an analogy with an erroneous circuit design.

The program must be reduced to a set of topological patterns, as for hardware SCA. Since a program of reasonable size is very difficult to reduce in this way, this step is usually computerized.

Six basic sneak patterns exist, as shown in Fig. 8.4. Note that most software sneak patterns are related to branching instructions, such as GOTO or THEN/ELSE statements. The conditions leading to and deriving from such statements, as well as the statements themselves, are important clues in the SCA.

Software sneak conditions are:

1. *Sneak output.* The wrong output is generated.
2. *Sneak inhibit.* Undesired inhibit of an input or output.
3. *Sneak timing.* The wrong output is generated because of its timing or incorrect input timing.
4. *Sneak message.* A program message incorrectly reports the state of the system.

Figure 8.1 illustrates a potential sneak message condition, since the program will not indicate that *C* has failed if *A* and *B* have failed. This failure is brought about by an incorrect line pattern. The program correctly identifies the correct *A* value and the incorrect *B* value, and proceeds to the output with no chance of testing *C*.

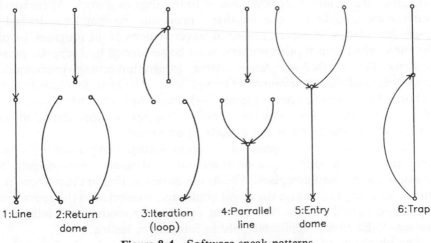

| 1:Line | 2:Return dome | 3:Iteration (loop) | 4:Parrallel line | 5:Entry dome | 6:Trap |

Figure 8.4 Software sneak patterns

SOFTWARE TESTING

Testing that a program will operate correctly over the range of input conditions is an essential part of the development process. Software testing must be planned and executed in a disciplined way since, even with the most careful design effort, errors are likely to remain in any reasonably large program, due to the impracticability of finding all errors by checking, as described above. Some features, such as timing, overflow conditions and module interfacing, are not easy to check.

There are limitations to software testing. It is not practicable to test exhaustively a reasonably complex program. The total number of possible paths through a program with n branches and loops is 2^n. It is not normally possible to plan a test strategy which will provide such coverage, and the test time would be exorbitant. Therefore the tests to be performed must be carefully selected to verify correct operation under the likely range of environments and input conditions, whilst being economical.

The software test process is iterative, whilst code is being produced. It is necessary to test code as soon as it is produced, to ensure that errors can be corrected quickly by the programmer who wrote it. It is also useful to test modules independently, since it is easier to devise effective tests for smaller, well-specified sections of code than for large programs. The detection and correction of errors is also much less expensive early in the development programme. As errors are corrected the software must be re-tested to confirm that the redesign has been effective and has not introduced any other errors. Later, when most or all modules have been written and tested, the complete program must be tested, errors corrected, and retested. Thus design and test proceed in steps, with test results being fed back to the programmers.

It is usual for programmers to test modules or small programs themselves. Given the specification and suitable instructions for conducting and reporting tests, they are usually in the best position to test their own work. Alternatively, programmers might test one another's programs, so that an independent approach is taken. However, testing of larger sections of the program, involving the work of several programmers, must be performed by a separate person or team. This is called *integration* testing. Integration testing covers module interfaces, and should demonstrate compliance with the system specification.

Formal configuration control (page 174) should be started when integration testing commences. Formal error reporting (see next section) should also be started at this stage, if it is not already in operation.

There are two main categories of program testing. *Verification* is the term used to cover all testing in a development or simulated environment, for example using a host computer. *Validation* covers testing in the real environment, including running on the target computer, connected to the operational input and output devices. Verification can include module and integration testing. Validation is applicable only to integration testing.

The objectives of software testing are to ensure that the system complies with the requirements and to detect as many errors as is practicable. Therefore the test plan must include:

1. Operation at extreme conditions (timing, input parameter values and rate of change, memory utilisation.)
2. Ranges of input sequence.
3. Fault tolerance (error recovery).

Since it is not practicable to test for the complete range of input conditions it is important to test for the most critical ones and for combinations of these. Random input conditions, possibly developed from system simulation, should also be used to provide assurance that a wide range of inputs is covered.

ERROR REPORTING

Reporting of software errors is an important part of the overall program documentation. The person who discovers an error is usually not the programmer or system designer, and therefore all errors, whether discovered during checking, testing or use, need to be written up with full details of program operating conditions at the time. The corrective action report should state the source of the error (specification, design, coding) and describe the changes made. Figure 8.5 shows an example of a software error reporting form. A software error reporting and corrective action procedure is just as important as a failure reporting system for hardware. The error reports and corrective action details should be retained with the module or program folder as part of the development record.

SOFTWARE RELIABILITY STATISTICS

Introduction

The statistics of software reliability are in a formative state at present. No models have yet gained widespread acceptance and there are no standard methods comparable with, say, MIL-STD-781 for reliability demonstration or MIL-STD-756 and MIL-HDBK-217 for reliability prediction, though attempts have been made to develop methods and standards.

Efforts to quantify software reliability usually relate to determining (or predicting) the probability of, or quantity of, errors existing in a program. Whilst this is a convenient starting point, there are practical difficulties. The reliability of a program depends not only upon whether or not errors exist but upon the probability that an existing error will affect the output, and the nature of the effect. Errors which are very likely to manifest themselves, e.g. those which cause a failure most times the program is run, are likely to be discovered and corrected during the development phase. An error which only causes a failure under very rare or unimportant conditions may not be a reliability

```
Program

Module

Error conditions:

        Input conditions:

    Description of failure:

Effect/importance:

Execution time since last failure:      Total run time:

Date:              Time:              Signed:

Program statement(s) involved:
    Line                                Statement


Error source:
        Code:        Design:        Specification:

Correction recommended:
    Code:

Design:

Specification:

Date:          Signed:          Approved:

    Correction made/tested:    Date:    Time:    Signed:

    Program master amended:    Date:    Time:    Signed:
```

Figure 8.5 Software error reporting form

problem, but the coding error that caused the total loss of a spacecraft, for example, was a disaster, despite all the previous exhaustive checking and testing.

Error generation, and the discovery and correction of errors, is a function of human capabilities and organization. Therefore, whilst theoretical models based upon program size might be constructed, the derivation of reliability values is likely to be contentious. For example, a well-structured modular program is much easier to check and test, and is less prone to error in the first place, than an unstructured program designed for the same function. A skilled and experienced programming team is less likely to generate errors than one which is less well endowed. A further difficulty in software reliability modelling is the fact that errors can originate in the specification, the design and the coding. With hardware, failure is usually a function of load, strength and time, and whether a weakness is due to the specification, the design or the production process, the physics of failure remain the same. With software, however, specification, design and coding errors are often different in nature, and the probability of their existence depends upon different factors. For example, the number of coding errors might be related to the number of code statements, but the number of specification errors might not have the same relationship. One specification error might lead to a number of separate program errors.

The following sections briefly outline some of the statistical models which have been proposed for software reliability.

The Poisson model (time-related)

It is assumed that errors can exist randomly in a code structure and that their appearance is a function of the time the program is run. The number of errors occurring in time t is $N(t)$. If the following conditions exist:

1. $N(0) = 0$,
2. not more than one error can occur in the time interval $(t, t + dt)$,
3. the occurrence of an error is independent of previous errors,

then the occurrence of errors is described by the non-homogeneous Poisson distribution:

$$P[N(t) = n] = \frac{[m(t)]^n}{n!} \exp[-m(t)] \qquad (n \geq 0) \tag{8.1}$$

where

$$m(t) = \int_0^t \lambda(s) \, ds$$

$m(t)$ is the mean (s-expected) number of errors occurring in the interval $(0, t)$:

$$m(t) = a[1 - \exp(-bt)]$$

where a is the total number of errors and b is a constant. The number of errors remaining after time t, assuming that each error which occurs is corrected without the introduction of others, is

$$\bar{N}(t) = a \exp(-bt) \tag{8.2}$$

The reliability function, after the most recent error occurs and is corrected at time s, is

$$R(t) = \exp[-a\{\exp(-bs) - \exp[-b(s+t)]\}] \tag{8.3}$$

In using a time-related model, the question arises as to what units of time should be used. The Poisson model has been tested against software error data using calendar time during which errors were detected and corrected and values for the parameters a and b derived. However, since software errors are not time-related in the way that physical (hardware) failure processes are, the use of time-related models for software errors is problematical.

The Musa model

The Musa model uses program execution time as the independent variable. A simplified version of the Musa model is

$$n = N_0 \left[1 - \exp \left(\frac{-Ct}{N_0 T_0} \right) \right]$$

(8.4)

where N_0 is the inherent number of errors, T_0 the MTTF at the start of testing (MTTF is mean time to failure) and C the 'testing compression factor' equal to the ratio of equivalent operating time to testing time.

The present MTTF:

$$T = T_0 \exp \left(\frac{Ct}{N_0 T_0} \right)$$

gives

$$R(t) = \exp \left(\frac{-t}{T} \right)$$

(8.5)

From these relationships we can derive the number of failures which must be found and corrected, or the program execution time necessary, to improve from T_1 to T_2:

$$\Delta_n = N_0 T_0 \left(\frac{1}{T_1} - \frac{1}{T_2} \right)$$

(8.6)

$$\Delta_t = \left(\frac{N_0 T_0}{C} \right) \ln \left(\frac{T_2}{T_1} \right)$$

(8.7)

Example 8.1

A large program is believed to contain about 300 errors and the recorded MTTF at the start of testing is 1.5 h. The testing compression factor is assumed to be 4. How much testing is required to reduce the remaining number of errors to ten? What will then be the reliability over 50 h of running?

From Eqns (8.6) and (8.7),

$$(300 - 10) = 300 \times 1.5 \left(\frac{1}{1.5} - \frac{1}{T_2} \right)$$

$$t = \left(\frac{300 \times 1.5}{4}\right) \ln\left(\frac{T_2}{1.5}\right)$$

Therefore

$$T_2 = 45 \text{ h}$$

and

$$\Delta_t = 382.6 \text{ h}$$

giving

$$R_{50} = \exp\left(\frac{-50}{45}\right) = 0.33$$

The Jelinski–Moranda and Schick–Wolverton models

Two other exponential-type models which have been derived are the Jelinski–Moranda (JM) model and the Schick–Wolverton (SW) model. In the JM and SW models, the hazard function $h(t)$ is given respectively by:

$$h(t_i) = \phi[N_0 - n_{i-1}] \tag{8.8}$$

$$h(t_i) = \phi[N_0 - n_{i-1}] t_i \tag{8.9}$$

where t_i is the length of the ith debugging interval, i.e. the time between the $(i-1)$th and the ith errors, and ϕ is a constant.

Littlewood models

Littlewood attempts to take account of the fact that different program errors have different probabilities of causing failure. If $\phi_1, \phi_2, \ldots, \phi_N$ are the rates of occurrence of errors $1, 2, \ldots, N$, the p.d.f. for the program time to failure, after the ith error has been fixed, is

$$f(t) = \lambda \exp(-\lambda t) \tag{8.10}$$

where λ is the program failure rate

$$\lambda = \phi_1 + \phi_2 + \cdots \phi_{N-i}$$

ϕ is assumed to be gamma-distributed, i.e. errors do not have constant rates of occurrence but rates which are dependent upon program usage. If the gamma distribution parameters are (α, β) (equivalent to (λ, a) in Eqn 2.34 then it can be shown, using a Bayes approach, that

$$f(t) = \frac{(N-1)\alpha(\beta + t')^{(N-i)\alpha}}{(\beta + t' + t)^{(N-1)(\alpha+1)}} \tag{8.11}$$

where t' is the time taken to detect and correct i errors. From this

$$R(t) = \left(\frac{\beta + t'}{\beta + t' + t}\right)^{(N-i)\alpha} \tag{8.12}$$

and

$$\lambda(t) = \frac{(N - i)\alpha}{\beta + t' + t} \tag{8.13}$$

At each error occurrence and correction, $\lambda(t)$ falls by an amount $\alpha/(\beta + t')$. It is assumed that all detected errors are corrected, without further errors being introduced.

Example 8.3

A large program is assumed to include a total of 300 errors, of which 250 have been detected and corrected in 20 h of execution time. Assuming the Littlewood model holds and the distribution parameters are $\alpha = 0.005$, $\beta = 4$, what is the expected reliability over a further 20 h?

From Eqn (8.12),

$$R(20) \doteq \left(\frac{4 + 20}{4 + 20 + 20} \right)^{(300 - 250)0.005}$$

$$= 0.86$$

Point process analysis

Since a program can be viewed as a repairable system, with errors being detected and corrected in a time continuum, the method of point process analysis described in Chapter 2 can be applied to software reliability measurement and analysis.

SOFTWARE RELIABILITY PREDICTION AND MODELLING

The utility of any statistical software reliability model depends upon acceptable values for the distribution parameters being available. Unlike hardware failure statistics, there is no physical basis for parameter estimation and the data that have been analysed to date are very limited compared with the wealth of available hardware failure data. Since software reliability is so dependent upon human performance and other non-physical factors, data obtained on one program or group of programs are unlikely to be accepted as being generally applicable, in the way that data on material properties are. Much work remains to be done to derive parameters applicable to various types of program, program structure and language, and the variability of programming teams, and indeed to develop the models already proposed to the point where there is concensus about practical applications.

The logical limitations inherent in the prediction of reliability as described in Chapter 5 apply equally to software. Indeed they are even more severe, since there are no physical or logical connections between past data and future expectation, as there are with many hardware failure modes.

HARDWARE/SOFTWARE INTERFACES

In software controlled systems, failures can occur which are difficult to diagnose to hardware or software causes, due to interactions between the two. We have already covered examples of such situations, where hardware elements provide program inputs. The software design can minimize these possibilities, as well as provide automatic diagnosis and fault indication. However, there are other types of failure which are more difficult, particularly when the hardware/software interface is less clearly defined.

Hardware meets software most closely within electronic devices such as processors and memories. A failure of a memory device, say of an individual memory cell which always indicates a logic state 1 (i.e. stuck at 1) regardless of the input, can cause failures which appear to be due to software errors. If the program is known to work under the input conditions, electronic fault-finding techniques can be used to trace the faulty device. There are times, particularly during development, when the diagnosis is not clear-cut, and the software and hardware both need to be checked. Timing errors, either due to device faults or software errors can also lead to this situation.

Memory devices of all types, from paper and magnetic tape to semiconductor memory devices, can cause system failures. Memory media and devices belong to the class of equipment sometimes called 'firmware', to indicate their interface status. Since many memory media are dynamic, i.e. the same data are handled in different locations at different times during program execution, firmware failures can lead to system failures which occur only under certain operating conditions, thus appearing to be due to software errors. Such failures can also be intermittent. Software and memory or microprocessor devices can be designed to protect the system against such failures. For example, the redundancy techniques described above could provide protection against some types of dynamic memory failure (other than a catastrophic failure of, say, a complete memory device). Redundancy can be provided to data or logic held in memory by arranging for redundant memory within the operating store or by providing independent, parallel memory devices. The program logic then has to be designed to store and access the redundant memory correctly, so the program becomes more complex.

CONCLUSIONS

Software is part of the operating system in a rapidly increasing range of modern engineered products, from large systems such as process plant, through more compact systems such as numerical control (NC) machine tools, to individual products such as domestic appliances and a wide range of electronic equipment. The versatility and economy offered by software control can lead to an under-estimation of the difficulty and cost of program generation. It is relatively easy to write a program to perform a simple defined function. To ensure that the program will operate satisfactorily under all conditions that

Figure 8.6 Software development for reliability

might exist, and which will be capable of being changed or corrected easily when necessary, requires an effort greater than that required for the basic design and first-program preparation. Careful groundwork of checking the specification, planning the program structure and assessing the design against the specification is essential, or the resulting program will contain many errors and will be difficult to correct. The cost and effort of debugging a large, unstructured program containing many errors can be so high that it is cheaper to scrap the whole program and start again.

Software that is reliable from the beginning will be cheaper and quicker to develop, so the emphasis must always be to minimize the possibilities of early errors and to eliminate errors before proceeding to the next phase. The essential elements of a software development project to ensure a reliable product are:

1. Specify the requirements completely and in detail.
2. Make sure that all project staff understand the requirements.
3. Check the specification thoroughly. Keep asking 'what if . . . ?'
4. Design a structured program and specify each module fully.
5. Check the design and the module specifications thoroughly against the system specification.
6. Check written programs for errors, line by line.
7. Plan module and system tests to cover important input combinations, particularly at extreme values.
8. Ensure full recording of all development notes, tests, checks, errors and program changes.

Figure 8.6 shows the sequence of development activities for a software project, with the emphasis on reliability.

BIBLIOGRAPHY

1. G. J. Myers, *Software Reliability: Principles and Practice.* Wiley, New York (1976).
2. M. Shooman, *Software Engineering — Design, Reliability, Management.* McGraw-Hill, New York (1983).
3. G. J. Myers, *The Art of Software Testing.* Wiley, New York (1979).
4. R. Longbottom, *Computer System Reliability.* Wiley (1980).
5. J. A. Thayer, M. Lipow and E. C. Nelson, *Software Reliability Study*, TRW series. North-Holland, Amsterdam (1976).
6. L. N. Harris and C. J. Dale, 'Approaches to software reliability prediction. *Procs. Annual Reliability and Maintainability Symposium*, IEEE (1982).
7. B. W. Kernighan and P. J. Plauger, *The Elements of Programming Style.* McGraw-Hill, New York (1974).
8. E. Yourdon, *Techniques of Program Structure and Design.* Prentice-Hall, Englewood Cliffs (1975).
9. J. E. Arsenault and J. A. Roberts, *Reliability and Maintainability of Electronic Systems.* Computer Science Press (1980).

232

10. US MIL-S-52779 (AD): *Software Quality Assurance Requirements*. Available from the National Technical Information Service, Springfield, Virginia.
11. UK Defence Standard 00-16: *Guide to the Achievement of Quality in Software*. HMSO.

Chapter 9

MAINTAINABILITY, MAINTENANCE AND AVAILABILITY

INTRODUCTION

Most systems are maintained, i.e. they are repaired when they fail, and work is performed on them to keep them operating. The ease with which repairs and other maintenance work can be carried out determines a system's maintainability. Maintainability can be quantified as the *mean time to repair* (MTTR). The time to repair, however, includes several activities, usually divided into three groups:

1. Preparation time: finding the person for the job, travel, obtaining tools and test equipment, etc.
2. Active maintenance time: actually doing the job.
3. Delay time (logistics time): waiting for spares, etc., once the job has been started.

Active maintenance time includes time for studying repair charts, etc., before the actual repair is started, and time spent in verifying that the repair is satisfactory. It might also include time for post-repair documentation when this must be completed before the equipment can be made available, e.g. on aircraft. Maintainability is often specified as a *mean active maintenance time* (MAMT), since it is only the active time (excluding documentation) that the designer can influence.

Maintained systems may be subject to *corrective* and *preventive* maintenance. Corrective maintenance includes all action to return a system from a failed to an operating or available state. The amount of corrective maintenance is therefore determined by reliability. Corrective maintenance action cannot be planned; it happens when we do not want it to.

Preventive maintenance seeks to retain the system in an operational or available state by preventing failures from occurring. This can be by servicing, such as cleaning and lubrication, or by inspection to find and rectify incipient failures, e.g. by crack detection or calibration. Preventive maintenance affects reliability directly. It is planned and should be performed when we want it to be. Preventive maintenance is measured by the time taken to perform the specified maintenance tasks and their specified frequency.

233

Maintainability affects availability directly. The time taken to repair failures and to carry out routine preventive maintenance removes the system from the available state. There is thus a close relationship between reliability and maintainability, one affecting the other and both affecting availability and costs. In the steady state, i.e. after any transient behaviour has settled down and assuming that maintenance actions occur at a constant rate:

$$\text{Availability} = \frac{\text{MTBF}}{\text{MTBF} + \text{MTTR} + \text{mean preventive maintenance time}}$$

Availability analysis and prediction were covered in Chapter 5.

The maintainability of a system is clearly governed by the design. The design determines features such as accessibility, ease of test and diagnosis, and requirements for calibration, lubrication and other preventive maintenance actions.

This chapter describes how maintainability can be optimized by design, and how it can be predicted and measured. It also shows how plans for preventive maintenance can be optimized in relation to reliability, to minimize downtime and costs.

MAINTENANCE TIME DISTRIBUTIONS

Maintenance times tend to be lognormally distributed (see Fig. 9.1). This has been shown by analysis of data. It also fits our experience and intuition that, for a task or group of tasks, there are occasions when the work is performed rather quickly, but it is relatively unlikely that the work will be done in much less time than usual, whereas it is relatively more likely that problems will occur which will cause the work to take much longer than usual.

In addition to the job-to-job variability, leading typically to a lognormal distribution of repair times, there is also variability due to learning. Depending upon how data are collected, this variability might be included in the job-to-

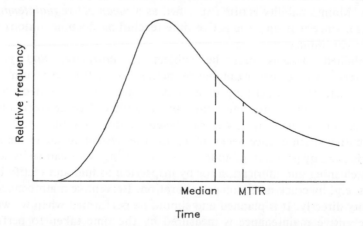

Figure 9.1 The lognormal distribution of maintenance times

job variability, e.g. if technicians of different experience are being used simultaneously. However, both the mean time and the variance should reduce with experience and training.

The properties of the lognormal distribution are described in Chapter 2. Lognormal or Weibull probability paper can be used for plotting maintenance time data (Chapter 3).

PREVENTIVE MAINTENANCE STRATEGY

The effectiveness and economy of preventive maintenance can be maximized by taking account of the time-to-failure distributions of the maintained parts and of the failure rate trend of the system.

In general, if a part has a decreasing hazard rate, any replacement will increase the probability of failure. If the hazard rate is constant, replacement will make no difference to the failure probability. If a part has an increasing hazard rate, then scheduled replacement at any time will in theory improve reliability of the system. However, if the part has a failure-free life (Weibull $\gamma > 0$), then replacement before this time will ensure that failures do not occur. These situations are shown in Fig. 9.2.

These are theoretical considerations. They assume that the replacement action does not introduce any other defects and that the time-to-failure distributions are exactly defined. As explained in Chapter 5, these assumptions must not be made without question. However, it is obviously of prime importance to take account of the time-to-failure distributions of parts in planning a preventive maintenance strategy.

In addition to the effect of replacement on reliability as theoretically determined by considering the time-to-failure distributions of the replaced parts, we must also take account of the effects of the maintenance action on reliability. For example, data might show that a high pressure hydraulic hose has an increasing hazard rate after a failure-free life, in terms of hose leaks. A sensible maintenance policy might therefore be to replace the hose after, say, 80 per cent. of the failure-free life. However, if the replacement action increases the probability of hydraulic leaks from the hose end connectors, it might be more economical to replace hoses on failure.

The effects of failures, both in terms of effects on the system and of costs of downtime and repair, must also be considered. In the hydraulic hose example, for instance, a hose leak might be serious if severe loss of fluid results, but end connector leaks might generally be only slight, not affecting performance or safety. A good example of replacement strategy being optimized from the cost point of view is the scheduled replacement of incandescent and fluorescent light units. It is cheaper to replace all units at a scheduled time before an expected proportion will have failed, rather than to replace each unit on failure.

In order to optimize preventive replacement, it is therefore necessary to

236

(m = scheduled replacement interval)

Decreasing hazard rate: scheduled replacement increases failure probability.

Constant hazard rate: scheduled replacement has no effect on failure probability.

Increasing hazard rate: scheduled replacement reduces failure probability.

Increasing hazard rate with failure—free life >m: scheduled replacement makes failure probability = 0.

Figure 9.2 Theoretical reliability and scheduled replacement relationships

know the following for each part:

1. The time-to-failure distribution parameters for the main failure modes.
2. The effects of all failure modes.
3. The cost of failure.
4. The cost of scheduled replacement.
5. The likely effect of maintainance on reliability.

We have considered so far parts which do not give any warning of the onset of failure. If incipient failure can be detected, e.g. by inspection, non-destructive testing, etc., we must also consider:

6. The rate at which defects propagate to cause failure.
7. The cost of inspection or test.

Note that, from 2, an FMECA is therefore an essential input to maintenance planning.

This systematic approach to maintenance planning, taking account of reliability aspects, is called *reliability centred maintenance* (RCM).

Example 9.2

A flexible cable on a robot assembly line has a time-to-failure distribution which is Weibull, with $\gamma = 150$ h, $\beta = 1.7$ and $\eta = 300$ h. If failure occurs whilst in use the cost of stopping the line and replacing the cable is \$5,000. The cost of replacement during scheduled maintenance is \$500. If the line runs for 5,000 hours a year and scheduled maintenance takes place every week (100 hours), what would be the annual expected cost of replacement at one-weekly or two-weekly intervals?

With no scheduled replacement the probability of a failure occurring in t hours will be

$$1 - \exp\left[\frac{-(t - 150)}{300}\right]^{1.7} \qquad \text{Eqn (2.37)}$$

With scheduled replacement after m hours, the scheduled maintenance cost in 5,000 h will be

$$\frac{5,000}{m} \times 500 = \frac{2.5 \times 10^6}{m}$$

and the expected failure cost in each scheduled replacement interval will be (assuming not more than one failure in any replacement interval):

$$5,000\left\{1 - \exp\left[\frac{-(m - 150)}{300}\right]^{1.7}\right\}$$

Then the total cost per year

$$C = \frac{2.5 \times 10^6}{m} + \frac{5,000 \times 5,000}{m} \left\{ 1 - \exp\left[\frac{-(m - 150)}{300} \right]^{1.7} \right\}$$

Results are as follows:

m	C	No. of scheduled replacements	Expected failures
100	$25,000	50	0
200	$18,304	25	1.2
400	$38,735	12	6.5

Therefore the optimum policy might be to replace the cables at alternate scheduled maintenance intervals, taking a slight risk of failure. (Note that the example assumes that not more than one failure occurs in any scheduled maintenance interval. If m is only a little more than γ this is a reasonable assumption.)

A more complete analysis could be performed using Monte Carlo simulation. We could then take account of more detailed maintenance policies; e.g. it might be decided that if a cable had been replaced due to failure shortly before a scheduled maintenance period, it would not be replaced at that time.

Practical implications

The time-to-failure patterns of components in a system therefore largely dictate the optimum maintenance policy. Generally, since electronic components do not wear out, scheduled tests and replacements do not improve reliability. Indeed they are more likely to induce failures (real or reported). Electronic equipment should only be subjected to periodic test or calibration when drifts in parameters or other failures can cause the equipment to operate outside specification without the user being aware. Built-in test and autocalibration can reduce or eliminate the need for periodic test.

Mechanical equipment subject to wear, corrosion, fatigue, etc., should be considered for preventive maintenance.

FMECA AND FTA IN MAINTENANCE PLANNING

The FMECA is an important prerequisite to effective maintenance planning and maintainability analysis. As shown earlier, the effects of failure modes (costs, safety implications, detectability) must be considered in determining scheduled maintenance requirements. The FMECA is also a very useful input for preparation of diagnostic procedures and checklists, since the likely causes of failure symptoms can be traced back using the FMECA results. When a fault tree analysis (FTA) has been performed, it can also be used for this purpose.

BUILT-IN TEST (BIT)

Complex electronic systems such as laboratory instruments, avionics and process control systems now frequently include built-in test (BIT) facilities. BIT consists of additional hardware (and often software) which is used for carrying out functional test on the system. BIT might be designed to be activated by the operator, or it might monitor the system continuously or at set intervals.

BIT can be very effective in increasing system availability and user confidence in the system. However, BIT inevitable adds complexity and cost and can therefore increase the probability of failure. Additional sensors might be needed as well as BIT circuitry and displays. In microprocessor-controlled systems BIT can be largely implemented in software, but even then additional memory might be required.

BIT can also adversely affect apparent reliability by falsely indicating that the system is in a failed condition. This can be caused by failures within the BIT, such as failures of sensors, connections, or other components. BIT should therefore be kept simple, and limited to monitoring of essential functions which cannot otherwise be easily monitored.

It is important to optimize the design of BIT in relation to reliability, availability and cost. Sometimes BIT performance is specified (e.g. '90 per cent. of failures must be detected and correctly diagnosed by BIT'). An FMECA can be useful in checking designs against BIT requirements since BIT detection can be assessed against all the important failure modes identified.

MAINTAINABILITY PREDICTION

Maintainability prediction is the estimation of the maintenance workload which will be imposed by scheduled and unscheduled maintenance. The standard method used for this work is US MIL-HDBK-472, which contains four methods for predicting the mean time to repair (MTTR) of a system. Method II is the most frequently used. This is based simply on summing the products of the expected repair times of the individual failure modes and dividing by the sum of the individual failure rates, i.e.:

$$\text{MTTR} = \frac{\sum (\lambda t_r)}{\sum \lambda} \tag{9.1}$$

The same approach is used for predicting the mean preventive maintenance time, with λ replaced by the frequency of occurrence of the preventive maintenance action.

MIL-HDBK-472 describes the methods to be used for predicting individual task times, based upon design considerations such as accessibility, skill levels required, etc. It also describes the procedures for calculating and documenting the analysis, and for selection of maintenance tasks when a sampling basis is to be used (method III), rather than by considering all maintenance activities, which is impracticable on complex systems.

MAINTAINABILITY DEMONSTRATION

The standard approach to maintainability demonstration is MIL-STD-471. The technique is the same as for maintainability prediction using method III of MIL-HDBK-472, except that the individual task times are measured rather than estimated from the design. Selection of task times to be demonstrated might be by agreement or by random selection from a list of maintenance activities.

DESIGN FOR MAINTAINABILITY

It is obviously important that maintained systems are designed so that maintenance tasks are easily performed, and that the skill level required for diagnosis, repair and scheduled maintenance are not too high, considering the experience and training of likely maintenance personnel and users. Features such as ease of access and handling, the use of standard tools and equipment rather than specials, and the elimination of the need for delicate adjustment or calibration are desirable in maintained systems. As far as is practicable, the need for scheduled maintenance should be eliminated. Whilst the designer has no control over the performance of maintenance people, he or she can directly affect the inherent maintainability of a system.

Design rules and checklists should include guidance, based on experience of the relevant systems, to aid design for maintainability and to guide design review teams.

Design for maintainability is closely related to design for ease of production. If a product is easy to assemble and test maintenance will usually be easier. Design for testability of electronic circuits is particularly important in this respect, since circuit testability can greatly affect the ease and accuracy of failure diagnosis, and thus maintenance and logistics costs. Design of electronic equipment for testability is covered in more detail in Chapter 6.

Interchangeability is another important aspect of design for ease of maintenance of repairable systems. Replaceable components and assemblies must be designed so that no adjustment or recalibration is necessary after replacement. Interface tolerances must be specified to ensure that replacement units are interchangeable.

BIBLIOGRAPHY

1. J. D. Patton, *Maintainability and Maintenance Management*. Instrument Society of America, Pittsburgh (1980).

2. A. S. Goldman and T. B. Slattery, *Maintainability: A Major Element of Systems Effectiveness*. Wiley, New York (1964.
3. B. S. Blanchard and E. E. Lowery, *Maintainability Principles and Practices*. McGraw-Hill, New York (1969).
4. C. E. Cunningham and W. Cox, *Applied Maintainability Engineering*. Wiley, New York (1972).
5. F. S. Nowlan and H. F. Heap, *Reliability Centred Maintenance*. Dolby Access Press (1979).
6. A. K. S. Jardine, *Maintenance, Replacement and Reliability*. Pitman 1973.
7. D. J. Smith, *Reliability and Maintainability in Perspective*. MacMillan 1981.
8. J. E. Arsenault and J. A. Roberts, *Reliability and Maintainability of Electronic Systems*. Computer Science Press (1980).
9. US MIL-STD-470: *Maintainability Program Requirements*. Available from the National Technical Information Service, Springfield, Virginia.
10. US MIL-STD-471: *Maintainability Demonstration*. Available from the National Technical Information Service, Springfield, Virginia.
11. US MIL-STD-472: *Maintainability Prediction*. Available from the National Technical Information Service, Springfield, Virginia.
12. UK Defence Standard 00-40: *The Management of Reliability and Maintainability*. HMSO.

Chapter 10

RELIABILITY IN MANUFACTURE

INTRODUCTION

It is common knowledge that a well-designed product can be unreliable in service because of poor quality of production. Control of production quality is therefore an indispensable facet of an effective reliability effort. Quality control in any production situation is a matter of ensuring conformance to specifications and tolerances. This involves controlling variability and identifying and solving problems. The most likely source of variability is nearly always human operations, particularly repetitive, boring or unpleasant tasks. Automation of such tasks therefore usually leads to quality improvement. Typical examples are paint spraying and spot welding by robots in automobile production, component placement and soldering in electronic production and CNC machining.

Variability can never be completely eliminated, since there will nearly always be some human operations and automatic processes are not without variation. A reliable design should cater for expected variation, so designers must be made aware of the variability inherent in the manufacturing processes to be used.

The production quality programme must be integrated with the design reliability programme to ensure that the objectives are harmonized. Not all the objectives of a production quality programme might enhance reliability. For example, undue emphasis might be placed on reducing production costs by saving test or inspection costs or on passing a final quality check without considering which outgoing tests would maximize reliability. In general, what is good for production quality control is good for reliability, but radical separation of the two programmes can reduce the effectiveness of both.

The design reliability and maintainability effort should also take account of the needs of production, and the quality team should use the information provided in stress analyses, FMECAs and reliability test reports. A reliable and easily maintained design will be cheaper to produce, in terms of reduced costs of scrap and rework.

The integration of reliability and quality programmes is covered in more detail in Chapter 14.

242

QUALITY COSTS

Quality costs are the costs of all activities specifically directed at quality control, and the costs of failure. Quality costs are usually considered in three categories:

1. Prevention costs
2. Appraisal costs
3. Failure costs

Prevention costs are those related to activities which prevent failures occurring. These include quality control of bought-in components and materials, training and management. In an integrated Q & R programme they also include the costs of the reliability activities during design and development testing.

Appraisal costs are those related to test and measurement, process control and quality audit.

Failure costs are the actual costs of failure. Internal failure costs are those incurred during manufacture. These cover scrap and rework costs (including costs of related work in progress, space requirements for scrap and rework, associated documentation, and related overheads). Failure costs also include external or post-delivery failure costs, such as warranty costs; these are the costs of unreliability.

Obviously it is necessary to minimize the sum of quality and reliability costs over a suitably long period. Therefore the immediate costs of prevention and appraisal must be related to the anticipated effects on failure costs, which might be affected over several years. Investment analysis related to Q & R is an uncertain business, because of the impossibility of accurately predicting and quantifying the results. Therefore the analysis should be performed using a range of assumptions to determine the sensitivity of the results to assumed effects, such as the yield at test stages and reliability in service.

When quality control is integrated with production activities, such as operator control of measurement and test, the quality costs will not be as easily separable from production costs. However, the true costs of the quality-related activities should be identified, in order to enable the overall quality costs to be minimized.

CONTROL OF PRODUCTION VARIABILITY

The main cause of production-induced unreliability is the variability inherent in production processes. In principle, a correct design, correctly manufactured, should not fail in normal use. However, all manual processes are variable. Automatic processes are also variable, but the variability is usually easier to control. Bought-in components and materials also have variable properties. Production quality control (QC) is primarily concerned with measuring and controlling these variations in the most cost-effective way.

Statistical quality control (SQC) is the term used for the measurement and control of production variability. In SQC, QC people rely heavily on the *s*-normal distribution. This is usually appropriate, because of the central limiting tendency to *s*-normality of the many sources of variability in production processes. QC people are also primarily concerned with the behaviour of variables about the central tendency, and therefore control is typically set about ±2 or 3 SDs of the measured parameter.

Control charts

Figure 10.1 (a) is an example of a process control chart. As measurements are made the values are marked as points on the control chart against the sample

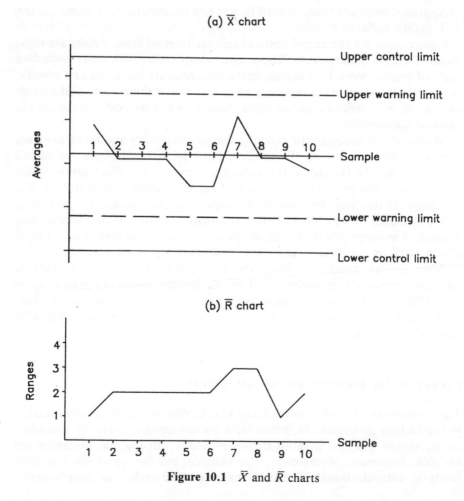

Figure 10.1 $\bar{X}$ and $\bar{R}$ charts

number on the horizontal scale. The data plotted can be individual values or sample averages; when sample averages are plotted the chart is called an $\bar{X}$ *chart*. The $\bar{X}$ chart shows very clearly how the process value varies. Upper and lower control limits can be drawn on the chart, as shown, to indicate when the process has exceeded preset limits.

The control limits on an $\bar{X}$ chart are based on the tolerance required of the process and from analysis of trial runs or past process data. For example, the control limits could be set at the ± 2 or 3σ values of the process distribution. Warning limits are also used. These are set within the control limits to provide a warning that the process might require adjustment. Usually 2 or more sample points must fall outside the warning limits before action is taken. However, any point falling outside the control limit indicates a need for immediate investigation and corrective action.

Figure 10.1 (*b*) is a *range chart* ($\bar{R}$ chart). The plotted points show the range of values within the sample. The chart indicates the repeatability of the process.

$\bar{X}$ and $\bar{R}$ charts (also called *Shewhart charts*) are the basic tools of SQC for control of manufacturing processes. Their ease of use and effectiveness make them very suitable for use by production operators for controlling their own work, and therefore they are commonly used in operator control and quality circles (see later in this chapter). Computer programs are available which produce $\bar{X}$ and $\bar{R}$ charts automatically when process data are input. Integrated measurement and control systems provide for direct input of measured values to the control chart program.

Control of human variability

Several methods have been developed for controlling the variability inherent in human operations in manufacturing, and these are well documented in the references on quality assurance. Psychological approaches, such as improving motivation by better work organization, exhortation and training, have been used since early industrialization, particularly since the 1940s. These were supported by the development of statistical QC methods, described earlier.

Inspection

One way of monitoring and controlling human performance is by independent inspection, and this was the standard QC approach until the 1950s, and is still used to a large extent. An inspector can be made independent of production requirements and can be given authority to reject work for correction or scrap. However, inspection is subject to three major drawbacks:

1. Inspectors are not perfect; they can fail to detect defects. On some tasks, such as inspecting large numbers of solder joints or small assemblies, inspection can be a very poor screen, missing 10 to 50 per cent. of defects. Inspector performance is also as variable as any other human process.

2. Independent inspection can reduce the motivation of production people to produce high quality work. They will be concerned with production quantity, relying on inspection to detect defects.
3. Inspection is expensive. It is essentially non-productive, the staff employed are often more highly paid than production people and output is delayed while inspection takes place. Probably worst, independent inspection can result in an overlarge QC department, unresponsive to the needs of the organization.

These drawbacks, particularly the last, have led increasingly to the introduction of operator control of quality.

Operator control

Under operator control, the production worker is responsible for monitoring and controlling his or her own performance. For example, in a machining operation the operator will measure the finished article, log the results and monitor performance on SQC charts. Inspection becomes part of the production operation and worker motivation is increased. The production people must obviously be trained in measurement and SQC methods, but this is usually found to present few problems. Operator control becomes even more relevant with the increasing use of production machinery which includes measuring facilities, such as self-monitoring computer numerically controlled (CNC) machines.

'Zero defects'

The *'zero defects'* (ZD) approach to QC was developed in the United States in the 1960s. ZD is based very much upon setting QC targets, publicizing results and exhortation by award presentations and poster campaigns. Successes were claimed for ZD but it has always been controversial. The motivational basis is evangelical and emotional, and the initial enthusiasm is hard to sustain. There are few managers who can set up and maintain a ZD programme as originally publicized, and consequently the approach is now seldom used.

Quality circles

The quality circles movement started in Japan in the 1950s, and is now used world wide. The idea is largely based on Drucker's management teaching, developed and taught for QC application by W. E. Deming. It uses the methods of operator control, consistent with Drucker's teaching that the most effective management is that nearest to the action; this is combined with basic SQC methods to identify and correct problems at their sources. The operator is often the person most likely to understand the problems of the process he or she operates, and how to solve them. However, the individual operator does not usually have the authority or the motivation to make changes. Also, he or she might not be able to influence problems elsewhere in the manufacturing system. The quality circles system gives workers this knowledge and influence, by organizing them into small work groups, trained to monitor quality performance, to analyse problems and to recommend solutions to management.

It is therefore a very different approach to that of ZD, since it introduces quality motivation as a normal working practice at the individual and team level, rather than by exhortation from above. Whilst management must be closely involved and supportive, it does not have to be continually active and visible in the way ZD requires.

Quality circles must be organized with care and with the right training, and must have full support from senior and middle management. In particular, quality circles recommendations must be carefully assessed and actioned whenever they will be effective.

The concept is really straightforward, enlightened management applied to the quality control problem, and in retrospect it might seem surprising that it has taken so long to be developed. It has proved to be highly successful in motivating people to produce better quality, and has been part of the foundation of the Japanese industrial revolution since the Second World War. The quality circles approach can be very effective when there is no formal quality control organization, for example in a small company.

Production FMECA

A production FMECA (page 161) is a useful method for identifying potential production-induced reliability problems.

ACCEPTANCE SAMPLING

Acceptance sampling provides a method for deciding whether to accept a particular production lot, based upon measurements of samples drawn at random from the lot. Sampling can be by *attributes* or by *variables*.

Sampling by Attributes

Sampling by attributes is applicable to go/no–go tests, using binomial and Poisson statistics. Sampling by attributes is covered in standard plans such as US MIL-STD-105 and BS 6001. These give accept and reject criteria for various sampling plans, based upon sample size and risk levels. The main criterion is the acceptable quality level (AQL), defined as the maximum percentage defective which can be accepted as a process average. The tables in the standards give accept and reject criteria for stated AQLs, related to sample size, and for 'tightened', 'normal' and 'reduced' inspection. These inspection levels relate to the consumer's risk in accepting a lot with a percentage defective higher than the AQL. Table 10.1 shows a typical sampling plan. MIL-S-19500 and MIL-M-38510 (see Chapter 7) give sampling plans based upon the lot tolerance percentage defective (LTPD). The plans provide the

Table 10.1 Master table for normal inspection—single sampling (MIL-STD-105D, Table II-A

Sample size code letter	Sample size	Acceptable quality
		0.010 0.015 0.025 0.040 0.065 0.10 0.15 0.25 0.40 0.65 1.0 1.5
		Ac Re Ac Re Ac Re Ac Re Ac Re Ac Re Ac Re Ac Re Ac Re Ac Re Ac Re Ac Re
A	2	
B	3	
C	5	
D	8	0 1
E	13	0 1
F	20	0 1
G	32	0 1 1 2
H	50	0 1 1 2 2 3
I	80	0 1 1 2 2 3 3 4
K	125	0 1 1 2 2 3 3 4 5 6
L	200	0 1 1 2 2 3 3 4 5 6 7 8
M	315	0 1 1 2 2 3 3 4 5 6 7 8 10 11
N	500	0 1 1 2 2 3 3 4 5 6 7 8 10 11 14 15
P	800	0 1 1 2 2 3 3 4 5 6 7 8 10 11 14 15 21 22
Q	1250	0 1 1 2 2 3 3 4 5 6 7 8 10 11 14 15 21 22
R	2000	1 2 2 3 3 4 5 6 7 8 10 11 14 15 21 22

⇩ = Use first sampling plan below arrow. If sample size equals, or exceeds, lot or batch

⇧ = Use first sampling plan above arrow.

minimum sample size to assure, with given risk, that a lot with a percentage defective equal to or more than the specified LTPD will be rejected. LTPD tests give lower consumers' risks that substandard lots will be accepted. LTPD sampling plans are shown in Table 10.2.

For any attribute sampling plan, an *operating characteristic* curve can be derived. The OC curve shows the power of the sampling plan in rejecting lots with a given percentage defective. For example, Fig. 10.2 shows OC curves for single sampling plans for 10 per cent. samples drawn from lots of 100, 200 and 1000, when one or more defectives in the sample will lead to rejection (acceptance number = 0). If the lot contains, say, 2 per cent. defective the probability of acceptance will be 10 per cent. for a lot of 1000, 65 per cent. for a lot of 200, and 80 per cent. for a lot of 100. Therefore the lot size is very important in selecting a sampling plan.

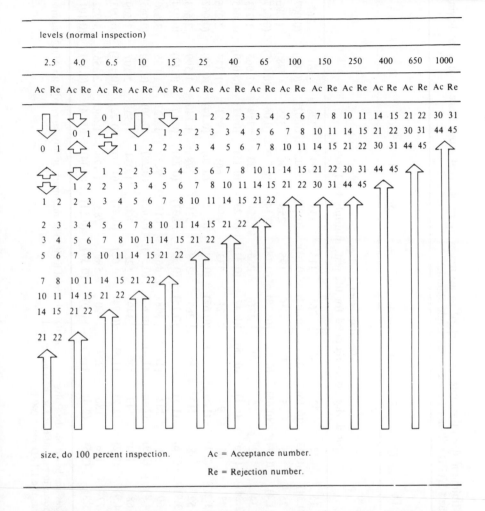

levels (normal inspection)

2.5	4.0	6.5	10	15	25	40	65	100	150	250	400	650	1000
Ac Re	Ac Re	Ac Re	Ac Re	Ac Re	Ac Re	Ac Re	Ac Re	Ac Re	Ac Re	Ac Re	Ac Re	Ac Re	Ac Re

size, do 100 percent inspection.

Ac = Acceptance number.
Re = Rejection number.

Double sampling plans are also used. In these the reject decision can be deferred pending the inspection of a second sample. Inspection of the second sample is required if the number of defectives in the first sample is greater than allowable for immediate acceptance but less than the value set for immediate rejection. Tables and OC curves for double (and multiple) sampling plans are also provided in the references quoted above.

Sampling by variables

Sampling by variables involves using actual measured values rather than individual attribute ('good or bad') data. The methods are based upon use of the *s*-normal distribution. They are described in US MIL-STD-414 and in reference 4 in the bibliography.

Table 10.2 LTPD sampling plans. [a] Minimum size of sample to be tested to assure, with 90 per cent. confidence, that a lot having percentage defective equal to the specified LTPD will not be accepted (single sample)

Acceptance number (c) ($r = c + 1$)	Max. percentage defective (LTPD) or λ												
	30	20	15	10	7	5	3	2	1.5	1	0.7	0.5	0.3
	Minimum sample sizes (for device-hours required for life test, multiply by 1,000)												
0	8 (0.64)	11 (0.46)	15 (0.34)	22 (0.23)	32 (0.16)	45 (0.11)	76 (0.07)	116 (0.04)	153 (0.03)	231 (0.02)	328 (0.02)	461 (0.01)	767 (0.007)
1	13 (2.7)	18 (2.0)	25 (1.4)	38 (0.94)	55 (0.65)	77 (0.46)	129 (0.28)	195 (0.18)	258 (0.14)	390 (0.09)	555 (0.06)	778 (0.045)	1,296 (0.027)
2	18 (4.5)	25 (3.4)	34 (2.24)	52 (1.6)	75 (1.1)	105 (0.78)	176 (0.47)	266 (0.31)	354 (0.23)	533 (0.15)	759 (0.11)	1,065 (0.080)	1,773 (0.045)
3	22 (6.2)	32 (4.4)	43 (3.2)	65 (2.1)	94 (1.5)	132 (1.0)	221 (0.62)	333 (0.41)	444 (0.31)	668 (0.20)	953 (0.14)	1,337 (0.10)	2,226 (0.062)
4	27 (7.3)	38 (5.3)	52 (3.9)	78 (2.6)	113 (1.8)	158 (1.3)	265 (0.75)	398 (0.50)	531 (0.37)	798 (0.25)	1,140 (0.17)	1,599 (0.12)	2,663 (0.074)
5	31 (8.4)	45 (6.0)	60 (4.4)	91 (2.9)	131 (2.0)	184 (1.4)	308 (0.85)	462 (0.57)	617 (0.42)	927 (0.28)	1,323 (0.20)	1,855 (0.14)	3,090 (0.085)
6	35 (9.4)	51 (6.6)	68 (4.9)	104 (3.2)	149 (2.2)	209 (1.6)	349 (0.94)	528 (0.62)	700 (0.47)	1,054 (0.31)	1,503 (0.22)	2,107 (0.155)	3,509 (0.093)
7	39 (10.2)	57 (7.2)	77 (5.3)	116 (3.5)	186 (2.4)	234 (1.7)	390 (1.0)	589 (0.57)	783 (0.51)	1,178 (0.34)	1,680 (0.24)	2,355 (0.17)	3,922 (0.101)

[a] MIL-S-19500 and MIL-M-38510. Sample sizes are based on the Poisson exponential binomial limit. The minimum quality (approximate AQL) required to accept (on the average) 19 of 20 lots is shown in parentheses for information only. The life test failure rate, λ, shall be defined as the LTPD per 1,000 h.

Figure 10.2 Operating characteristic (OC) curves for single sampling plans (10 per cent. sample, acceptance number = 0)

Sampling by variables is not as popular as sampling by attributes, since it is a more complex method. However, it can be useful when a particular production variable is important enough to warrant extra control.

General comment on sampling

Whilst standard sampling plans can be useful for controlling manufacturing processes, and for providing some assurance that the proportion defective is below a specified figure, they do not provide the high assurance necessary for many modern products. For example, if an electronic assembly consists of 100 components, all with an AQL of 0.1 per cent., there is on average a probability of about 0.9 that an assembly will be free of defective components. If 10,000 assemblies are produced, about 1,000 on average will be defective. The costs involved in diagnosis and repair or scrap during manufacture would obviously be very high. With such quantities typical of much modern manufacturing, higher assurance than can realistically be obtained from statistical sampling plans is obviously necessary. Also, standard QC sampling plans often do not provide assurance of long term reliability.

Electronic component manufacturers often quote outgoing quality levels in parts defective per million (p.p.m.). The figures are quoted as averages, derived from their own 100 per cent. outgoing inspection and test. Typical figures for commercial integrated circuits are 100 to 300 p.p.m. (0.01 to 0.03 per cent. defective), and lower for simpler components such as transistors, resistors and passive components. Incoming quality is assured either by monitoring production test results or, more effectively, by testing or screening incoming components before assembly. The competitive pressures on component manufacturers are such that even when demand is very high, quality levels are maintained. Governments and other purchasing organizations have also imposed standards (see Chapter 14) for controlling electronic component quality, and component manufacturers stress the quality of their products in their advertising.

Because of the much wider range of non-electronic component types, there is more diversity of approach. Generally, competition ensures high quality standards. Equipment manufacturers, particularly larger companies such as automobile or domestic equipment makers, often impose severe quality requirements on suppliers. These typically include provision for the replacement of all defective items, payment of costs incurred, transfer of warranty costs and audit of the supplier's QC systems. As with electronic components, SQC has become insufficiently effective by itself to ensure the high quality levels necessary for mass production in competitive markets.

QUALITY CONTROL IN ELECTRONICS PRODUCTION

Test methods

Electronic equipment production is characterized by very distinct assembly stages, and specialist test equipment has been developed for each of these. Since electronic production is so ubiquitous, and since test methods can greatly affect quality costs and reliability, it is appropriate to consider test methods for electronics in this context. Automatic test equipment (ATE) for electronic production falls into the following main categories:

1. *Component testers.* These range from simple bench-top testers to very complex automatic testers for high speed testing of large-scale integrated circuits (e.g. microprocessors, memories).
2. *Bare board testers.* These test for shorts and opens on bare printed circuit boards (PCBs) before components are mounted to them.
3. *In-circuit testers.* In-circuit testers (ICT) test individual components, sequentially, after they have been mounted (soldered) on to PCBs, by making contact via an array of test pins to test points on the PCB surface.
4. *Functional testers.* Functional testers, of either assembled PCBs or higher levels of assembly, involve applying power, operating the unit under test through its specified functions and measuring the outputs. Functional testers may either be general purpose or designed for testing particular systems.

Nearly all electronic test equipment is software-controlled, and so can be configured to test, for example, any PCB pattern or assembled PCB. There is also some overlap of functions: e.g. modern PCB functional testers often include ICT facilities. ATE also includes diagnostic capabilities. ICT equipment will indicate which components are defective and the nature of the fault (e.g. shorted capacitors, IC faulty logic). Functional testers have guided probes, which the operator uses under machine instruction to isolate fault locations.

The rapid development of more complex circuits has led to a corresponding effort on the part of manufacturers to provide ATE which can adequately test and diagnose modern electronic systems. Generally speaking, ATE has not managed to keep up-to-date, and testing of electronic systems containing modern ICs can be difficult and expensive.

It is very important to design digital circuits to be testable by providing test access points and by including additional circuitry, or it will not be possible to test all functions or to diagnose the causes of certain failures. Design for testability can have a major impact on quality costs and on future maintenance costs.

The optimum test strategy for a particular electronic product will depend upon factors such as component quality, quality of the assembly processes, test equipment capability, costs of diagnosis and repair at each stage, and production throughput. Generally, the earlier a defect is detected, the lower will be the cost. For example, if a defective component is detected and rejected at the component test, the cost will be the cost of the component. If the defective component is not detected until the PCB functional test, the cost will include time for diagnosis and repair, interruption to production, retest and additional documentation. A common rule of thumb states that the cost of detecting a defect increases by a factor of 10 at each test stage (component/PCB, ICT, functional test stages, in-service). Therefore any test strategy must be based on detecting and correcting defects as early as practicable in the production cycle.

No test is 100 per cent. effective in detecting defects. For example, a defective component might survive the component test and ICT, but fail at the functional test. Sometimes it is argued that defects detected at early test stages (e.g. a marginally out-of-tolerance component) would not cause later failure, and therefore early test stages might be too effective. This, however, ignores the fact that a marginal component is more likely to fail later, possibly in service, when the cost of failure is much higher.

Since defects can be introduced at any manufacturing stage it is not sufficient to rely upon early testing alone.

The only practical way in which the many variables involved in electronic equipment testing can be assessed is to run computer models of the test options, under the range of assumed input conditions. Figure 10.3 shows some typical test flow arrangements. In Fig. 10.3 (a), the manufacturer performs all test stages. In (b), the manufacturer relies on his suppliers to provide tested components and PCBs, applies only functional test and uses ICT only for diagnosis of PCBs which fail the functional test. Depending upon the circumstances, either strategy might be optimal.

Test results must be monitored continuously to ensure that costs are

254

(a)

(b)

Figure 10.3 Electronics equipment test strategies

controlled and minimized. There will inevitably be variation over time. It is important to analyse the causes of failures at later test stages, to determine whether they could have been detected and corrected earlier. Throughout, the causes of defects must be rapidly detected and eliminated, and this requires a very responsive data collection and analysis system (see page 258). Some ATE systems provide data logging and analysis, and data networking between test stations is also available.

Reliability of connections

Solder connections for electronic components on PCBs are not usually a reliability problem, but for equipment subject to long periods in storage, or which must operate in severe environments of vibration, temperature cycling or corrosive atmospheres, solder joint quality can be critical, particularly as the number involved is always large. Complete opens and shorts will nearly always be detected during functional test, but open-circuit failures can occur in use due to corrosion or fatigue of joints which pass initial visual and functional tests. The main points to watch to ensure solder joint reliability are:

1. *Process control.* Solder temperature, solder mix (when this is variable, e.g. in wave solder machines), soldering time, fluxes, PCB cleaning.
2. *Component mounting.* The solder joint should not be used to provide structural support, particularly in equipment subject to vibration or shock, or for components of relatively large mass, such as trimpots and some capacitors.
3. *Component preparation.* Component leads must be clean and wettable by the solder. Components which have been stored unpackaged for more than a few days, or packaged for more than six months, require special attention, as oxide formation on the leads may inhibit solderability, particularly on wave solder machines, where all joints are subject to the same time in the solder wave. If necessary, such components should have their leads cleaned and retinned prior to assembly. All components should be subject to sampling tests for solderability, as near in time to the assembly stage as practicable.
4. *Solder joint inspection.* Inspectors performing visual inspection of solder joints are typically about 80 per cent effective in seeing joints which do not pass visual inspection standards. Also, it is possible to have unreliable joints which meet appearance criteria. If automatic testing for opens and shorts is used instead of 100 per cent. visual inspection, it will not show up marginal joints which can fail later.

Burn-in (Stress Screening)

'*Burn-in*' is the name given to processes to stimulate failure in defective electronic components and assemblies by accelerating the stresses that will cause defective items to fail without damaging good items. The term '*stress screening*' is also used, usually in relation to assemblies. Component burn-in is an effective method for reducing production quality costs and for improving reliability. It is particularly necessary for LSI and VLSI integrated circuits, and for equipment which must operate in severe environments. It is mandatory on most high-reliability procurement contracts, such as for defence, space and telecom systems.

The justification for component burn-in is based on the evidence that mass production of electronic components generates three distinct categories. Most components are 'good', and are produced to specification. Some are initially defective and fail on test, and are then rejected. However, a proportion will be defective but will pass on test, the defects being potential causes of failure at some future time. Typical defects of this type are weak wire bond connections, impurities, imperfections, inclusions and non-hermetic seals. These components are called '*freaks*'. Figure 10.4 shows the lifetime distribution pattern of the three categories.

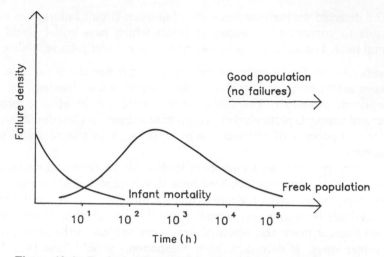

Figure 10.4 Typical failure density functions of electronic components when no component burn-in has been carried out

Standard methods have been developed for electronic component burn-in as part of the screen test and inspection sequence, particularly for semiconductor devices (See Chapter 7). The usual burn-in conditions for ICs are 168 hours at 125 °C, whilst being electrically stimulated and monitored. IC burn-in can be *static* (reverse bias, power, switching) or *dynamic* (all modes exercised, using special burn-in circuit drivers). Dynamic burn-in is a more effective screen, but it is more expensive. It should be considered for LSI and VLSI components, particularly for custom and semi-custom ICs.

Burn-in duration for commercial applications is often less than the 168 h required for military applications, 30–48 h being typical. Burn-in temperature can also be altered (MIL-STD-883 incudes a burn-in test at 250 °C). Optimum burn-in conditions, from the cost or effectiveness viewpoints, can be derived by assessment of the main failure modes and their activation energies, and from analysis of failure data during burn-in.

Equipment manufacturers sometimes perform their own component burn-in, tailored to their applications of the components. This has the advantage of close control by the equipment manufacturer, but to be effective he must have staff who are competent to operate the equipment and processes. In-house burn-in can only be justified when component throughput is high, since the cost of a full burn-in facility is high. Burn-in can also be performed by an independent test house which will have the facilities and expertise to ensure that burn-in of smaller quantities becomes cost-effective.

A less expensive form of high-temperature, short-duration test is often performed, particularly by commercial manufacturers, in which ICs are functionally tested individually after passing over a heated rail.

Burn-in of assemblies

Since electronic assemblies are collections of components which are subject mainly to a decreasing hazard rate and since the assembly processes generate similar failure patterns, burn-in of electronic assemblies is an effective way of improving reliability of delivered equipment. Generally it is cheaper and most effective to burn-in the complete equipment, unless it is a large system made up of several separate testable assemblies, but sometimes burn-in of PCBs may be performed as well, particularly for equipment where PCBs are wired in, making repair expensive at the full assembly stage.

Electronic equipment is often burned-in using test cycles given in MIL-STD-781 (see Chapter 11). Burn-in ovens are available which provide the environmental cycling. The equipment under test may be stimulated and monitored manually, but automatic systems are more often used, coupled with automatic data logging. Most evidence supports burn-in of 50–150 h as being effective in showing up 80–90 per cent. of component and production-induced defects (e.g. solder joints, component drift or mismatch, weak components) and reducing the instantaneous failure rate by factors ranging from 2 to 10. The effectiveness of burn-in depends upon the severity of the environmental cycles, and on the adequacy of the preceding QC methods. However, since equipment burn-in temperatures seldom exceed about 70 °C, burn-in at this level should not be relied upon to screen out defective ICs.

If burn-in shows up very few defects, it is either insufficiently severe or the product being burned-in is already highly reliable. Obviously the latter situation is preferable, and all failures during burn-in should obviously be analysed to determine if QC methods should have prevented them or discovered them earlier, particularly prior to assembly. At this stage, repairs of failures are expensive, so eliminating the need for them by using high quality components and tight QC during production is a worthwhile objective. However, for relatively complex equipment or when reliability is important, burn-in of assemblies is usually necessary and is often stipulated in government contracts.

Equipment burn-in is expensive, so its application must be carefully monitored. Burn-in failure data should be analysed continuously to enable the process to be optimized in terms of operating conditions and duration. Times between failures should be logged to enable the burn-in operation to be optimized.

Component and equipment burn-in must be considered in the development of the production test strategy. The costs and effects in terms of reduced failure costs in service, and the relationships with other test methods and QC methods, must be assessed as part of the integrated quality assurance approach. Much depends upon the reliability requirement and the costs of failures in service. Also, burn-in will ensure that manufacturing quality problems are detected before shipment, so that they can be corrected before they affect much production ouput. With large scale production, particularly of commercial and domestic equipment, burn-in of samples is sometimes applied for this purpose.

FAILURE REPORTING AND ANALYSIS

Internal failure reporting and analysis is an important part of the QA function. The system must provide for:

1. Reporting of all test and inspection failures with sufficient detail to enable investigation and corrective action to be taken.
2. Reporting the results of investigation and action.
3. Analysis of failure patterns and trends, and reporting on these.

The data system should be computerized for economy and accuracy. Modern ATE sometimes includes direct test data recording and inputting to a central system by networking. Manual input to computer terminals connected to the network also speeds up the system, but data input using paper forms is still common. The data analysis system should provide Pareto analysis (Chapter 13), probability plots (Chapter 3) and trend analyses, for management reporting.

Production defect data reporting and analysis must be very quick to be effective. Trends should be analysed daily, or weekly at most, particularly for high rates of production, to enable timely corrective action to be taken. Production problems usually arise quickly, and the system must be able to detect a bad batch of components or an improperly adjusted process as soon as possible. Many problems become immediately apparent without the aid of a data analysis system, but a change of, say, 50 per cent. in the proportion of a particular component in a system failing on test might not be noticed otherwise. The data analysis system is also necessary for indicating areas for priority action, using the Pareto principle of concentrating action on the few problem areas that contribute the most to quality costs. For this purpose longer term analysis, say monthly, is necessary.

Defective components should not be scrapped immediately, but should be labelled and stored for a period, say 1 to 2 months, so that they are available for more detailed investigation if necessary.

Production defect data should not be analysed in isolation by people whose task is primarily data management. The people involved (production, supervisors, QC engineers, test operators, etc.) must participate to ensure that the data are interpreted by those involved and that practical results are derived. The quality circles approach provides very effectively for this.

Production defect data are important for highlighting possible in-service reliability problems. Many in-service failure modes manifest themselves during production inspection and test. For example, if a component or process generates failures on the final functional test, and these are corrected before delivery, it is possible that the failure mechanism exists in products which pass test and are shipped. Metal surface protection and soldering processes present such risks, as can almost any component in electronic production. Therefore production defects should always be analysed to determine the likely effects on reliability and on external failure costs, as well as on internal production quality costs.

BIBLIOGRAPHY

1. J. M. Juran, *Quality Control Handbook* (3rd Edition). McGraw-Hill (1974).
2. A. V. Feigenbaum, *Total Quality Control* (3rd Edition). McGraw-Hill (1983).
3. J. M. Juran, F. M. Gryna, *Quality Planning and Analysis* (2nd Edition). McGraw-Hill (1980).
4. E. L. Grant, R. S. Leavenworth, *Statistical Quality Control* (5th Edition). McGraw-Hill Kogakusha Ltd. (1980).
5. J. M. Groocock, *The Cost of Quality*. Pitman (1974).
6. B. Davis, *The Economics of Automatic Test Equipment*. McGraw-Hill (1982).
7. *The Reliability Handbook*, National Semiconductor Corporation (1982).
8. F. Jensen, N. E. Peterson, *Burn-in: An Engineering Approach to the Design and Analysis of Burn-in Procedures*. J. Wiley (1983).
9. US Military Standard (Q) 9858–Quality Program Requirements.
10. British Standard BS 5750, Quality Systems.
11. UK Defence Standard 05–21, *Quality Control System Requirements for Industry*. HMSO.
12. US Military Standard 105D, *Sampling Procedures and Tables for Inspection by Attributes*.
13. US Military Standard 414, *Sampling Procedures and Tables for Inspection by variables*.
14. British Standard BS 6001, *Sampling procedures and tables for inspection by attributes*.
15. *Quality Progress,* Journal of the American Society for Quality Control.
16. *Quality Assurance,* Journal of the Institute of Quality Assurance (UK).

Chapter 11

RELIABILITY TESTING

INTRODUCTION

Testing is an essential part of any development programme. If the development risks are high the test programme becomes a major component of the overall development effort, in terms of time and other resources. For example, a new type of hydraulic pump or a new model of a video recording system will normally undergo exhaustive tests to determine that the design is reliable under the expected operating environments and for the expected operating life. Reliability testing is necessary because designs are seldom perfect and because designers cannot usually be aware of, or be able to analyse, all the likely causes of failure of their designs in service. The disciplines described in Chapters 6, 7 and 8, when systematically applied, can contribute to a large extent to inherently reliable design. They can also result in fewer failures during testing, and thus reduce the time and cost of the test programme.

Reliability testing should be considered as part of an integrated test programme, which usually includes:

1. Functional testing, to ensure that the design will meet the basic performance requirements.
2. Environmental testing, to ensure that the design is capable of operating under the expected range of environments.
3. Reliability testing.
4. Safety testing.

It is obviously impracticable to separate entirely the various categories of test. All testing will provide information on performance and reliability, and there will be common requirements for expertise, test equipment and other resources. The different categories of test do have certain special requirements. In particular, statutory considerations often determine safety tests, some of which may have little in common with other tests.

To provide the basis for a properly integrated development test programme, the design specification should cover all criteria to be tested (function, environment, reliability, safety). The development test programme should be drawn up to cover assurance of all these design criteria. It is important to avoid competition between people running the different categories of test, with the resulting arguments about allocation of models, facilities, and priorities. An integrated test programme reduces the chances of conflict.

The development test programme should include:

1 Model allocations
2 Requirements for facilities such as test equipment
3 A common test and failure reporting system
4 Test schedule

One person should be put in charge of the entire programme, with the responsibility and authority for ensuring that all specification criteria will be demonstrated.

There is one conflict inherent in reliability testing as part of an integrated test programme, however. To obtain information about reliability in a cost-effective way, i.e. quickly, it is necessary to generate failures. Only then can safety margins be ascertained. On the other hand, failures interfere with functional and environmental testing. The development test programme must address this dilemma. It can be very tempting for the people running the development test programme to minimize the chance of failure occurring, in order to make the programme run smoothly and at least cost. However, weaknesses in the design (or in the way it is made) must be detected and corrected before the production phase. This can only be achieved realistically by generating failures. An ideal test programme will show up every failure mode which might occur in service.

The development test dilemma should be addressed by dividing tests into two main categories:

1. Tests in which failures are undesirable.
2. Tests which deliberately generate failures.

Functional testing and most environmental testing are in category 1. Most reliability testing (and some safety testing) are in category 2. (The next chapter describes reliability demonstration tests in which failures are expected but are not deliberately generated.) The two categories of test must be run in parallel, with good communication between them and, wherever practicable, using common approaches. In particular, there must be a common reporting system for test results and failures, and for action to be taken to analyse and correct failure modes. Test and failure reporting and corrective action are covered in more detail later.

The category 2 testing should be started as soon as hardware (and software, when appropriate) is available for test. The effect of failures on schedule and cost increases progressively, the later they occur in the development programme. Therefore tests should be planned to show up failure modes as early as is practicable.

PLANNING RELIABILITY TESTING

Using design analysis data

The design analyses performed during the design phase (reliability prediction, FMECA, stress analysis, parameter variation analysis, sneak circuit analysis, FTA) should be used in preparing the reliability test plan. These analyses should have highlighted the risks and uncertainties in the design, and the reliability test programme should specifically address these. For example, if the FMECA shows a particular failure mode to be highly critical, the reliability test programme should confirm that the failure is very unlikely to occur within the use environment and lifetime. Inevitably the test programme will also show up failure modes and effects not perceived during the design analyses, otherwise there would be little point in testing. Therefore, the test programme must cover the whole range of use conditions, including storage, handling, testing, repair and any other aspect which might affect reliability.

Considering variability

We have seen (Chapter 3) how variability affects the probability of failure. A major source of variability is the range of production processes involved in converting designs into hardware. Therefore the reliability test programme must cover the effects of variability on the expected and unexpected failure modes. If parameter variation analyses have been performed, these can be very useful in planning reliability tests to confirm the effects of variation. However, to ensure that the effects of variability are covered as far as is practicable, it is important to carry out reliability testing on several items. The number of systems to be tested must be determined by considering:

1. The extent to which the key variables can be controlled
2. The criticality of failure
3. The cost of test hardware and of testing

Only rarely will fewer than four items be adequate. For fairly simple systems (transistors, fasteners, hydraulic check valves) it might be relatively easy to control the few key variables and the criticality of failures might be relatively low. However, it is not expensive to test large quantities. For systems of moderate complexity (e.g. LSI circuits, TV sets, machine tools) it is much harder to control key variables, since there are so many. Every interface within the system introduces further sources of variability which can affect reliability. Therefore it is very important to test a relatively large number, 5 to 20 being a typical range. For complex systems (areo engines, large computers) hardware and test cost tend to be the major constraints, but at least four items should be subjected to a reliability test. There are only two types of systems for which reliability testing of fewer than four might be appropriate:

1. Very complex, expensive systems, such as power stations and spacecraft launchers.
2. Systems which will be manufactured in very small quantities (e.g. spacecraft, ships, power stations), of which the items tested will be used operationally.

The effects of known sources of variability can sometimes be assessed by testing items in which variable parameters (e.g. dimensions, process variables) have been deliberately set at worst case values. Analysis of variance (ANOVA) and other statistical engineering optimization techniques, as described in Chapter 13, can be used for analysing results in this type of test.

Time effects

The reliability test programme must take account of the pattern of the main failure modes with respect to time (or cycles, distances, etc., with which the time dimension is associated).

If the failure modes have increasing hazard rates, testing must be directed towards assuring adequate reliability during the expected life. Therefore reliability tests must be of sufficient duration to demonstrate this, or they must be accelerated. Accelerated testing is covered later. Generally speaking, mechanical components and assemblies are subject to increasing hazard rates, when wear, fatigue, corrosion or other deterioration processes can cause failure. Systems subject to repair and overhaul can also become less reliable with age, due to the effects of maintenance, so the appropriate maintenance actions must be included in the test plan.

For items and systems which are not subject to an increasing hazard or failure rate, endurance testing is less important. Electronic components and systems are in this category. There is usually little to be gained by subjecting such hardware to long duration reliability testing. Typically 80 to 90 per cent. of failures will have occurred in the first 200 hours of operation of each item, and it is far more effective to test several items for relatively short periods. Since such items are not prone to wearout, the tested items can usually be used for other development work without having to be overhauled.

TEST ENVIRONMENTS

The reliability test programme must cover the range of environmental conditions which the product is likely to have to endure. The main reliability-affecting environmental factors, affecting most products, are:

Temperature
Vibration

Shock
Humidity
Power input and output
Dirt
People

In addition, electronic equipment might be subjected to:

Electromagnetic effects
Voltage transients, including static electrical discharge

Certain other environments can affect reliability in special cases. Examples are:

Radiation (ultraviolet, cosmic, X-rays)
Lubricant age or contamination
High altitude
Space vacuum
Industrial pollution
Electromagnetic pulse (lightning, nuclear)
Salt spray
Fungus
High intensity noise

US MIL-STD-810 and UK Defence Standard 07-55 provide test methods appropriate to most of these environmental conditions. However, these standards do not address reliability directly, since the objective is to show that the product will not fail or incur damage under the test conditions. Also, most of the tests do not require that the equipment be operating during the tests, and the tests are single-environment, not combined.

The environmental test programme will address the formal environmental test requirements, particularly when these are necessary in order to comply with legal or contractual requirements. The environmental aspects of the reliability test programme must take account of the environmental requirements stated in the design specification and of the planned environmental test. However, to be effective as a means of ensuring a reliable product, the environmental aspects of reliability testing must be assessed in much greater detail.

The environmental aspects of reliability testing must be determined by considering which environmental conditions, simply and in combination with others, are likely to be the most critical from the reliability point of view. In most cases, past experience and codes of practice will provide adequate guidelines. For example, US MIL-STD-781 provides information on how to assess environmental conditions and to design the tests accordingly. Typically, a reliability test environment for an electronic system to be used in a vehicle

Figure 11.1 Typical CERT environmental cycles:
electronic equipment in a vehicle application

or aircraft might be as shown in Fig. 11.1. Such testing is known as *combined environmental reliability testing* (CERT).

Test chambers are available for CERT testing, particularly for electronic systems. These include facilities for temperature cycling and for vibration input to the unit under test by locating the chamber over a floor-mounted vibrator, with a moveable or flexible floor for the chamber. Electrical signals, (power, control and monitoring) can be fed through connectors in the chamber wall. Special chambers can be provided with other facilities, such as humidity and reduced pressure. Control of the chamber conditions can be programmed, and the unit under test can be controlled and monitored using external equipment, such as programmable power supplies, data loggers, etc. Figure 11.2 shows a typical CERT test facility.

When past experience on standard methods is inappropriate, e.g. for a high risk product to be used in a harsh environment, the test environments must be carefully evaluated, particularly:

1. Rate of change of conditions, not just maximum and minimum values. For example, a high rate of change of temperature can cause fracture or fatigue due to thermal mismatch and conductivity effects.
2. Operating and dormant conditions in relation to the outside environment. For example, moisture-assisted corrosion might cause more problems when equipment is idle than when it is operating.
3. The effects of combined environments, which might be much more severe than any one condition. ANOVA methods (Chapter 13) applied to test data can be used to evaluate these effects.

Figure 11.2 CERT test facility
(Courtesy Thermotron Industries Inc.)

4. Direction and modes of vibration and shock. This is dealt with in more detail later.
5. Particular environmental conditions applicable to the product, such as handling, storage, maintenance and particular physical conditions.

Vibration testing

Adequate vibration testing is particularly important for products which must survive vibration conditions. However, specifying and obtaining the right conditions can be difficult, and it is easy to make expensive mistakes.

The main principles of effective vibration testing are:

1. Vibration should be input to the unit under test (UUT) through more than one axis, preferably simultaneously.
2. Vibration input should cover the complete range of expected frequencies and intensities, so that all resonances will be excited.
3. Vibration input should be random, rather than swept frequency, so that different resonances will be excited simultaneously (see below).
4. Test fixtures to mount the UUT to the vibration tables should be designed so that they do not alter the vibration output (no fixture resonances or damping). Whenever practicable, the UUT should be mounted directly on to the vibrator platform.

The simplest vibration test is a fixed frequency 'shake', usually with a sine wave input. However, this is of little value in reliability testing, but can be

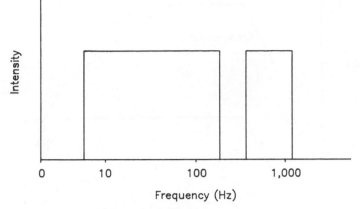

Figure 11.3 Simple vibration test specification

useful in development testing. Modern vibrators can be programmed to generate any desired profile within the operating range. For example, the test profile might continuously sweep a sine wave vibration at a prescribed intensity, across a specified frequency range, but with a gap over a particular frequency band (Fig. 11.3).

Swept frequency sine testing is useful for resonance searches, to enable the design to be modified if unacceptable resonances are detected.

Peak acceleration for a given frequency of sine wave vibration can be calculated using the formula:

$$A = 0.002\, f^2\, D \qquad\qquad (11.1)$$

where
A = peak acceleration (g)
f = frequency (Hz)
D = peak-to-peak displacement (mm)

for example, if $f = 50$ Hz and $D = 2$ mm
then $A = 10$ g

Alternatively, the spectrum could be a random input within a specified range and density function. Random vibration testing in which the input contains many frequencies is more effective than swept frequency for reliability testing, to show up vibration-induced failure modes, since it simultaneously excites all resonances. It is also more representative of real life.

The unit of measurement for random vibration inputs with continuous spectra is *acceleration spectral density* (ASD). The units are $g^2\,Hz^{-1}$. Typically inputs of up to $0.1\ g^2\,Hz^{-1}$ are used for equipment which must be shown to withstand fairly severe vibration, or for screening tests on assemblies such as electronic equipment. A typical random vibration spectrum is shown in Fig. 11.4.

Figure 11.4 Road transport vibration levels

It is important to apply power to electrical or electronic equipment and to monitor its performance while it is being vibrated, so that intermittent failures can be detected and investigated.

Since dynamic responses are usually affected more by resonances within the product, due to the design and to production variation, than by the input spectrum from the vibrator, it is seldom cost-effective to simulate accurately the operating environment, even if it is known in detail. Since the objective in vibration reliability testing is to excite resonances simultaneously, and since almost any random spectrum will do this, test costs can be minimized by permitting large spectral tolerances, e.g. ± 6 dB.

Temperature testing

Temperature testing for reliability is a less complex subject than vibration testing. The only aspects that need to be considered are:

Extreme values
Rate of change

As for any other environmental condition, these should be selected by considering the expected product environment and the likely effects. Temperature testing for electrical and electronic equipment is particularly important, since reliability is very much affected by operating temperature (Chapter 7). Equipment should be powered during temperature testing, otherwise the tests will be unrepresentative of the thermal patterns and gradients in use. It should also be monitored continuously to ensure that intermittent failures are detected.

Electromagnetic compatibility (EMC) testing

EMC testing is very important for microprocessor-controlled systems, since data corruption due to electromagnetic interference (EMI) or from voltage transients in power supplies can have serious consequences. The equipment must be subjected to EMI and transients to confirm that it will perform without failure under these conditions. The levels of EMI and transient waveforms must be ascertained by evaluating or measuring the operating environment, or they might be specified.

Internally induced EMI and transients must also be protected against, and tests to ensure that transmitted EMI is within limits might also be necessary.

Customer simulation testing

Functional, environmental, reliability and safety tests are all designed to demonstrate that the equipment will meet its design parameters. In general these tests are carried out by people with extensive engineering backgrounds and they are, with the best will in the world, far removed from the average user of the equipment. It is therefore important, particularly in the field of consumer products (televisions, copiers, washing machines, etc.), that some reliability testing be conducted using people who are more nearly representative of typical customers. This approach is very useful in highlighting failure modes that do not show up when the equipment is used by experienced personnel.

ACCELERATED TESTS

For some products the test time necessary to provide adequate reliability assurance under normal operating conditions might be inordinately long, and therefore very expensive. Reliability data-gathering should not hold up development, and should be as economical as practicable, so it is important to be able to accelerate reliability tests.

Reliability tests can be accelerated by increasing the sample size, provided that the life distribution does not show a wearout characteristic during the anticipated life. Increasing the sample size is appropriate for small, cheap items which can be produced in quantity, such as electronic components, bearings, hydraulic seals and mechanical fastening devices. Being able to use a large sample reduces the error in the reliability estimate for the population due to part-to-part variability. However, large-sample reliability tests, to provide a high total operating time, should be supported by some long duration testing if there is reason to suspect that failure modes exist which have high times to first failure. For example, a reliability test of a mechanical fastener must provide data on endurance, and a short duration test of a very large number of items would not do this, since such a device would not have a constant failure rate or a zero time to first failure under typical load conditions. Extrapolation of reliability data over long periods of time must be treated with caution, and therefore whenever practicable long duration tests should be considered.

A particular type of large sample test is sudden death testing, in which the sample is split into subgroups and the time to first failure in each group is plotted to provide the distribution parameters. Sudden death probability plotting was covered in Chapter 3.

Increasing the severity of the test is an obvious approach when large samples cannot be provided. However, we are then faced with two problems:

1. What is the equivalent operating time under normal stress?
2. Are the failures induced under the accelerated test conditions the same as those which might occur under normal conditions?

Equivalent time

The degree of test acceleration can be determined in single-stress situations, such as in fatigue testing by using Miner's rule (Eqn 4.9), or in electronic device testing by using the Arrhenius law (Eqn 7.1). These relationships are usually empirical or based upon previous test data, so exact correlations between stress acceleration and reliability should not be expected.

Example 11.1

A reliability test of a CMOS integrated circuit memory device gave the following results:

Sample size = 300
Test time = 5000 h
Test temperature = 168 °C
Failures = 5

Assuming that the Arrhenius relationship holds for the failure modes experienced, what is the failure rate at 50 °C? (Assume constant hazard rate).

The Arrhenius equation (7.1) is

$$\lambda = K \exp\left(\frac{-A}{T}\right)$$

The value of $A \, (= E/k)$ for CMOS $= 7{,}532$ (from MIL-HDBK-217). At $168\,^\circ C$ (441 K),

$$\lambda_{168^\circ C} = \frac{5}{(300 \times 5{,}000)} = 3.3 \times 10^{-6} \text{ h}$$

Therefore

$$K = \frac{3.3 \times 10^{-6}}{\exp\,(-7{,}532/441)} = 86.3$$

and

$$\lambda_{50^\circ C} = 86.3 \, \exp\,(-7{,}532/323)$$
$$= 0.0064 \times 10^{-6} \text{ h}$$

This result should be compared with Fig. 7.3.

Accelerated test results can be evaluated using probability plots. A higher stress level than expected is applied, and time-to-failure data are plotted in the usual way. The failure distribution parameters (e.g. failure-free life, B_{10} life) must then be related to the expected operating conditions using the types of relationship mentioned above.

Step-stress testing

The accelerated tests covered so far involve the test being performed at a fixed but elevated stress level. Step-stress testing is a technique whereby the item is tested initially at normal stress, but after a certain time the stress is increased, and stepwise increases are continued until failure. Therefore instead of a time axis, we have a stress-time axis for the probability plot of failure data. The probability of failure at a stress-time value can then be determined. Multiple stresses can be combined, each stress value being incremented simultaneously. Example 11.2 provides an illustration of a combined step-stress test.

Example 11.2

A rocket engine component must operate under fatigue loading conditions at high temperature. A reliability of 0.995 must be demonstrated under the extreme conditions of completely reversed cyclic (150 Hz) stress of $2 \times 10^8 \, \text{N m}^{-2}$ at $450\,^\circ C$, over a 10 min. engine run-time. Design an accelerated test to demonstrate this.

The specified load and temperature stresses are fixed as the values for the start of the test. These values are maintained for 10 min, after which load is increased by $0.1 \, \text{N m}^{-2}$ and temperature by $10\,^\circ C$ and the test continued for 30 s. These increments are repeated after each subsequent 30 s period. Ten samples are tested to failure with results as shown in Table 11.1. The results

Table 11.1 Results for ten samples tested to failure (Example 11.2).

Failure no.	Mean rank $x_i/(10+1)$ (%)	Time (min)	Stress Load ($\mathrm{N\,m^{-2} \times 10^8}$)	Temperature (°C)
1	9.1	13.1	3.1	520
2	18	13.2	3.1	520
3	27	14.4	3.4	550
4	36	15.0	3.4	550
5	45	15.0	3.4	550
6	54	15.0	3.4	550
7	64	15.5	3.5	560
8	73	15.6	3.6	570
9	82	15.9	3.6	570
10	91	15.9	3.6	570

are plotted in Fig. 11.5. This shows the line of plotted failure data lying to the right of the design point (0.5 per cent. failure, 10 min). Therefore the test demonstrates compliance with the requirement.

It is not meaningful to estimate distribution parameters or s-confidence limits for this type of data, since additional variables have been added. A failure-free life will probably exist in many test situations of this type. However, this will not normally be detectable from curvature of the plotted data, since step-stress failure data plot as a steeper slope than ordinary life data. Normally the design would provide for a failure-free stress life greater than the design life. The step-stress test can then be used to confirm economically that the design is correct.

Failure modes

It is important in accelerated testing to ensure that unrealistic failure modes are not introduced by the higher stresses. The physics of the materials being tested and analysis or failures should indicate whether or not new failure modes are likely to be or have been stimulated. Obviously failure modes which occur only at stresses above the maximum operating stress will not be of interest. For example, increasing temperature beyond a certain level may change the strength of a material, so it is important that temperature increments are kept within such limits. The occurrence of a new failure mode can sometimes be detected by an upward curvature of the plotted data.

It is also possible that interactions may occur between separate stresses, so that the combined weakening effect is greater than would be expected from a simple additive process. Interactive effects are covered in Chapter 13, which describes experiments which can quantify the effects of variables and of their interactions. Accelerated tests should be supported by experiments to provide

Figure 11.5 Step-stress Test (Example 11.2)

the insights necessary for planning the tests when combined environmental stressing is to be applied. The amounts by which the separate stresses are increased should be determined in relation to their separate and combined effects.

MIL-STD-2068

US MIL-STD-2068 (*Reliability Development Tests*) provides an excellent description of how development reliability tests should be managed and integrated with other development tasks. The standard was developed to place more emphasis on testing to detect weaknesses in the product, rather than on formal reliability demonstration methods in which the test objective is (from the supplier's point of view) to have as few failures as possible. Formal reliability demonstration methods and the quantification of the results are covered in the next chapter.

FAILURE REPORTING AND CORRECTIVE ACTION SYSTEMS (FRACAS)

FRACAS is an apt acronym for the task of failure reporting and corrective action. It is essential that all failures which occur during development testing are carefully reported and investigated. It can be very tempting to categorize a failure as irrelevant, or not likely to cause problems in service, especially when engineers are working to tight schedules and do not want to be delayed by filling in failure reports. However, time and cost will nearly always be saved in the long run if the first occurrence of every failure mode is treated as a problem to be investigated and corrected. Failure modes which affect reliability in service can often be tracked back to incidents during development testing, when no corrective action was taken.

A failure review board should be set up with the task of assessing failures, instigating and monitoring corrective action, and monitoring reliability growth. An important part of the board's task is to ensure that the corrective action is effective in preventing any recurrence of failure. The board should consist of:

The project reliability engineer
The designer
Others who might be able to help with the solutions, such as the quality engineer, production or test engineer

The failure review board should operate as a team which works together to solve problems, not as a forum to argue about blame or to consign failure reports to the 'random, no action required' category. Its recommendations should be actioned quickly or reported to the project management if the board cannot decide on immediate action, e.g. if the solution to the problem requires more resources. This aproach has much in common with the quality circles method described in Chapter 10.

US MIL-STD-781 (*Reliability Qualification and Production Approval Tests*) provides a good description of failure reporting methods. Since a consistent reporting system should be used throughout the programme MIL-STD-781 can be recommended as the basis for this.

These data should be recorded for each failure:

1. Description of failure symptoms, and effect of failure.
2. Immediate repair action taken.

3. Equipment operating time at failure (e.g. elapsed time indicator reading, mileage).
4. Operating conditions.
5. Date/time of failure.
6. Failure classification (e.g. design, maintenance-induced, QC).
7. Report of investigation into failed component and reclassification, if necessary.
8. Recommended action to correct failure mode.
9. Corrective action follow-up (test results, etc.).

Failure report forms should be designed to allow these data to be included. They should also permit easy input to computer files, by inclusion of suitable coding boxes for use by the people using the forms.

Corrective Action Effectiveness

When a change is made to a design or to a process to correct a cause of failure, it is important to repeat the test which generated the failure to ensure that the corrective action is effective. Corrective action sometimes does not work. For example, it can have the effect of transferring the problem to the next weakest item in the sequence of stress-bearing items, or the true underlying cause of failure might be more complex than initial analysis indicates. Therefore re-test is important to ensure that no new problems have been introduced and that the change has the desired effect.

Analysis of test results must take account of the expected effectiveness of corrective action. Unless the causes of a failure are very well understood, and there is total confidence that the corrective action will prevent recurrence, 100 per cent. effectiveness should not be assumed.

BIBLIOGRAPHY

1. US MIL-STD-810: *Environmental Test Methods*. Available from the National Technical Information Service, Springfield, Virginia.
2. UK Defence Standard 07-55: *Environmental Testing*.
3. US MIL-STD-2068: *Reliability Development Tests*. Available from the National Technical Information Service, Springfield, Virginia.
4. US MIL-STD-781: *Reliability Qualification and Production Approval Tests*. Available from the National Technical Information Service, Springfield, Virginia.
5. C. E. Harris and C. E. Crede (Eds), *Shock and Vibration Handbook*. McGraw-Hill (1976).
6. C. T. Morrow, *Shock and Vibration Engineering*. Wiley (1963).
7. D. S. Steinberg, *Vibration Analysis for Electronic Equipment*. Wiley (1973).
8. W. T. Thomson, *Theory of Vibration with Applications*. Prentice-Hall (1981).
9. *The Journal of Environmental Sciences*. Institute of Environmental Sciences (USA). Published monthly.

Chapter 12

MEASURING AND IMPROVING RELIABLITY

INTRODUCTION

It is often necessary to measure the reliability of equipment and systems during development, production and in use. Demonstration of reliability might be required as part of a development and production contract, or prior to release to production, to ensure that the requirements have been met. Two basic forms of reliability measurement are used. A sample of equipments may be subjected to a formal reliability test, with the conditions specified in detail. Reliability may also be monitored during development and use, as test and utilization proceed, without tests being set up specifically for reliability measurement. Both approaches have common features, such as the need to define failures and to collect and analyse data, and there are advantages in practice if both approaches are used as parts of an integrated reliability programme.

In Chapter 3 we covered statistical reliability analysis methods using probability and hazard plots. In this chapter we describe standard methods of test and analysis which are used to demonstrate compliance with reliability requirements. We will also describe techniques for monitoring the growth of reliability during a programme of development or utilization.

The standard methods are not substitutes for the statistical analysis methods described in Chapter 3. The standard methods may be referenced in procurement contracts, particularly for government equipment, but they may not provide the statistical engineering insights given, for example, by Weibull probability plotting of failure data. Therefore the standards should be seen as complementary to the statistical engineering methods and useful (or mandatory) for demonstrating and monitoring reliability of products which are past the experimental development phase.

SEQUENTIAL TESTING: US MIL-STD-781

The standard method for formal reliability demonstration testing for repairable equipment which operates for periods of time, such as electronic equipment, motors, etc., is US MIL-STD-781: *Reliability Qualification and*

Figure 12.1 Typical probability ratio sequential test (PRST) plan

Production Approval Tests. MIL-STD-781 testing is based on *probability ratio sequential testing* (PRST), the results of which (failures and test time) are plotted as in Fig. 12.1. Testing continues until the 'staircase' plot of failures versus time crosses a decision line. The reject line (dotted) indicates the boundary beyond which the equipment will have failed to meet the test criteria. Crossing the accept line denotes that the test criteria have been met. The decision lines are truncated to provide a reasonable maximum test time. Test time is stated as multiples of the specified MTBF.

Test plans

MIL-STD-781 contains a number of test plans, which allow a choice to be made between the statistical risks involved (i.e. the risk of rejecting an equipment with true reliability higher than specified or of accepting an equipment with true reliability lower than specified) and the ratio of minimum acceptable to target reliability. The risk of a good equipment (or batch) being rejected is called the *producer's risk* and is denoted by α. The risk of a bad equipment (or batch) being accepted is called the *consumer's risk,* denoted by β. MIL-STD-781 test plans are based upon the assumption of a constant failure rate, so MTBF is used as the reliability index. Therefore MIL-STD-781 tests are appropriate for equipment where a constant failure rate is likely to be encountered, such as fairly complex maintained electronic equipment, after an initial burn-in period. Such equipment in fact was the original justification for the development of the precursor to MIL-STD-781, the AGREE report (see page 9). If predominating failure modes do not occur at a constant rate, tests based on the methods described in Chapter 3 should be used. In any case it is a good idea to test the failure data for trend, as described in Chapter 3.

The criteria used in MIL-STD-781 are:

1. Design MTBF, θ_0.
2. Low limit MTBF, θ_1. This is the specified, or contractually agreed, MTBF to be demonstrated.
3. Design ratio, $d = \theta_0/\theta_1$.

4. Producer's risk, α (the probability that equipment with higher than design MTBF will be rejected).
5. Consumer's risk, β (the probability that equipment with lower than low limit MTBF will be accepted).

The PRST plans available in MIL-STD-781 are shown below in Table 12.1. In addition, a number of fixed-length test plans are included (plans IX–XVI), in which testing is required to be continued for a fixed multiple of design MTBF. These are listed in Table 12.2. A further test plan (XVII) is provided, for production reliability acceptance testing (PRAT), when all production items are to be tested. The plan is based on test plan III. The test time is not truncated by a multiple of MTBF but depends upon the number of equipments produced.

Table 12.1 MIL-STD-781C PRST plans.

Test plan	Decision risks (%)		Design ratio, $d = \theta_0/\theta_1$
	α	β	
I	10	10	1.5
II	20	20	1.5
III	10	10	2.0
IV	20	20	2.0
V	10	10	3.0
VI	20	20	3.0
VII[a]	30	30	1.5
VIII[a]	30	30	2.0

[a]Test plans VII and VIII are known as short-run high risk PRST plans.

Table 12.2 MIL-STD-781C fixed length test plans.

Test plan	Decision risks (%)		Design ratio, d	Test duration, $X\theta_1$	Reject $\geqslant$ failures	Accept $\leqslant$ failures
	α	β				
IX	10	10	1.5	45.0	37	36
X	20	20	1.5	21.1	18	17
XI	10	10	2.0	18.8	14	13
XII	20	20	2.0	7.8	6	5
XIII	30	30	2.0	3.7	3	2
XIV	10	10	3.0	9.3	6	5
XV	20	20	3.0	4.3	3	2
XVI	30	30	3.0	1.1	1	0

Statistical basis for PRST plans

MIL-STD-781 PRST testing is based on the assumption of a constant failure rate. The decision risks are based upon the risks that the estimated MTBF will not be less than the lower limit MTBF (for acceptance) or not more than the design MTBF (for rejection).

We thus set up two null hypotheses:

For H_0: $\hat{\theta} \geqslant \theta_1$

For H_1: $\hat{\theta} \leqslant \theta_0$

The probability of accepting H_0 is $(1 - \alpha)$, if $\hat{\theta} = \theta_1$; the probability of accepting H_1 is β, if $\hat{\theta} = \theta_0$. The time at which the ith failure occurs is given by the exponential distribution function $f(t_i) = (1/\theta) \exp (- t_i/\theta)$. The *sequential probability ratio,* or ratio of the expected number of failures given $\theta = \theta_0$ or θ_1, is

$$\prod_{i=1}^{n} \frac{(1/\beta_1) \exp (- t_i/\theta_1)}{(1/\theta_0) \exp (- t_i/\theta_0)}$$

where n is the number of failures.

The upper and lower boundaries of any sequential test plan specified in terms of θ_0, θ_1, α and β can be derived from the sequential probability ratio. However, arbitrary truncation rules are set in MIL-STD-781 to ensure that test descisions will be made in a reasonable time. The truncation alters the accept and reject probabilities somewhat compared with the values for a non-truncated test, and the α and β rules given in MIL-STD-781 are therefore approximations. The exact values can be determined from the operating characteristic (OC) curve appropriate to the test plan and are given in MIL-STD-781. The OC curves are described in the next section.

Operating characteristic curves and expected test time curves

An operating characteristic (OC) curve can be derived for any sequential test plan to show the probability of acceptance (or rejection) for different values of true MTBF. Similarly, the s-expected test time (ETT — time to reach an accept or reject decision) for any value of θ can be derived. OC and ETT curves are given in MIL-STD-781 for the specified test plans. Typical curves are shown in Figs 12.2 and 12.3.

Selection of test criteria

Selection of which test plans to use depends upon the degrees of risk which are acceptable and upon the cost of testing. For example, during development of a new equipment, when the MTBF likely to be achieved might be uncertain, a test plan with 20 per cent. risks may be selected. Later testing, such as production batch acceptance testing, may use 10 per cent. risks. A higher design

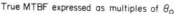

Figure 12.2 Operating characteristic (OC) curve. Test plan 1: $\alpha = 10\%$, $\beta = 10\%$ and $d = 1.5$

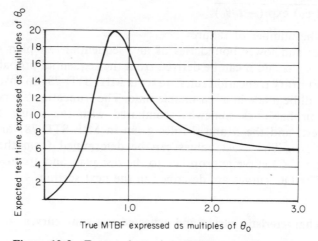

Figure 12.3 Expected test time (ETT) curve. Test plan 1: $\alpha = 10\%$, $\beta = 10\%$ and $d = 1.5$

ratio would also be appropriate for early development reliability testing. The higher the risks (i.e. the lower the values of α and β) and the lower the design ratio, the longer will be the s-expected test duration and therefore the expected cost.

The design MTBF should be based upon reliability prediction, development testing and previous experience. MIL-STD-781 requires that a reliability prediction be performed at an early stage in the development programme and updated as development proceeds. We discussed the uncertainties of reliability prediction in Chapter 5. For many types of equipment the MTBF values to be used in a test might have to be based upon an 'educated' estimate, rather than on a stress analysis prediction. However, this should only apply to the first reliability test, since the results of this can be used for setting criteria for subsequent tests.

The low limit MTBF may be a figure specified in a contract, as is often the case with military equipment, or it may be an internally generated target, based upon past experience or an assessment of market requirements.

Guidelines on the selection of test plans are given in Appendix C to MIL-STD-781C.

Test sample size

MIL-STD-781 provides recommended sample sizes for reliability testing. For a normal development programme, early reliability testing (qualification testing) should be carried out on at least two equipments. For production reliability acceptance testing the sample size should be based upon the production rate, the complexity of the equipment to be tested and the cost of testing. Normally at least three equipments per production lot should be tested.

Burn-in

If equipment is burned-in prior to being submitted to a production reliability acceptance test, MIL-STD-781 requires that all production equipments are given the same burn-in prior to delivery.

Test environmental conditions

The test conditions specified in MIL-STD-781 are based upon the expected conditions of use most likely to generate failure. The conditions used are called combined environment reliability testing (CERT), and normally include:

1. Operating stress (usually electrical stress).
2. Temperature cycling.
3. Vibration cycling (fixed frequency or swept frequency).
4. Humidity cycling.

The environmental stresses (type, severity, cyclic pattern) to be used for a particular test depend upon the anticipated usage environmental conditions. Appendix B to MIL-STD-781 C describes typical conditions, incuding the analyses necessary to establish these. The equipment's environmental specification is usually derived from the 'mission profile', which is often specified for military equipment. Where it is not specified, typical or worst case profiles should be established by analysis and measurement. For example, if the equipment will be subjected to shock or vibration during transport or use, the conditions should be measured and used as the basis for planning the environmental test conditions. A typical CERT test profile was shown in Fig. 11.1. The reliability test operating conditions are not usually worst case or design limit values. Environmental qualification tests (see Chapter 11) should have been performed prior to reliability demonstration to prove the ability of the design to withstand these conditions. The purpose of MIL-STD-781 testing is to show, within specified levels of risk, that the equipment's MTBF is greater than the low limit MTBF and possibly also to provide contractual milestones. Therefore the objective is not to induce failures by applying unrealistic stresses.

Practical problems of MIL-STD-781 testing

Reliability demonstration testing to MIL-STD-781 is subject to some severe practical problems and limitations, which cause it to be a controversial method. We have already covered one fundamental limitation: the assumption of a constant failure rate. However, it is also based upon the implication that MTBF is an inherent parameter of a system which can be experimentally demonstrated, albeit within s-confidence limits. In fact, reliability measurement is subject to the same fundamental constraint as reliability prediction: *reliability is not an inherent physical property of a system, as is mass or electric current.* The mass or power consumption of a system is measurable (also within statistical bounds, if necessary). Anyone could repeat the measurement with any copy of the system and would expect to measure the same values. However, if we measure the MTBF of a system in one test, *it is unlikely* that another test will demonstrate the same MTBF, quite apart from considerations of purely statistical variability. In fact there is no logical or physical reason to expect repeatability of such experiments. This can be illustrated by an example.

Suppose a computer is subjected to an MIL-STD-781 reliability demonstration. Four computers are tested for 400 hours and show failure patterns as follows:

No.1   2 memory device failures (20 h, 48 h)
        1 connector intermittent (150 h)
        1 capacitor short circuit (60 h)

No.2   1 open-circuit PCB track (40 h)
        1 IC socket failure (200 h)

No.3   No failures

No.4   1 shorted connector (trapped on assembly) (0 h)

Total failures: 6
Total running time: 1,600 h. Demonstrated MTBF (θ) = 267 h

Note that these failures are quite typical. However, if the experiment were repeated with another four computers, there would be no reason to expect the same number or pattern of failures. If the same four computers were tested for another 1,600 hours the pattern of failures would almost certainly be different. The pattern of failures and their likelihood can be influenced by quality control of manufacture and repair. Therefore the MTBF measured in this test is really no more than historical data, related to those four computers over that period of their lives. It does not predict the MTBF of other computers or of those four over a subsequent period, any more than four sales in one day would be a prediction of the next day's sales.

Of course, if a large number of computers were tested we would be able to extrapolate the results with rather greater credibility and to monitor trends (e.g. average failures per computer). However, MIL-STD-781 testing cannot be extended to such large quantities because of the costs involved.

MIL-STD-781 testing is often criticized on the grounds that in-service experience of MTBF is very different to the demonstrated figure. From the discussion above this should not surprise anyone. In addition, in-service conditions are almost always very different to the environments of MIL-STD-781 testing, despite any attempts to simulate realistic conditions in CERT.

MIL-STD-781 testing is not consistent with the reliability test philosophy described in the last chapter, since the objective is to count failures and to hope that few occur. An effective reliability test programme should aim at generating failures, *since they provide information on how to improve the product.* Failure-counting should be a secondary consideration to failure analysis and corrective action. Also, a reliability test should not be terminated solely because more than a predetermined number of failures occur. MIL-STD-781 testing is very expensive, and the benefit to the product in terms of improved reliability is sometimes questionable.

MIL-STD-781 testing can be a useful method for providing reliability milestones and motivation in development projects where the customer specifies the product and pays for its development, particularly if contractual incentives are involved. However, it should not replace the test philosophy described in Chapter 11. MIL-STD-781 methods are not used in projects in which the supplier specifies the product and funds its development, since that type of motivation is not normally necessary. This 'commercial' approach is becoming more widely adopted in military and other government contracting. The introduction of MIL-STD-1635 (Reference 6) is an indication of this trend.

RELIABILITY DEMONSTRATION FOR ONE-SHOT ITEMS: US MIL-STD-105

For equipments which operate only once, or cyclically, such as pyrotechnic devices, missiles, fire warning systems and switchgear, the sequential method of testing based on operating time may be inappropriate. MIL-STD-105 — *Sampling Procedures and Tables for Inspection by Attributes* — and BS6001 provide test plans based on success ratio for such items. These plans are described in Chapter 10. Alternatively, a MIL-STD-781 test could be adapted for items which operate cyclically, using a baseline of mean cycles to failure, or MTBF assuming a given cycling rate.

COMBINING RESULTS USING BAYESIAN STATISTICS

It can be argued that the result of a reliability demonstration test is not the only information available on a product, but that information is available prior to the start of the test, from component and subassembly tests, previous tests on the product and even intuition based upon experience. Why should this information not be used to supplement the formal test result? Bayes theorem (Chapter 2) states (Eqn 2.9):

$$P(B \mid A) = \frac{P(A \mid B)P(B)}{P(A)}$$

enabling us to combine such probabilities. Equation (2.9) can be extended to cover probability distributions:

$$p(\lambda \mid \phi) = \frac{f(\phi \mid \lambda)p(\lambda)}{f(\phi)} \qquad (12.1)$$

where λ is a continuous variable and ϕ represents the new observed data: $p(\lambda)$ is the *prior* distribution of λ; $p(\lambda \mid \phi)$ is the *posterior* distribution of λ, given ϕ; and $f(\phi \mid \lambda)$ is the sampling distribution of ϕ, given λ.

Let λ denote failure rate and t denote successful test time. Let the density function for λ be gamma distributed with

$$p(\lambda) = \frac{t}{\Gamma(a)} (\lambda t)^{a-1} \exp(-\lambda t)$$

If the prior parameters are a_0, t_0, then the prior mean failure rate is $\mu = a_0/t_0$ and the prior variance is $\sigma^2 = a_0/t_0^2$ (appropriate symbol changes in Eqn 2.34). The posterior will also be gamma distributed with parameters a_1 and t_1 where $a_1 = a_0 + n$ and $t_1 = t_0 + t$ and n is the number of events in the interval from 0 to t. The s-confidence limits on the posterior mean are $[\chi^2_{\alpha,\beta}(\nu = 2a_1)]/2t_1$.

Example 12.1

The prior estimate of the failure rate of an item is 0.02, with a standard deviation of 0.01. A reliability demonstration test results in $n = 14$ failures in $t = 500$ h. What is the posterior estimate of failure rate and the 90 per cent. lower s-confidence limit?

The prior mean failure rate is

$$\mu = \frac{a_0}{t_0} = 0.02 \text{ h}^{-1}$$

$$\sigma^2 = \frac{a_0}{t_0^2} = 10^{-4} \text{ h}^{-2}$$

Therefore,

$$a_0 = \frac{\mu^2}{\sigma^2} = 4.0 \text{ failures}$$

$$t_0 = \frac{\mu}{\sigma^2} = \frac{0.02}{10^{-4}} = 200 \text{ h}$$

$$a_1 = 4 + 14 = 18 \text{ failures}$$

$$t_1 = 200 + 500 = 700 \text{ h}$$

The posterior estimate for failure rate is

$$\lambda_1 = \frac{a_1}{t_1} = \frac{18}{700} = 0.0257 \text{ h}^{-1}$$

This compares with the traditional estimate of the failure rate from the test result of $14/500 = 0.028 \text{ h}^{-1}$.

The 90 per cent. lower s-confidence limit on the mean is

$$\frac{\chi^2_{10}(\nu = 2 \times 18)}{2t_1} = \frac{26}{2 \times 700} = 0.0186 \text{ h}^{-1}$$

compared with the traditional estimate (from Eqn 2.43) of

$$\frac{1}{\lambda_1} = \frac{2 \times 500}{\chi^2_{10}(\nu = 2 \times 14 + 1)}$$

$$\lambda_1 = 0.0198 \text{ h}^{-1}$$

In Example 12.1 use of the prior information has resulted in a failure rate estimate lower than that given by the test, and closer s-confidence limits.

The Bayesian approach is very controversial in reliability engineering, particularly as it has been argued that it provides a justification for less reliability testing. Choosing a prior distribution based on subjective judgement or other test experience can also be very contentious. Combining subassembly test results in this way also ignores the possibility of interface problems. The Bayesian approach is not normally recommended and it has not been approved in any national standards.

RELIABILITY GROWTH MONITORING (DUANE METHOD)

It is common for new products to be less reliable during early development than later in the programme, when improvements have been incorporated as a result of failures observed and corrected. Similarly, products in service often display reliability growth. This was first analysed by J. T. Duane, who derived an empirical relationship based upon observation of the MTBF improvement of a range of items used on aircraft. Duane observed that the cumulative MTBF θ_c (total time divided by total failures) plotted against total time on log–log paper gave a straight line. The slope (α) gave an indication of

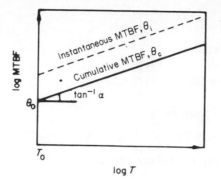

Figure 12.4 Duane reliability growth

reliability (MTBF) gowth, i.e.

$$\log \theta_c = \log \theta_0 + \alpha(\log T - \log T_0)$$

where θ_0 is the cumulative MTBF at the start of the monitoring period T_0. Therefore,

$$\theta_c = \theta_0 \left(\frac{T}{T_0}\right)^{\alpha} \tag{12.2}$$

The relationship is shown plotted in Fig. 12.4.

The slope α gives an indication of the rate of MTBF growth and hence the effectiveness of the reliability programme in correcting failure modes. Duane observed that typically α ranged between 0.2 and 0.4, and that the value was correlated with the intensity of the effort on reliability improvement.

The Duane method is applicable to a population with a number of failure modes which are progressively corrected, and in which a number of items contribute different running times to the total time. Therefore it is not appropriate for monitoring early development testing, and it is common for early test results to show a poor fit to the Duane model.

We can derive the instantaneous MTBF θ_i of the population by differentiation of Eqn (12.2):

$$\theta_c = \frac{T}{n}$$

where n is the number of failures. Therefore,

$$n = \frac{T}{\theta_c} = \frac{T}{\theta_0(T/T_0)^{\alpha}}$$

$$= T^{(1-\alpha)} \left(\frac{T_0^{\alpha}}{\theta_0}\right)$$

(T_0^α/θ_0) is a constant. Differentiation gives

$$\frac{dn}{dT} = (1-\alpha)\, T^{-\alpha}\,\left(\frac{T_0^\alpha}{\theta_0}\right) = (1-\alpha)\,\left(\frac{T_0}{T}\right)^\alpha \frac{1}{\theta_0}$$

$$= \frac{1-\alpha}{\theta_c}$$

$$\frac{dn}{dT} = \frac{1}{\theta_i}$$

So

$$\theta_i = \frac{\theta_c}{1-\alpha} \tag{12.3}$$

θ_i is shown in Fig. 12.4. The plot of θ_i is parallel to that for θ_c. A reliability monitoring programme may be directed towards a target either for cumulative or instantaneous MTBF.

After the end of a development programme in which MTBF growth is being managed, the anticipated MTBF of production items is θ_i, measured at the end of the programme. This assumes that the development testing accurately simulated the expected in-use stresses of the production items and that the standard of items being tested at the end of the development programme fully represents production items. Of course, these assumptions are often not valid and extrapolations of reliability values from one set of conditions to another must always be considered to be tentative approximations. Nevertheless, the empirical Duane method provides a reasonable approach to monitoring and planning MTBF growth for items such as electronic equipment.

The Duane method can also be used to assess the amount of test time required to attain a target MTBF. If the MTBF is known at some early stage, the test time required can be estimated if a value is assumed for α. The value chosen must be related to the expected effectiveness of the programme in detecting and correcting causes of failure. Knowledge of the effectiveness of past reliability improvement programmes operated by the organization can provide guidance in selecting a value for α. The following may be used as a guide:

$\alpha = 0.4-0.6$. Programme dedicated to the elimination of failure modes as a top priority. Use of accelerated (overstress) tests. Immediate analysis and effective corrective action for all failures.

$\alpha = 0.3-0.4$. Priority attention to reliability improvement. Normal (typical expected stresses) environment test. Well-managed analysis and corrective action for important failure modes.

$\alpha = 0.2$. Routine attention to reliability improvement. Testing without applied environmental stress. Corrective action taken for important failure modes.

$\alpha = 0.2$–0. No priority given to reliability improvement. Failure data not analysed. Corrective action taken for important failure modes, but with low priority.

Example 12.2

The first reliability qualification test on a new electronic test equipment generates 11 failures in 600 h, with no one type of failure predominating. The requirement set for the production standard equipment is an MTBF of not less than 500 h in service. How much more testing should be planned, assuming values for α of 0.3 and 0.5?

$$\hat{\theta}_0 = \frac{600}{11} = 54.4 \text{ h}$$

When $\theta_i = 500$,

$$\theta_c = 500(1 - \alpha)$$

$$\begin{cases} = 350 \quad \text{(for } \alpha = 0.3) \\ = 250 \quad \text{(for } \alpha = 0.5) \end{cases}$$

Using $\theta_0 = 54.4$, from Eqn (12.2),

$$\theta_c = \theta_0 \left(\frac{T}{T_0}\right)^\alpha$$

$$T = T_0 \left(\frac{\theta_c}{\theta_0}\right)^{1/\alpha}$$

$$= 600 \left(\frac{350}{54.4}\right)^{1/0.3} = 297{,}200 \text{ h} \quad \text{(for } \alpha = 0.3)$$

$$= 600 \left(\frac{250}{54.4}\right)^{1/0.5} = 12{,}670 \text{ h} \quad \text{(for } \alpha = 0.5)$$

Graphical construction can be used to derive the same result, as shown in Fig. 12.5 [$\tan^{-1}(0.3) = 17°$; $\tan^{-1}(0.5) = 27°$; $\theta_i = \theta_c/(1 - \alpha)$].

Obviously nearly 300,000 h of testing is unrealistic, and therefore in this case a value for α of 0.5 would have to be the objective to achieve the MTBF requirement of 500 h in a further $(12{,}670 - 600) \approx 12{,}000$ h of testing.

Example 12.2 shows that the results of a Duane analysis are very sensitive to the starting assumptions. If θ_0 was 54.4 h at $T_0 = 200$ h, the test time required for a 500 h MTBF would be 4,200 h. The initial reliability figure is usually uncertain, since data at the early stage of the programme are limited. It might be more appropriate to use a starting reliability based upon a combination of data and engineering judgement. If in the previous example immediate corrective action was being taken to remove some of the causes of

Figure 12.5 Duane plot for Example 12.2

earlier failures, a higher value of θ_0 could have been used. It is important to monitor early reliability growth and to adjust the plan accordingly as test results build up.

The Duane model is criticized as being empirical and subject to wide variation. It is also argued that reliability improvement in development is not usually progressive but occurs in steps as modifications are made. However, the model is simple to use and it can provide a useful planning and monitoring method for reliability growth. Difficulties can arise when results from different types of test must be included, or when corrective action is designed but not applied to all models in the test programme. These can be overcome by commonsense approaches. A more fundamental objection arises from the problem of quantifying and extrapolating reliability data. The comments made earlier about the realism of reliability demonstration testing apply equally to reliability growth measurement. As with any other failure data, trend tests as described in Chapter 3 should be performed to ascertain whether the assumption of a constant failure rate is valid.

Other reliability gowth models are also used, some of which are described in Ref.5 in the Bibliography, which also describes the management aspects of reliability growth monitoring. Reliability growth monitoring for one-shot items can be performed similarly by plotting cumulative success rate. Statistical tests for MTBF or success rate changes can also be used to confirm reliability growth, as described in Chapter 2. Example 12.3 shows a typical reliability growth plan and record of achievement.

Example 12.3

Reliability growth plan:
Office Copier Mk 4

Specification:
In-use call rate: 2 per year max. (at end of development)
      1 per year max. (after first year)
Average copies per machine per year: 40,000

Assumptions:
$\theta_0 = 1,000$ copies per failure at 10,000 copies on prototypes

$$\alpha \begin{cases} = 0.5 \text{ during development} \\ = 0.3 \text{ in service} \end{cases}$$

Notes to Duane plot (Fig. 12.6)
1. Prototype reliability demonstration: models 2 and 3: 10,000 copies each
2. First interim reliability demonstration: models 6–8, 10: 10,000 copies each
3, 5. Overstress and ageing tests (data not included in θ_c)
4. Second interim reliability demonstration: models 8, 10, 12, 13: 10,000 copies each
6. Final reliability demonstration: models 12, 13: 10,000 copies each. Models 8, 10: 20,000 copies each

Reliability demonstration test results give values for θ_i.

Note that in Example 12.3 the overstress test results are plotted separately, so that the failures during these tests will not be accumulated with those encountered during tests in the normal operating environment. Therefore, overstress failure data will be obtained to enable potential in-use failure modes to be highlighted, without confusing the picture as far as measured reliability achievement is concerned. The improvement from the first to the second overstress test has a higher Duane slope than the main reliability growth line, indicating effective improvement. The example also includes a longevity test, to show up potential wearout failure modes, by having two of the first reliability demonstration units continue to undergo test in the second interim and final reliability demonstrations. These units will have been modified between tests to include all design improvements shown to be necessary. A lower value for α is assumed for the in-service phase, as improvements are more difficult to implement once production has started.

Figure 12.6 Duane plot for Example 12.3

NON-PARAMETRIC METHODS

Non-parametric statistical techniques (see page 53) can be applied to reliability measurement. They are arithmetically very simple and so can be useful as quick tests in advance of more detailed analysis, particularly when no assumption is made of the underlying failure distribution.

The C-rank method

If n items are tested and k fail, the reliability of the sample is

$$R_C \approx 1 - [C\text{-rank of the } (k+1)\text{th ordered value in } (n+1)] \qquad (12.4)$$

where C denotes the confidence level required, using the appropriate median rank table (Appendix 6).

Example 12.4

Twenty items were subjected to a 100 h test in which three failed. What is the reliability at the 50 and 95 per cent. lower confidence levels?

$$k + 1 = 3 + 1 = 4$$
$$n + 1 = 20 + 1 = 21$$

From Appendix 6, from the median and 95 per cent. rank tables, the C-rank of four items in a sample of 21 is:

At 50%:   0.172   (i.e. $R_{50} \approx 0.828$)

At 95%:   0.329   (i.e. $R_{95} \approx 0.671$)

(cf. $17/20 = 0.85$ for $\hat{R}$).

The success-run method

When tests are run without failure, reliability can be estimated using the equation

$$R_C \approx (1 - C)^{1/(n+1)} \qquad (12.5)$$

Example 12.5

Twenty items are tested, without failure. What is the 90 per cent. lower confidence limit of reliability?

$$R_{0.9} \approx (1 - 0.9)^{1/21}$$
$$= 0.897$$

RELIABILITY GROWTH ESTIMATION BY FAILURE DATA ANALYSIS

Reliability growth can be estimated by considering the failure data and the planned corrective action. No empirical model is used and the method takes direct account of what is known and planned, so it can be easier to sell. However, it can only be applied when sufficient data are available, well into a development programme or when the product is in service.

If we know that 20 per cent. of failures are caused by failure modes for which corrective action is planned and we are sure that the changes will be effective, we can simply estimate that the improvement in failure rate will be 20 per cent. Alternatively, we could assign an effectiveness value to the changes, say 80 per cent, in which case the failure rate improvement will be 16 per cent.

This approach should be used whenever failure data and failure investigations are comprehensive enough to support a Pareto analysis, as described in Chapter 13. The method can be used in conjunction with a Duane plot. If known failure modes can be corrected, reliability growth can be anticipated. However, if reliability is below target and no corrective action is planned, the reliability growth forecast will not have much meaning.

MAKING RELIABILITY GROW

In this chapter we have covered methods for measuring reliability achievement and growth. Of course it is not enough just to measure performance. Effort must be directed towards achieving the objective. In reliability engineering this means taking positive action to unearth design and production shortfalls which can lead to failure, to correct these deficiencies and to prove that the changes are effective. In earlier chapters we have covered the methods of stress analysis, design review, testing and failure data analysis which can be used to ensure a product's reliability. To make these activities as effective as possible, it is necessary to ensure that they are all directed towards the quantified reliability achievement which has been specified. If reliability growth is being measured at stages in the development and early in-use phases, the programme must be related to reliability requirements to be demonstrated at key points.

There is a dilemma in operating such a programme. There will be a natural tendency to try to demonstrate that the reliability requirements have been met. This can lead to reluctance to induce failures by increasing the severity of tests and to a temptation to classify failures as non-relevant or unlikely to recur. On the other hand, reliability growth is maximized by deliberate and aggressive stress-testing, analysis and corrective action as described in Chapter 11. Therefore, the objective should be to stimulate failures during development testing and not to devise tests and failure-reporting methods which will maximize the chances of demonstrating that a specification has been satisfied. Such an open and honest programme makes high demands on teamwork and integrity, and emphasizes the importance of the project manager understanding the aims and being in control of the reliability programme.* The reliability milestones should be stated early in programme and achievement should be monitored against this statement.

Test, analyse and fix

Reliability growth programmes as described above have come to be known as test, analyse and fix (TAAF) programmes. It is very important in such programmes that:

1. All failures are analysed fully, and action taken in design or production to ensure that they should not recur. No failure should be dismissed as being 'random' or 'non-relevant' during this stage, unless it can be demonstrated conclusively that such a failure cannot occur on production units in service.
2. Corrective action must be taken as soon as possible on all units in the development programme. This might mean that drawings have to be altered more often, and can cause programme delays. However, if faults are not corrected reliability growth will be delayed, potential failure modes at the 'next weakest link' may not be highlighted, and the effectiveness of the corrective action will not be adequately tested.

Whenever failures occur, the investigation should refer back to the reliability predictions, stress analyses and FMECAs to determine if the analyses were correct. Discrepancies should be noted and corrected to aid future work of this type.

Reliability growth in service

The same principles as described above should be applied to reliability growth in service. However, there are three main reasons why in-service reliability growth is more difficult to achieve than during the development phase.

1. Failure data are often more difficult to obtain. Warranty or service contract repair reports are a valuable source of reliability data, but they are often harder to control, investigation can be more difficult with equipment in the users' hands and data often terminate at the end of the warranty period. Some companies make arrangements with selected dealers to provide comprehensive service data. Military and other government customers often operate their own in-use failure data systems. However, in-use data very rarely match the needs of a reliability growth programme.
2. It is much more difficult and much more expensive to modify delivered equipment or to make changes once production has started.

---

*The politics of test planning to ensure accept decisions for contractual or incentive purposes is an aspect of reliability programme management which will not be covered here.

3. A product's reputation is made by its early performance. Reliance on reliability growth in use can be very expensive in terms of warranty costs and markets.

Nevertheless, a new product will often have in-service reliability problems despite an intensive development programme. Therefore reliability data must be collected and analysed, and improvements designed and implemented. Most products which deserve a reliability programme have a long life in service and undergo further evolutionary development, and therefore a continuing reliability improvement programme is justifiable. When evolutionary development takes place in-use data can be a valuable supplement to further reliability test data, and can be used to help plan the follow-on development and test programme. Appendix 9 describes the features of a data collection and analysis system.

Another source of reliability data is that from production test and inspection. Many products are tested at the end of the production line and this includes burn-in for many types of electronic equipment. Whilst data from production test and inspection are collected primarily to monitor production quality costs and vendor performance, they can be a useful supplement to in-use reliability data. Also, as the data collection and the product are still under the manufacturer's control, faster feedback and corrective action can be accomplished.

The manufacturer can run further tests of production equipment to verify that reliability and quality standards are being maintained. Such tests are often stipulated as necessary for batch release in government production contracts. As in-house tests are under the manufacturer's control, they can provide early warning of incipient problems and can help to verify that the reliability of production units is being maintained or improved.

BIBLIOGRAPHY

1. US MIL-STD-781: *Reliability Qualification and Production Acceptance Tests—Exponential Distribution,* Issue C. Available from the National Technical Information Service, Springfield, Virginia.
2. British Standard, BS 5760: *Reliability of Systems, Equipments and Components,* Part 2. British Standards Institution, London.
3. IEC Publication 605: *Equipment Reliability Testing.* International Electrotechnical Commission, Geneva (1978).
4. C. Lipson and M. J. Sheth, *Statistical Design and Analysis of Engineering Experiments.* McGraw-Hill, New York (1973).
5. US MIL-HDBK-189: *Reliability Growth Management.* Available from the National Technical Information Service, Springfield, Virginia.
6. US MIL-STD-1635: *Reliability Growth Testing.* Available from the National Technical Information Service, Springfield, Virginia.

Chapter 13

SOLVING RELIABILITY PROBLEMS

INTRODUCTION

The causes of many reliability problems are obvious and the solutions can be deduced fairly easily. However, we sometimes come across problems which are not straightforward, or which defy the apparently obvious solutions. This chapter presents analytical methods which can be used to help track down the causes of failure.

PARETO ANALYSIS

As a first step in reliability problem-solving, we can use the Pareto principle of the 'significant few and the insignificant many'. It is often found that a large proportion of failures in a product are due to a small number of causes. Therefore, if we analyse the failure data, we can determine how to solve the largest proportion of the overall reliability problem with the most economical use of resources. We can often eliminate a number of failure causes from further analysis by creating a Pareto plot of the failure data. For example, Fig. 13.1 shows failure data on a domestic washing machine, taken from warranty records. These data indicate that attention paid to the programme switch, the outlet pump, the high level switch and leaks would be likely to show the greatest payoff in warranty cost reduction. However, before committing resources it is important to make sure than the data have been fully analysed, to obtain the maximum amount of information contained therein. The data in Fig. 13.1 show the parts replaced or adjusted.

In this case further analysis of records reveals:

1. For the programme switch: 77 failures due to timer motor armature open-circuit, 18 due to timer motor end bearing stiff, 10 miscellaneous. Timer motor failures show a decreasing hazard rate during the warranty period.
2. For the outlet pump: 79 failures due to leaking shaft seal allowing water to reach motor coils, 21 others. Shaft seal leaks show an increasing hazard rate.
3. For the high level switch: 58 failures due to failure of spot weld, allowing contact assembly to short to earth (decreasing hazard rate), 10 others.

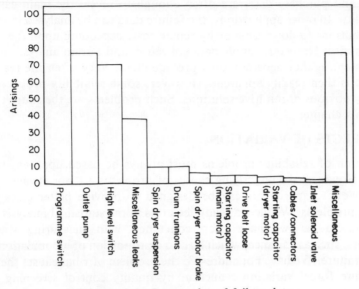

Figure 13.1 Pareto plot of failure data

These data reveal definite clues for corrective action. The timer motor and high level switch appear to exhibit manufacturing quality problems (decreasing hazard rate). The outlet pump shaft leak is a wear problem (increasing hazard rate). However, the leak is made more important because it damages the pump motor. Two types of corrective action might be considered: reorientation of the pump so that the shaft leak does not affect the motor coils and attention to the seal itself. Since this failure mode has an increasing rate of occurrence relative to equipment age, improving the seal would appreciably reduce the number of repair calls on older machines. Assuming that corrective action is taken on these four failure modes and that the improvements will be 80 per cent. effective, future production should show a reduction in warranty period failures of about 40 per cent.

The other failure modes should also be considered, since, whilst the absolute payoff in warranty cost terms might not be so large, corrective action might be relatively simple and therefore worth while. For example, some of the starting capacitor failures on older machines were due to the fact that they were mounted on a plate, onto which the pump motor shaft leak and other leaks dripped. This caused corrosion of the capacitor bodies. Therefore, re-arrangement of the capacitor mounting and investigation into the causes of leaks would both be worth considering.

This example shows the need for good data as the basis for decision-making on where to apply effort to improve reliability, by solving the important problems first. The data must be analysed to reveal as much as possible about the relative severity of problems and the likely causes. Even quite simple data, such as a brief description of the cause of failure, machine and item part number, and purchase date, are often enough to reveal the main causes of unreliability. In other applications, the failure data can be analysed in relation to contributions to down-time or by repair cost, depending upon the criteria of importance. However, simple data collection and Pareto analysis are only the first step, as they generally only provide clues. Very often the real cause of failure is then readily apparent. However, some reliability problems defy seemingly obvious or intuitive solutions. Such problems are the subject of the rest of this chapter.

THE EFFECTS OF VARIATION

The solution of reliability problems must always be based upon engineering realities. However, the underlying causes of problems are not always immediately obvious. Experiments can be designed to provide better clues to the causes of problems. We have already seen this in the statistical analysis of life test data. So far we have been concerned with measuring statistical parameters, such as variance, which lead to a proportion of a population being prone to failure. We have not addressed the problem of what causes the variation. 'Brute force' variation reduction by quality control screening is not always the best or most economical approach. We need to be able to find the causes of variation in processes which might involve several variables. These variables might have different effects upon the total variation, and interactions between the variables might also be significant. For example, if our problem is the reliability of electrical reed relays, the causes of the variation leading to failure might be switching frequency, operating temperature, switching current, relay manufacturer or interactions such as the combined effect of current and temperature. Wear in a plain bearing might be affected by lubricant type, pressure, temperature, clearances and by interactions. Deciding which of the several variables and interactions are the main causes of failure can be determined by tests in which the variables are altered. The design and application data on components such as reed relays, bearings and lubricants are based upon such experiments, and in most cases proper use of the component will lead to a reliable design. Problems can arise when sufficient data are not available, when for some reason the component does not comply with the application data or when failure is due to the effects of interactions not covered by the data.

To emphasize the importance of identifying and controlling the variables which have the greatest effect upon the desired response, it is necessary to remember that the SDs of s-independent random variations add by the sums of their squares. Therefore, if two variables have SDs of 5 units and 2 units, the combined SD of their sum or difference is $(5^2 + 2^2)^{1/2} = 5.4$. If we could totally eliminate the lesser source of variation, the combined SD would be reduced to 5. However, reducing the greater source by the same amount, to 3, would result in a combined SD of $(3^2 + 2^2)^{1/2} = 3.6$, a much greater reduction. The importance of distributed variables as causes of unreliability or process difficulty is often not appreciated by engineers taught to consider problems from a deterministic point of view. In the next section we will consider the analysis of variance and the way in which the technique can lead to solutions to problems which, because of their statistical nature, often lead to a great deal of frustration in the search for corrective action and in maintaining control afterwards.

ANALYSIS OF VARIANCE

The analysis of variance (ANOVA) technique was developed by R. A. Fisher and is a very elegant, economical and powerful method for determining the s-significant variances and interactions in multivariable situations. Analysis of variance is used widely in such fields as market research, optimization of chemical and metallurgical processes, agriculture and medical research. It can provide the insights necessary for solving quality and reliability problems. Whilst the methods described below might appear tedious computer programs are available which greatly ease the analytical task.

Analysis of single variables

The variance of a set of data (sample) is equal to

$$\frac{1}{n} \sum_{i=1}^{n} (x_i - \bar{x})^2$$

(Eqn 2.15), where n is the sample size and $\bar{x}$ is the mean value. The population variance estimate is derived by dividing the sum of squares, $\Sigma(x_i - \bar{x})^2$, not by n but by $(n-1)$, where $(n-1)$ denotes the number of degrees of freedom (DF). Then $\hat{\sigma} = \Sigma_i(x_i - \bar{x})^2/(n-1)$ (Eqn 2.16).

Example 13.1

To show how the variance of a group of samples can be analysed, consider a simple experiment in which 20 bearings, 5 each from 4 different suppliers, are run to failure. Table 13.1 shows the results.

We need to know if the observed variation between the samples is s-significant or is only a reflection of the variations of the populations from which the samples were drawn. Within each sample of five there are quite large variations. We must therefore analyse the difference between the 'between

Table 13.1 Times to failure of 20 bearings.

Sample	Times to failure, [a]x_i(h)					Sample totals	Sample means, x_i
1	4	1	3	5	7	20	4
2	6	6	5	10	3	30	6
3	3	2	5	7	8	25	5
4	7	8	8	12	10	45	9

[a]$\Sigma x_i = 120$.

Table 13.2 Values of $(x_i - \bar{x})^2$ for the data of Table 13.1.

Sample	$(x_i - \bar{x})^2$					$\Sigma(x_i - \bar{x})^2$ [a]
1	4	25	9	1	1	40
2	0	0	1	16	9	26
3	9	16	1	1	4	31
4	1	4	4	36	16	61

[a] Overall $\Sigma(x_i - \bar{x})^2 = 158$.

sample' (BS) and the 'within sample' (WS) variance and relate this to the populations.

The next step is to calculate the sample totals and sample means, as shown in Table 13.1. The overall average value

$$\bar{x} = \frac{\Sigma x_i}{n} = \frac{120}{20} = 6$$

Since n is 20, the total number of degrees of freedom (DF) is 19. We then calculate the values of $(x_i - \bar{x})^2$. These are shown in Table 13.2.

Having derived the total sum of squares and the DF, we must derive the WS and BS values. To derive the BS effect, we assume that each item value is equal to its sample mean (x_i'). The sample means for each item in samples 1 to 4 are 4, 6, 5 and 9, respectively. The BS sums of squares $(x_i' - \bar{x})^2$ are then, for each item in samples 1 to 4, 4, 0, 1 and 9, respectively, giving sample totals $\Sigma(x_i' - \bar{x})^2$ of 20, 0, 5 and 45 and an overall BS $\Sigma(x_i' - \bar{x})^2$ of 70. The BS DF is $4 - 1 = 3$.

Table 13.3 Values of $(x_i - x_i')$ for the data of Table 13.1.

Sample	x_i''					WS $\Sigma x_i''$
1	0	-3	-1	1	3	0
2	0	0	-1	4	-3	0
3	-2	-3	0	2	3	0
4	-2	-1	-1	3	1	0

Table 13.4 Values of WS$(x_i'' - \bar{x}'')^2$ for the data of Table 13.1.

Sample	WS$(x_i'' - \bar{x}'')^2$					WS $\Sigma(x_i'' - \bar{x}'')^2$ [a]
1	0	9	1	1	9	20
2	0	0	1	16	9	26
3	4	9	0	4	9	26
4	4	1	1	9	1	16

[a] Overall $\Sigma(x_i'' - \bar{x}'')^2 = 88$.

Table 13.5 Sources of variance for the data in Table 13.1.

Source of variance	$\Sigma(.)$	DF	σ^2
BS	70	3	23.33
WS (residual)	88	16	5.50
Total	158	19	8.32

Now we derive the equivalent values for the WS variance, by removing the BS effect. We achieve this by subtracting x_i' from each item in the original table to give a value x_i''. The result is shown in Table 13.3. Table 13.4 gives the values of the WS sums of squares, derived by squaring the values as they stand, since now $\bar{x}'' = 0$. The number of WS DF is $(5 - 1) \times 4 = 16$ (4 DF within each sample, for a total of four samples). We can now tabulate the analysis of variance (Table 13.5).

If we can assume that the variables are s-normally distributed and that all the variances are equal, we can use the F-test (variance ratio test) (page 53) to test the null hypothesis that the two variance estimates (BS and WS) are estimates of the same common (population) variance. The WS variance represents the experimental error, or *residual* variance. F is the ratio of the variances, and a table showing the s-significance levels of variance ratios in relation to the number of degrees of freedom is given in Appendix 4. In this case

$$F = \frac{23.33}{5.50} = 4.24$$

For 3 DF in the greater variance estimate and 16 DF in the smaller variance estimate, Appendix 4 shows that the 5 per cent. s-significance level of F is 3.24. Since our value of F is greater than this, we conclude that the variance ratio is s-significant at the 5 per cent. level, and the null hypothesis that there is no difference between the samples is therefore rejected at this level.

It is possible to simplify the analysis of variance procedure described above. Having set up the initial table of values, we produce Table 13.6 showing the squares of the values. We then use a *correction factor* equal to $(\Sigma x_i)^2/n$. In

Table 13.6 Squares of the data values in Table 13.1 (Example 10.1).

Sample			x_i^2			Totals
1	16	1	9	25	49	100
2	36	36	25	100	9	206
3	9	4	25	49	64	151
4	49	64	64	144	100	421

this case, $\Sigma x_i = 120$; therefore the correction factor $CF = 720$. The total sum of squares is then

$$\Sigma(x_i^2) - CF = 158$$

(as derived earlier). The between sample sum of squares is

$$\frac{\text{Sum of squares of sample totals}}{\text{Number of items in sample}} - CF = 70$$

The within sample sum of squares is the difference between the total sum of squares and the between sample sum of squares:

$$158 - 70 = 88$$

The degrees of freedom are derived as before.

If the sample sizes are not all the same, it is necessary to derive the between sample sum of squares by squaring each sample total, dividing by the numbers in the samples, adding these values and then subtracting the correction factor.

A further simplification can be applied when the values are not small numbers. If a constant value is subtracted from all the values, the variances (which form the subject of the analysis) are unaffected, but the arithmetic can be simplified. Since the first step in the analysis is to square the values, we can subtract a number which results in negative values. Subtracting a number close to the overall mean value provides the greatest simplification. Of course this trick will not always be necessary if an electronic calculator is being used, but even then it can speed up the calculation as it reduces the number of keystrokes, especially when dealing with the squares of fairly large numbers. The analysis results are similarly unaffected by division by a constant value.

Analysis of multiple variables (factorial experiments)

The method described above can be extended to analyse more than one source of variance. When there is more than one source of variance, interactions may occur between them, and the interactions may be more s-significant than the individual sources of variance. The following example will illustrate this for a three-factor situation.

Example 13.2

On a hydraulic system using 'O' ring seals, leaks occur apparently randomly. Three manufacturer's seals are used, and hydraulic oil pressure and temperature vary through the system. Argument rages as to whether the high temperature locations are the source of the problem, or the manufacturer, or even pressure. Life test data show that seal life may be assumed to be s-normally distributed at these stress levels, with variances which do not change with stress. A test rig is therefore designed to determine the effect on seal performance of oil temperature, oil pressure and type of seal. We set up

Table 13.7 Results of experiments on 'O' ring seals.

Temperature	Type (T) 1		T2		T3	
	p_1	p_2	p_1	p_2	p_1	p_2
t_1	104	209	181	172	157	178
	196	132	129	151	187	211
t_2	136	140	162	133	141	164
	97	122	108	114	174	128
t_3	108	96	99	112	122	135
	121	110	123	109	130	118

Table 13.8 The data of Table 13.7 after subtracting 100 from each datum.

Temperature	Type (T) 1		T2		T3	
	p_1	p_2	p_1	p_2	p_1	p_2
t_1	4	109	81	72	57	78
	96	32	29	51	87	111
t_2	36	40	62	33	41	64
	− 8	22	8	14	74	28
t_3	8	− 4	− 1	12	22	35
	21	10	23	9	30	18

$\Sigma x_i (t)$

$\Sigma x_i (T)$

the experiment in which we apply two values of pressure, 15 and 18 $MN\,m^{-2}$ (denoted p_1 and p_2, in increasing order), and three values of temperature, 80, 100 and 120 °C (t_1, t_2 and t_3, also in increasing order). We then select seals for application to the different test conditions, with two tests at each test combination. The results of the test are shown in Table 13.7, showing the time (h) to show a detectable leak.

To simplify Table 13.7, subtract 100 from each value. The coded data then appear as in Table 13.8.

The sums of squares for the three sources of variation are then derived as follows:

1. The correction factor:

$$CF = \frac{(\sum x_i)^2}{n} = \frac{(1,409)^2}{36} = 55,147$$

2. The between types sum of squares:

$$SS_T = \frac{\sum x_i^2(T)}{2 \times 3 \times 2} - CF$$

$$= \frac{371^2 + 393^2 + 645^2}{12} - 55,147 = 3,863$$

Between three types there are two DF.

3. The between pressures sum of squares:

$$SS_p = \frac{\sum x_i^2(p)}{3 \times 3 \times 2} - CF$$

$$= \frac{675^2 + 734^2}{18} - 55,147 = 96$$

Between two pressures there is one DF.

4. The between temperatures sum of squares:

$$SS_t = \frac{\sum x_i^2(t)}{2 \times 3 \times 2} - CF$$

$$= \frac{807^2 + 419^2 + 183^2}{12} - 55,147 = 16,544$$

Between three temperatures there are two DF.

5. The types–temperature interaction sum of squares:

$$SS_{Tt} = \frac{\sum x_i^2(Tt)}{2 \times 2} - CF - SS_T - SS_t$$

$$= \frac{241^2 + 95^2 + 35^2 + 233^2 + 117^2 + 43^2 + 333^2 + 207^2 + 105^2}{4}$$

$$- 55,147 - 3863 - 16,544$$

$$= 176$$

The T and the t effects each have 2 DF, so the interaction has $2 \times 2 = 4$ DF.

6. The types–pressure interaction sum of squares:

$$SS_{Tp} = \frac{\sum x_i^2(Tp)}{6} - CF - SS_T - SS_p$$

$$= \frac{162^2 + 209^2 + 202^2 + 191^2 + 311^2 + 334^2}{2 \times 3} - 55,147 - 3,863 - 96$$

$$= 142$$

The T effect has 2 DF and the p effect has 1 DF, so the Tp interaction has $2 \times 1 = 2$ DF.

7. The temperature–pressure interaction sum of squares:

$$SS_{tp} = \frac{\sum x_i^2(tp)}{2 \times 3} - CF - SS_t - SS_p$$

$$= \frac{354^2 + 453^2 + 218^2 + 201^2 + 103^2 + 80^2}{6} - 55,147 - 16,544 - 96$$

$$= 789$$

The t effect has 2 DF and the p effect 1 DF. Therefore the tp interaction has $2 \times 1 = 2$ DF.

8. The type–temperature–pressure interaction sum of squares:

$$SS_{Ttp} = \frac{\sum x_i^2(Ttp)}{2} - CF - SS_T - SS_t - SS_p - SS_{Tt} - SS_{tp} - SS_{Tp}$$

$$= \frac{154,613}{2} - 55,147 - 3,863 - 16,544 - 96 - 176 - 789 - 142$$

$$= 550$$

There are $2 \times 1 \times 2 = 4$ DF.

9. The total sum of squares:

$$SS_{tot} = \sum x_i^2 - CF$$

$$= 91,427 - 55,147 = 36,280$$

Total DF $= 36 - 1 = 35$ DF.

10. Residual (experimental error) sum of squares is

$$SS_{tot} - \text{(all other SS)}$$
$$= 36,280 - 3,863 - 96 - 16,544 - 376 - 142$$
$$- 789 - 550$$
$$= 13,920$$

The residual DF:
Total DF − (all other DF) $= 35 - 2 - 1 - 2 - 4 - 2 - 2 - 4 = 18$ DF

Examination of the analysis of variance table (Table 13.9) shows that all the interactions show variance estimates much less than the residual variance, and therefore they are clearly not s-significant.

Having determined that none of the interactions are s-significant, we can assume that these variations are also due to the residual or experimental variance. We can therefore combine these sums of squares and degrees of freedom to provide a better estimate of the residual variance. The revised residual variance is thus:

$$\frac{176 + 142 + 789 + 550 + 13,920}{4 + 2 + 2 + 4 + 18} = \frac{15,577}{30}$$

$$= 519 \text{ with 30 DF}$$

Table 13.9 Analysis of variance table.

| | Main factors | | | Interactions | | | | |
| | | | | First order | | | Second order | |
Effect	Type (T)	Pressure (p)	Temperature (t)	Tt	Tp	tp	Ttp	Residual
SS	3,863	96	16,544	176	142	789	550	13,920
DF	2	1	2	4	2	2	4	18
SS/DF[a]	1,932	96	8,272	44	71	395	138	773
SD	44	10	91	10	8	20	9	28

[a]Variance estimate.

We can now test the s-significance of the main factors. Clearly the effect of pressure is not s-significant. For the 'type' (T) main effect, the variance ratio $F = 1,932/519 = 3.72$ with 2 and 30 DF. The F-distribution table (Appendix 4) shows that this is s-significant at the 5 per cent. level. For the temperature (t) main effect, $F = 8,272/519 = 15.94$ with 2 and 30 DF. This is s-significant, even at the 1 per cent. level.

Therefore the experiment shows that the life of the seals is s-significantly dependent upon operating temperature and upon type. Pressure and interactions show effects which, if important, are not discernible within the experimental error; the effects are 'lost in the noise'. In other words, no type is s-significantly better or worse at higher pressures. Referring back to the original table of results, we can calculate the mean lives of the three types of seal, under the range of test conditions: type 1, 130 h; type 2, 132 h; and type 3, 154 h. Assuming that no other aspects such as cost predominate, we should therefore select type 3, which we should attempt to operate at low oil temperatures.

Non-normally distributed variables

In Example 13.2 above it is important to note that the method described is statistically correct only for s-normally distributed variables. As shown in Chapter 2, in accordance with the central limit theorem many parameters in engineering are s-normally distributed. However, it is prudent to test variables for s-normality before performing an analysis of variance. If any of the key variables are substantially non-s-normally distributed, non-parametric analysis methods should be used, as described later.

Two-level factorial experiments

It is possible to simplify the analysis of variance method if we adopt a a two-level factorial design for the experiment. In this approach, we take only two values for each main effect, high and low, denoted by + and − . Example 13.3 below shows the results of a three-factor non-replicated experiment (such an experiment is called a 2^3 factorial design, i.e. three factors, each at two levels).

Example 13.3

From the results of Table 13.10, the first three columns of Table 13.11 represent the *design matrix* of the factorial experiment. The response value is the mean of the values at each test combination or in this case, as there is no replication, the single test value.

The main effects can be simply calculated by averaging the difference between the response values for each high and low factor setting, using the appropriate signs in the 'factor' columns. Thus:

Main effect $A = [(15 − 13) + (13 − 12) + (14 − 11) + (14 − 12)] \div 4 = 2$

This is the same as: $(15 + 13 + 14 + 14 − 13 − 12 − 11 − 12) \div 4 = 2$
Likewise,
Main effect B $= (15 + 13 − 14 − 14 + 13 + 12 − 11 − 12) \div 4 = 0.5$
and,
Main effect C $= (15 − 13 + 14 − 14 + 13 − 12 + 11 − 13) \div 4 = 0.5$

Table 13.10 Results of a three-factor non-replicated experiment.

A	B			
	−	−	+	+
		C		
	−	+	−	+
−	12	11	12	13
+	14	14	13	15

Table 13.11 Response table and interactions of effects A, B, C.

Factor			Interactions				
A	B	C	AB	AC	BC	ABC	Response
+	+	+	+	+	+	+	15
+	+	−	+	−	−	−	13
+	−	+	−	+	−	−	14
+	−	−	−	−	+	+	14
−	+	+	−	−	+	−	13
−	+	−	−	+	−	+	12
−	−	+	+	−	−	+	11
−	−	−	+	+	+	−	12

The sum of squares of each effect is then calculated using the formula

$$SS = 2^{k-2} \times (\text{effect estimate})^2$$

where k is the number of factors. Therefore,

$$SS_A = 2 \times 2^2 = 8$$
$$SS_B = 2 \times 0.5^2 = 0.5$$
$$SS_C = 2 \times 0.5^2 = 0.5$$

We can derive the interaction effects by expanding Table 13.11. Additional columns are added, one for each interaction. The signs under each interaction column are derived by algebraic multiplication of the signs of the constituent main effects. The ABC interaction signs are derived from AB × C or AC × B or BC × A.

The AB interaction is then

$$(15 + 13 - 14 - 14 - 13 - 12 + 11 + 12) \div 4 = -0.5$$

and

$$SS_{AB} = 2 \times (-0.5)^2 = 0.5$$

The other interaction sums of squares can be calculated in the same way and the analysis of variance table (Table 13.12) constructed.

Whilst the experiment in Example 13.3 indicates a large A main effect, and possibly an important BC interaction effect, it is not possible to test these statistically since no residual variance is available. To obtain a value of residual variance further replication would be necessary, or a value of experimental error might be available from other experiments. Alternatively, if the interactions, particularly higher order interactions, are insignificant, they can be combined to give a residual value. Also, with so few degrees of freedom, the F test requires a very large value of the variance ratio in order to give high confidence that the effect is significant. Therefore, an unreplicated 2^3 factorial experiment may not always be sufficiently sensitive. However, the example will serve as an introduction to the next section, dealing with situations where several variables need to be considered.

Table 13.12 Analysis of variance table.

	Effect							
	A	B	C	AB	AC	BC	ABC	Total
SS	8	0.5	0.5	0.5	0.5	2	0	12
DF	1	1	1	1	1	1	1	7
SS/DF[a]	8	0.5	0.5	0.5	0.5	2	0	

[a] Variance estimate

Fractional factorial experiments

So far we have considered experiments in which all combinations of factors were tested, i.e. *full factorial* experiments. These can be expensive and time-consuming, since if the number of factors to be tested is f, the number of levels is L and the number of replications is r, then the number of tests to be performed is rL^f, i.e. in the hydraulic seal example $2 \times 3^2 = 18$, or in a three-level four-factor experiment with two replications $2 \times 3^4 = 162$. In a *fractional factorial* experiment we economize by eliminating some test combinations. Obviously we then lose information, but if the experiment is planned so that only those effects and interactions which are already believed to be unimportant are eliminated, we can make a compromise between total information, or experiment cost, and experimental value.

A point to remember is that higher order interactions are unlikely to have any engineering meaning or to show statistical significance, and therefore the full factorial experiment with several factors can give us information not all of which is meaningful.

We can design a fractional factorial experiment in different ways, depending on which effects and interactions we wish to analyse. Selection of the appropriate design is made starting from the full factorial design matrix. Table 13.13 gives the full design matrix for a 2^4 factorial experiment.

We select those interactions which we do not consider worth analysing. In the 2^4 experiment, for example, the ABCD interaction would not normally be considered significant. We therefore select all the rows in the table in which the ABCD column shows 1 . Thus we eliminate half of our test combinations, leaving a *half factorial* experiment. What else do we lose as a result? Table 13.14 shows what is left of the experiment.

Examination of this table shows the following pairs of identical columns

A, BCD; B, ACD; C, ABD; D, ABC; AB, CD; AC, BD; AD, BC

This means that the A main effect and the BCD interaction effect will be indistinguishable in the results. In fact in any experiment to this design we will not be able to distinguish response values for these effects; we say that they are *aliased* or *confounded*. If we can assume that the first-and second-order interactions aliased in this case are insignificant, then this will be an appropriate fractional design, reducing the number of tests from $2^4 = 16$ to $\frac{1}{2} \times 2^4 = 8$. We will still be able to analyse all the main effects, and up to three first-order interactions if we considered that they were likely to be significant. For example, if engineering knowledge tells us that the AB interaction is likely to be significant but the CD interaction not, we can attribute the variance estimate to the AB interaction.

A similar breakdown of a 2^3 experiment would show that it is not possible to produce a fractional factorial without aliasing main effects with first-order interactions. This is unlikely to be acceptable, and fractional factorial designs are normally only used when there are four or more factors to be analysed.

Table 13.13 The full design matrix for a 2^4 factorial experiment.

Test no.	1 A	2 B	3 C	4 D	5 AB	6 AC	7 AD	8 BC	9 BD	10 CD	11 ABC	12 ABD	13 ACD	14 BCD	15 ABCD
1	−	−	−	−	+	+	+	+	+	+	−	−	−	−	+
2	−	−	−	+	+	+	−	+	−	−	−	+	+	+	−
3	−	−	+	−	+	−	+	−	+	−	+	−	+	+	−
4	−	−	+	+	+	−	−	−	−	+	+	+	−	−	+
5	−	+	−	−	−	+	+	−	−	+	+	+	−	+	−
6	−	+	−	+	−	+	−	−	+	−	+	−	+	−	+
7	−	+	+	−	−	−	+	+	−	−	−	+	+	−	+
8	−	+	+	+	−	−	−	+	+	+	−	−	−	+	−
9	+	−	−	−	−	−	−	+	+	+	+	+	+	−	−
10	+	−	−	+	−	−	+	+	−	−	+	−	−	+	+
11	+	−	+	−	−	+	−	−	+	−	−	+	−	+	+
12	+	−	+	+	−	+	+	−	−	+	−	−	+	−	−
13	+	+	−	−	+	−	−	−	−	+	−	−	+	+	+
14	+	+	−	+	+	−	+	−	+	−	−	+	−	−	−
15	+	+	+	−	+	+	−	+	−	−	+	−	−	−	−
16	+	+	+	+	+	+	+	+	+	+	+	+	+	+	+

Table 13.14 Table 13.13 omitting rows where ABCD gives minus.

A	B	C	D	AB	AC	AD	BC	BD	CD	ABC	ABD	ACD	BCD	ABCD
+	+	+	+	+	+	+	+	+	+	+	+	+	+	+
+	−	−	+	−	−	+	+	−	−	+	−	−	+	+
+	−	+	−	−	+	−	−	+	−	−	+	−	+	+
+	+	−	−	+	−	−	−	−	+	−	−	+	+	+
−	−	−	−	+	+	+	+	+	+	−	−	−	−	+
−	+	+	−	−	−	+	+	−	−	−	+	+	−	+
−	+	−	+	−	+	−	−	+	−	+	−	+	−	+
−	−	+	+	+	−	−	−	−	+	+	+	−	−	+

Table 13.15 Sixteenth fractional factorial layout for seven main effects.

Test no.	A	B	C	D	E	F	G
1	+	+	+	+	+	+	+
2	+	+	−	+	−	−	−
3	+	−	+	−	+	−	−
4	+	−	−	−	−	+	+
5	−	+	+	−	−	+	−
6	−	+	−	−	+	−	+
7	−	−	+	+	−	−	+
8	−	−	−	+	+	+	−
Alias				AB	AC	BC	ABC

Quarter factorial designs can be used when appropriate, following similar logic to that described above. The value of using fractional factorial designs increases rapidly when large numbers of effects must be analysed. For example, if there are seven main effects, a full factorial experiment would analyse a large number of high level interactions which would not be meaningful, and would require $2^7 = 128$ tests for no repeats. We can design a sixteenth fractional factorial layout with only eight tests, which will analyse all the main effects and most of the first-order interactions, as shown in Table 13.15. The aliases are deliberately planned, e.g. by evaluating the signs for the D column by multiplying the signs for A and B, and so on. We can select which interactions to alias by engineering judgement. It should be noted that other effects are also aliased. The full list of aliased effects can be derived by multiplying the aliased effects. For example, if D and AB are aliased, then the ABD interaction effect is also aliased, and so on. (If a squared term arises, let the squared term equal unity, e.g. $AB \times AC = A^2BC = BC$.)

Example 13.4

A jet engine fuel control system contains a number of components, seven of which are considered likely to affect the desired fuel flow ($30 \text{ dm}^3 \text{min}^{-1}$)

Table 13.16 Results of a sixteenth fractional factorial experiment for seven jet engine components affecting fuel flow.

Test no.	Component							Response $(dm^3\ min^{-1})$	Response -30
	A	B	C	D	E	F	G		
1	+	+	+	+	+	+	+	36	6
2	+	+	−	+	−	−	−	34	4
3	+	−	+	−	+	−	−	34	4
4	+	−	−	−	−	+	+	28	−2
5	−	+	+	−	−	+	−	28	−2
6	−	+	−	−	+	−	+	30	0
7	−	−	+	+	−	−	+	32	2
8	−	−	−	+	+	+	−	26	−4
Alias				AB	AC	BC	ABC		

under specified conditions. A sixteenth fractional factorial experiment is carried out with each of the seven components at high and low values within their tolerances. The results are given in Table 13.16. (The experiment is designed on the basis that components D, E, F, G and the aliased interactions are not considered likely to be significant.)

The main effects are:

A: $(6 + 4 + 4 - 2 - 2 - 0 - 2 + 4) \div 4 = 3$

B: $(6 + 4 - 4 + 2 + 2 + 0 - 2 + 4) \div 4 = 3$

C: $(6 - 4 + 4 + 2 + 2 - 0 + 2 + 4) \div 4 = 4$

D: $(6 + 4 - 4 + 2 - 2 - 0 + 2 - 4) \div 4 = 1$

E: $(6 - 4 + 4 + 2 - 2 + 0 - 2 - 4) \div 4 = 0$

F: $(6 - 4 - 4 - 2 + 2 - 0 - 2 - 4) \div 4 = -2$

G: $(6 - 4 - 4 - 2 - 2 + 0 + 2 + 4) \div 4 = 0$

$SS = 2^{(7-2)} \times (\text{effect estimate})^2 = 32 \times (\text{effect estimate})^2$. Therefore

$SS_A = 288$

$SS_B = 288$

$SS_C = 512$

$SS_D = 32$

$SS_E = 0$

$SS_F = 64$

$SS_G = 0$

Therefore the variance of component C is the most critical factor in the design, with components A and B also requiring control.

NON-PARAMETRIC METHODS

Non-parametric methods may also be used for the two-way analysis of variance. Non-parametric methods can be useful when the variables cannot be assumed to be s-normally distributed since the methods imply no distribution for the data. They are also quite easy to apply. We will describe only the

Kruskal–Wallis test, since other tests require fairly large numbers of observations in each group and this is often not realistic in reliability work.

In the Kruskal–Wallis test we replace the observed values by their rank order numbers. If the null hypotheses (that there is no difference in variances) is true, then the rank order numbers $1 \ldots N$ should be evenly distributed among the groups, so the mean of the rank order numbers in groups should be about the same. If T is total of rank order numbers and T_i is the total of rank order numbers in the ith group, then

$$T = \frac{N(N+1)}{2}$$

The average rank in group i is T_i/n_i (where n_i is the number of observations in group i).

The sum of squares of the rank order numbers is

$$\sum \frac{T_i^2}{n_i} - \frac{T^2}{N} = \sum \frac{T_i^2}{n_i} - \frac{N(N+1)^2}{4}$$

The Kruskal–Wallis criterion is

$$\chi^2_{\nu=(n_i-1)} = \frac{12}{N(N+1)} \times (\text{sum of squares of rank order numbers})$$

$$= \frac{12}{N(N+1)} \sum \frac{T_i^2}{n_i} - 3(N+1)$$

Example 13.5

Test data on five groups are given in Table 13.17 and the rank order numbers in Table 13.18.

$$N = 30$$
$$n_i = 6$$
$$\chi^2_{\nu=4} = \frac{12}{30 \times 31} \left(\frac{80.83^2 + 112.33^2 + 128.5^2 + 53.5^2 + 87.83^2}{6} \right) - (3 \times 31)$$

$$= 99.4 - 93 = 6.4$$

From Appendix 2, this is s-significant at about 20 per cent.

Table 13.17 Test data on five groups.

			Test			
Group	1	2	3	4	5	6
1	4	17	21	341	35	39
2	17	26	29	34	40	42
3	25	25	33	41	41	43
4	1	5	109	22	27	34
5	14	24	31	32	33	32

Table 13.18 Rank order numbers (Example 13.5).

Group	Test 1	2	3	4	5	6	T_i
1	2	$6\frac{1}{2}$	8	$15\frac{1}{3}$	24	25	80.83
2	$6\frac{1}{2}$	13	$15\frac{1}{3}$	$22\frac{1}{2}$	26	29	112.33
3	$11\frac{1}{2}$	$11\frac{1}{2}$	$20\frac{1}{2}$	$27\frac{1}{2}$	$27\frac{1}{2}$	30	128.5
4	1	3	4	9	14	$22\frac{1}{2}$	53.5
5	5	10	$15\frac{1}{3}$	$18\frac{1}{2}$	$20\frac{1}{2}$	$18\frac{1}{2}$	87.83

RANDOMIZING THE DATA

At this stage it is necessary to point out an essential aspect of any statistically designed experiment. s-Significant sources of variance must be made to show their presence not only against the background of experimental error but against other sources of variation which might exist but which might not be tested for in the experiment. For instance, in Example 13.2, a source of variation might be the order in which seals are tested, or the batch from which seals are drawn. To eliminate the effects of extraneous factors and to ensure that only the effects being analysed will affect the results, it is imperative that the experiment be randomized. Thus, the items selected for test and the sequence of tests must be selected at random, using random number tables or another suitable randomizing process. In example 13.2, test samples should have been drawn at random from several batches of seals of each type and the test sequences should also have been randomized. It is also very important to eliminate human bias from the experiment, by hiding the identity of the items under test, if practicable.

If the items under test undergo a sequence of processes, e.g. heat treatment, followed by machining, then plating, it is essential that the items undergo each process in random order, i.e. separately randomized for each process.

Because of the absolute necessity for randomizing the data, it is nearly always necessary to design and plan an experiment to provide the data for analysis of variance. Data collected from a process as it usually occurs is unlikely to be valid for this purpose. Careful planning is important, so that once the experiment starts unforeseen circumstances do not cause disruption of the plan or introduce unwanted sources of variation or bias. A dummy run can be useful to confirm that the experiment can be run as planned.

ENGINEERING INTERPRETATION OF RESULTS

A statistical experiment can always, by its nature, produce results which conflict with the physical or chemical basis of the situation. The probability of a variance estimate being s-significant in relation to the experimental error is

Figure 13.2 Temperature–pressure interactions

determined in the analysis, but we must always be on the lookout for the occurrence of chance results which do not fit our knowledge of the processes being studied. For example, in the hydraulic seal experiment we could study further the temperature–pressure interaction. Since the variance estimate for this interaction was higher than for the other interactions we might be tempted to suspect some interaction. In another experiment the seals might be selected in such a way that this variance estimate showed s-significance.

Figure 13.2 shows the interaction graphically, in two forms. If there were no interaction, i.e. the two effects acted quite independently upon seal life, the lines would be parallel between t_1t_2, t_2t_3 or p_1p_2. In this case an interaction appears to exist between 80(t_1) and 100 °C(t_2), but not between 100 and 120 °C(t_3). This we can dismiss as being unlikely, taking into account the nature of the seals and the range of pressures and temperatures applied. That is not to say that such an interaction would always be dismissed, only that we can legitimately use our engineering knowledge to help interpret the results of a statistical experiment. The right balance must always be struck between the statistical and engineering interpretations. If a result appears highly s-significant, such as the temperature main effect, then it is conversely highly unlikely that it is a perverse result. If the engineering interpretation clashes with the statistical result and the decision to be made based on the result is important, then it is wise to repeat the experiment, varying the plan to emphasize the effects in question. In the hydraulic seal experiment, for example, we might perform another experiment, using type 3 seals only, but at three values of pressure as well as three values of temperature, and making four replications of each test intead of only two.

EVOLUTIONARY OPERATION

Evolutionary operation (EVOP) is a method used to optimize a process or system. It depends upon the vital few variables having been identified by methods such as those described above, and allows the optimum values to be determined in relation to the response required.

Consider an item in which two s-significant variables X and Y have been isolated. The results of a two-factor two-level experiment can be depicted as shown in Fig. 13.3. These results indicate that the response value might be increased by further increasing the values of X and Y. We therefore plan a further experiment, using the previous high values of X and Y as the new low values. Continuing in this way, we might generate a series of responses as shown in Fig. 13.4. The experiments are continued until no further improvement is achieved. In this case the combination of X and Y values which give the response of 23 appear to be close to the optimum.

The method can be extended to include further variables. For three variables a three-dimensional chart can be drawn to show the results, but computer analysis is required for more than three variables. Further refinement can be achieved by running a further experiment, as indicated by the dotted square in Fig. 13.4. If none of the corners indicate an improvement over the previous optimum, then we have achieved the objective, unless even higher optimization is required.

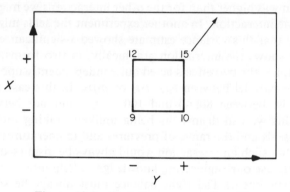

Figure 13.3 Evolutionary operation (1)

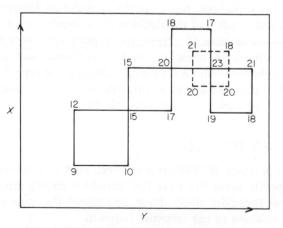

Figure 13.4 Evolutionary operation (2)

In performing this type of optimization it must be remembered that we are looking for s-significant changes in the response. The amount of change which is s-significant will have become apparent from the first analysis of variance. A further point to bear in mind is that in some cases the *response surface,* considered as a contour of the response values projected perpendicular to the plane of the variables in the two-variable case, may have more than one maximum value. Evolutionary operation is also called *response surface analysis*. Computer programs are used to analyse situations in metallurgy, component matching, etc., in which several variables affect the response.

CONCLUSIONS

Identifying the causes of unreliability or, conversely, optimizing an item or process to maximize reliability is an essential part of a reliability programme. Good data collection and straightforward engineering interpretation will provide the answers to most problems, especially if the cause of the problem is mainly deterministic. However, more detailed analyses and statistical engineering might be necessary to solve others, particularly when the underlying causes are statistically distributed variables or when interactions occur. Discovering and controlling the key variables is often not as straightforward as might appear intuitively. Intuitive corrective action often takes longer to provide a solution, as it is uncontrolled and does not always provide an answer to what really caused the problem. A properly planned statistical experiment should take no longer to implement, and is more likely to identify the source of the problem, with known statistical confidence. It also provides the basis for optimizing the solution.

The statistical approach to problem-solving also leads to better understanding of which variables need to be controlled, and by how much, in order to keep the overall response under control. Statistical experiments should be considered as part of development and manufacturing programmes, not only to optimize a product or process but also to provide the data necessary to anticipate problems. Control can then be concentrated on the key variables, and future problems can be solved by bringing these variables back under control, rather than by a late search for answers. Obviously it will not always be possible to anticipate every problem. Since such an approach will have to provide reference data for future action it will be necessary to arrange for careful recording and to have procedures which ensure that the data are used subsequently. Also, the data may become invalidated by changes in materials, processes, etc., so repeat experiments should be carried out when necessary.

This chapter has described analytical methods for solving engineering problems. The analysis must always be supported by engineering knowledge, or if necessary by other tests or experiments, to give engineering confidence in the results. The techniques described do not explain why a result occurs. However, they can provide very good clues in some difficult situations and can therefore

318

indicate the likely physical or other basis of problems, and thus the correct engineering approach to better understanding and control.

BIBLIOGRAPHY

1. M. J. Moroney, *Facts from Figures*. Penguin, Harmondsworth (1965).
2. G. E. Box, W. G. Hunter and J. S. Hunter, *Statistics for Experimenters*. Wiley (1978).
3. C. Lipson and N. J. Sheth, *Statistical Design and Analysis of Engineering Experiments*. McGraw-Hill, New York (1973).
4. W. G. Cochran and G. M. Cox, *Experimental Designs,* 2nd edn. Wiley, New York (1957).
5. A. H. Bowker and G. J. Lieberman, *Engineering Statistics*. Prentice-Hall, Englewood Cliffs (1972).
6. A. J. Duncan, *Quality Control and Industrial Statistics,* 4th edn. Irwin, Homewood, Illinois (1974).
7. M. Hollander and D. A. Wolfe, *Non-parametric Statistical Methods*. Wiley, New York (1963).

Chapter 14

RELIABILITY MANAGEMENT

CORPORATE POLICY FOR RELIABILITY

A really effective reliability function can exist only in an organization where the achievement of high reliability is recognized as part of the corporate strategy and is given top management attention. If these conditions are not fulfilled, and if it receives only lip service, reliability effort will be cut back whenever cost or time pressures arise. Reliability staff will suffer low morale and will not be accepted as part of project teams. Therefore, quality and reliability awareness and direction must start at the top and must permeate all functions and levels where reliability can be affected. It is significant that in Japan the government has made quality a national objective, with a requirement that a product attains set quality standards as a condition of being granted an export licence. It is common in Japanese corporations for reliability training to start at senior executive level and to be extended downwards to include every designer. The effects of this high level attention have been dramatic in leading Japanese industry to the forefront in many areas of technological competition.

Several factors of modern industrial business make such high level awareness essential. The high costs of repairs under warranty, and of those borne by the user, even for relatively simple items such as domestic electronic and electrical equipment, make reliability a high value property. Other less easily quantifiable effects of reliability, such as customer goodwill and product reputation, and the relative reliability of competitive products, are also important in determining market penetration and retention.

INTEGRATED RELIABILITY PROGRAMMES

The reliability effort should always be treated as an integral part of the product development and not as a parallel activity unresponsive to the rest of the development programme. This is the major justification for placing responsibility for reliability with the project manager. Whilst specialist reliability services and support can be provided from a central department in a matrix management structure, the responsibility for reliability achievement must not be taken away from the project manager, who is the only person who can ensure that the right balance is struck in allocating resources and time between the various competing aspects of product development.

The elements of a comprehensive reliability programme are shown, related to the overall development and production programme, in Fig. 1.6. This shows the continuous feedback of information, so that design iteration can be most effective.

Since production quality will be the final determinant of reliability, quality control is an integral part of the reliability programme. Quality control cannot make up for design shortfalls, but poor quality can negate much of the reliability effort. The quality control effort must be responsive to the reliability requirement and must not be directed only at reducing production costs and the passing of a final test or inspection. Quality control can be made to contribute most effectively to the reliability effort if:

1. Quality procedures, such as test and inspection criteria, are related to factors which can affect reliability, and not only to form and function. Examples are tolerances, inspection for flaws which can cause weakening, and the need for adequate burn-in when appropriate.
2. Quality control test and inspection data are integrated with the other reliability data.
3. Quality control personnel are trained to recognize the relevance of their work to reliability, and trained and motivated to contribute.

An integrated reliability programme must be disciplined. Whilst creative work such as design is usually most effective when not constrained by too many rules and guidelines, the reliability (and quality) effort must be tightly controlled and supported by mandatory procedures. The disciplines of design analysis, test, reporting, failure analysis and corrective action must be strictly imposed, since any relaxation can result in a reduction of reliability, without any reduction in the cost of the programme. There will always be pressure to relax the severity of design analyses or to classify a failure as non-relevant if doubt exists, but this must be resisted. The most effective way to ensure this is to have the agreed reliability programme activities written down as mandatory procedures, with defined responsibilities for completing and reporting all tasks, and to check by audit and during programme reviews that they have been carried out.

RELIABILITY AND LIFE CYCLE COSTS (LCC)

Achieving high reliability is expensive, particularly when the product is complex or involves relatively untried technology. The techiques described in earlier chapters require the resources of trained engineers, management time, test equipment and products for testing, and it often appears difficult to justify the necessary expenditure in the quest of an inexact quantity such as reliability. It can be tempting to trust to good basic design and production and to dispense with specific reliability effort, or to provide just sufficient effort to placate a customer who insists upon a visible programme without interfering with the 'real' development activity. However, experience points to the fact that all well-managed reliability efforts pay off. There are usually practical limits to how much can be spent on reliability during a development programme. However, the author is unaware of any programme in which experience indicated that too much effort was devoted to reliability or that the law of diminishing returns was observed to be operating to a degree which indicated that the programme was saturated. This is mainly due to the fact that nearly every failure mode experienced in service is worth discovering and correcting during development, owing to the very large cost disparity between corrective action taken during development and similar action (or the cost of living with the failure mode) once the equipment is in service. The earlier in a development programme that the failure mode is identified and corrected the cheaper it will be, and so the reliability effort must be instituted at the outset and as many failure modes as possible eliminated during the early analysis, review and test phases.

The cost of achieving reliability

It is dangerous to generalize about the cost of achieving a given reliability value, or of the effect on reliability of stated levels of expenditure on reliability programme activities. Some texts show a relationship as in Fig. 14.1, with the lowest total cost (life cycle cost — LCC) indicating the 'optimum reliability' point. However, whilst such a relationship might hold in practice, it is usually very difficult to determine the optimum expenditure and the optimum reliability. The in-service costs can usually be estimated fairly accurately, related to assumed reliability levels. However, the question of how much money buys how much reliability is much more difficult to answer, since so much depends upon factors that are hard to quantify.

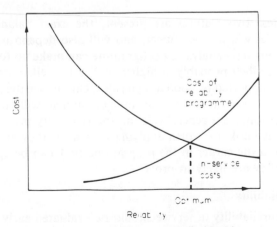

Figure 14.1 Reliability–cost relationship

For example, two similar products, developed with similar budgets, may have markedly different reliabilities, due to differences in quality control in production, differences in the quality of the initial design or differences in the way the reliability aspects of the development programme were managed. It is even harder to say by how much a particular reliability activity will affect reliability. $20,000 spent on FMECA might make a large or a negligible difference to achieved reliability, depending upon whether the failure modes uncovered would have manifested themselves and been corrected during the development phase, or the extent to which the initial design was free of shortcomings. In earlier chapters we mentioned the effects of testing and of using higher quality parts (particularly electronic) on achieved reliability. The costs of testing and of procurement of higher quality parts can be estimated accurately and the reliability impact assessed (Duane method for testing, π_Q for electronic components). However, it must never be assumed that these estimating or parametric approaches can be used with precision. They must be tempered with experience related to the type of product and the particular experience of the design and development team.

The value gained from a reliability programme must, to a large extent, be a subjective judgement based upon experience and related to the way the programme is managed. The reliability programme will usually be constrained by the resources which can be usefully applied within the development time-scale. Allocation of resources to reliability programme activities should be based upon an assessment of the risks. For a complex new design, design analysis must be thorough and searching, and performed early in the programme. For a relatively simple adaptation of an existing product, less emphasis may be placed on analysis. In both cases the test programme should be related to the reliability requirement, the risks assessed in achieving it and the costs of non-achievement. The two most important features of the programme are (1) the statement of the reliability aim in such a way that it is understood, feasible, mandatory and demonstrable, (2) dedicated, integrated management of the programme.

Provided these two features are present, the exact balance of resources between activities will not be critical, and will also depend upon the type of product. A strong test–analyse-fix programme can make up for deficiencies in design analysis, albeit probably at higher cost; an excellent design team, well controlled and supported by good design rules, can reduce the need for testing. The reliability programme for an electronic equipment will not be the same as for a power station. As a general rule, all the reliability programme activities described in this book are worth applying insofar as they are appropriate to the product, and the earlier in the programme they can be applied the more cost-effective they are likely to prove.

Costs of unreliability

The costs of unreliability in service should be evaluated early in the development phase, so that the effort on reliability can be justified and requirements can be set, related to expected costs. The analysis of unreliability costs takes different forms, depending on the type of development programme and how the product is maintained. Two examples are given to illustrate these differences in typical programmes.

Example 14.1

A commercial electronic communication equipment is to be developed as a risk venture. The product will be sold outright, with a 2 year parts and labour warranty. Outline the LCC analysis approach and comment on the support policy options.

The analysis must take account of direct and indirect costs. The direct costs can be related directly to failure rate (or removal rate, which is likely to be higher).

The *direct costs* are:

1. Warranty repair costs.
   The annual warranty repair cost will be:

   (Number of warranted units in use) × (annual call rate per unit) × (cost per call)

   The number of warranted units will be obtained from market projections. The call rate will be related to MTBF and expected utilization.
2. Spares production and inventory costs for warranty support.
   Spares costs: to be determined by analysis (e.g. Poisson model, simulation) using inputs of call rate, proportion of calls requiring spares, spares costs, probability levels of having spares in stock, repair time to have spares back in stock, repair and stockholding costs.
3. Net of profits on post-warranty repairs and spares.
   Annual profit on post-warranty spares and repairs: analysis to be similar to warranty costs analysis, but related to post-warranty equipment utilization.

Indirect costs (not directly related to failure or removal rate):

1. Service organization (training manuals, overheads). (Warranty period contribution.)
2. Production reputation.

These costs cannot be derived directly. A service organization will be required in any case and its performance will affect the product's reputation. However, a part of its costs will be related to servicing the warranty. A parametric estimate should be made under these headings, for example:

324

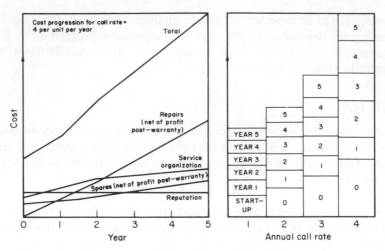

Figure 14.2 Reliability cost progression

Service organization: 50% of annual warranty cost in first 2 years, 25% thereafter.

Product reputation: agreed function of call rate.

Since these costs will accrue at different rates during the years following launch, they must all be evaluated for, say, the first 5 years. The unreliability costs progression should then be plotted (Fig. 14.2) to show the relationship between cost and reliability.

The net present values of unreliability cost should then be used as the basis for planning the expenditure on the reliability programme.

This situation would be worth analysing from the point of view of what support policy might show the lowest cost for varying call rates. For example, a very low call rate might make a 'direct exchange, no repair' policy cost-effective, or might make a longer warranty period worth considering, to enhance the product's reputation. Direct exchange would result in service department savings, but a higher spares cost.

Example 14.2

A cabin air-conditioning unit is to be developed for a military aircraft. The planned deployment is 400 aircraft, each utilizing one unit. The failure rate of a similar unit in a previous application was 2.4 per 1,000 flight hours, and the average flight time per aircraft per year is 350 h. It is decided to carry out an additional reliability test programme with the objective of forcing the MTBF to 1,000 h for the new unit. Assuming a Duane growth slope of 0.3, a cost per test hour of $80 and a starting MTBF of 417 h (1,000/2.4) after 1,500 h of development testing, comment on the cost-effectiveness of the reliability growth programme, taking into account the first 4 years of operation. The cost of a unit is $6,000 and the cost of replacing a unit is $900. Make the simplifying assumptions:

(a) The fleet starts at 400, i.e. there is no build-up period.
(b) Use constant money values.
(c) There will be no reliability growth in service.
(d) Consider only removals for failure.

The total cost of operating with a 417 h MTBF, over 4 years, would be:

$$\frac{400 \times 350 \times 4 \times 900}{417} = \$1,210,000$$

The total cost for a 1,000 h MTBF is $500,000. Therefore the LCC saving on repair costs would be about $700,000.

The instantaneous MTBF θ_i is equal to $\theta_c/(1 - \alpha)$, where θ_c is the cumulative MTBF on the Duane MTBF growth plot (Chapter 12) and α is the Duane slope. Therefore,

$$\theta_c = \theta_i (1 - \alpha)$$
$$= 1,000(1 - 0.3) = 700 \text{ h}$$

We must determine the number of test hours required to increase the cumulative MTBF from an initial value of 417 h to 700 h, at a slope of 0.3. With $T_0 = 1,500$ h and using the formula

$$\theta_c = \theta_0 \left(\frac{T}{T_0}\right)^\alpha$$

we obtain

$$T = 8,435 \text{ h}$$

At a cost of $80 per test hour, the test cost would be $670,000. Therefore on repair cost considerations alone there would be an approximate breakeven. However, we should also consider the value of reduced spares procurement.

If the aircraft are to be deployed on ten bases of 40 aircraft each, and we desired at least a 90 per cent. probability that a spare unit was available when needed, assuming Poisson arisings and a 1 month repair cycle, we can calculate the base spares requirements as

$$\text{Flying hours per month per base} = \frac{40 \times 350}{12}$$

$$= 1,167 \text{ h}$$

The expected monthly arising rate with a failure rate of 2.4 per 1,000 h is 2.8, and with a failure rate of 1 per 1,000 h is 1.17.

Referring to Fig. 2.8, the cumulative Poisson probability curves, five spare units are required per base for a 2.8 monthly arising rate, but only two per base for a 1.17 monthly arising rate. Therefore a total saving of $10(5 - 2) = 30$ units could be made in the spares buy. This represents a further saving of \$180,000.

The two examples given above involve very simple analysis and simplifying assumptions. In both cases a Monte Carlo simulation would be a more suitable approach if we needed to consider more complex dynamic effects, such as distributed repair times and costs, multiechelon repair and progressive increase in units at risk. However, simple analysis is often sufficient to indicate the magnitude of costs, and in many cases this is all that is needed, as the input variables in logistics analysis are usually somewhat imprecise, particularly failure (removal) rate values. Simple analysis is adequate if relatively gross decisions are required. However, if it is necessary to attempt to make more precise judgements or to perform sensitivity analyses, then more powerful methods such as simulation should be considered.

There are of course other costs which can be incurred as a result of a product's unreliability. Some of these are hard to quantify, such as goodwill and market share, though these can be very large in a competitive situation and where consumer organizations are quick to publicize failure. In extreme cases unreliability can lead to litigation, especially if damage or injury results. An unreliability cost often overlooked is that due to failures in production due to unreliable features of the design. A reliable product will usually be cheaper to manufacture, and the production quality cost monitoring system should be able to highlight those costs which are due to design shortfalls.

PRODUCT LIABILITY

Recent product liability legislation in the United States and in Europe adds a new dimension to the importance of eliminating safety-related failure modes, as well as to the total quality assurance approach in product development and manufacture. Before product liability (PL), the law relating to risks in using a product was based upon the principle of *caveat emptor* ('let the buyer beware'). PL introduced *caveat venditor* ('let the supplier beware'). PL was an outgrowth of the Ralph Nader campaigns in the United States, and it makes the manufacturer of a product liable for injury or death sustained as a result of failure of his product. A designer can now be held liable for a failure of his design, even if the product is old and the user did not operate or maintain it correctly. Claims for death or injury in many product liability cases can only be defended successfully if the producer can demonstrate that he has taken all practical steps towards identifying and eliminating the risk, and that the injury was entirely unrelated to failure or to inadequate design or manufacture. Since these risks may extend over ten years or even indefinitely, depending upon the law in the country concerned, long term reliability of safety-related features becomes a critical requirement. The size of the claims, liability being unlimited in the United States, necessitates top management involvement in minimizing these risks, by ensuring that the organization and resources are provided to manage and execute the quality and reliability tasks which will ensure reasonable protection. PL insurance is a new business area for the insurance companies, who naturally expect to see a suitable reliability and safety programme being operated by the manufacturers they insure.

RELIABILITY STANDARDS

Reliability programme standard requirements are issued by several large agencies which place development contracts on industry. The best known of these is US MIL-STD-785 — *Reliability Programs for Systems and Equipments, Development and Production* — which covers all development programmes for the US Department of Defense. Most reliability programme standards issued by other agencies in the United States and in Europe cover the same requirements as MIL-STD-785, and many of them go into more detail on specific requirements. MIL-STD-785 is supported by other military standards, handbooks and specifications, and these have been referenced in earlier chapters. It is necessary for personnel involved in a reliability programme, whether from the customer or supplier side, to be familiar with the appropriate standards. This book has referred mainly to US military documents, since in many cases they are the most highly developed and best known.

In the United Kingdom, Defence Standards 00–40 and 00–41 cover reliability programme management and methods for defence equipment. BS 5760 has also been prepared to provide a standard which can be referenced by any organization in development contracts. At the time of writing a NATO standard on reliability and maintainability is being prepared. Other national standards exist, and some large agencies such as NASA, major utilities and corporations issue their own reliability standards.

CUSTOMER MANAGEMENT OF RELIABILITY

When a product is being developed under a development contract, as is often the case with military and other public purchasing, the purchasing organization plays an important role in the reliability and quality programme. As has been shown, such organizations often produce standards for application to development contracts, covering topics such as reliability programme management, design analysis methods and test methods.

There is often a tendency for reliability (and quality, safety and other 'ilities') to become bureaucratized and inflexible in large purchasing organizations. It is important for such organizations to set up effective, responsive reliability management, with the main effort devoted to projects rather than to policies.

A reliability manager should be assigned to each project. The reliability manager might have other responsiblities, such as for production quality control. Project reliability management by a centralized reliability department, not responsible to the project manager, is just as likely to result in lower effectiveness, as is this approach in the contractor's organization. A central reliability department is necessary to provide general standards, training and advice, but should not be relied upon to manage reliability programmes across a range of projects. If there is a tendency for this to happen it is usually an indication that inadequate standards or training have been provided for project staff, and these problems should then be corrected.

The prime responsibilities of the purchaser in a development reliability programme are to:

1. Specify the reliability requirements (page 330).
2. Specify the standards and methods to be used.
3. Set up the financial and contractural framework (page 333).
4. Specify the reporting requirements.
5. Monitor contract performance.

The purchaser should not attempt to run the programme, since this would usurp the supplier's management responsibilities. Proper attention to the first three items above should ensure that the supplier is effectively directed and motivated, so that the purchaser has visibility of activities and progress without having to become too deeply involved.

It is usually necessary to negotiate aspects of the specification and contract. During the specification and negotiation phases it is usual for a central reliability organization to be involved, since it is important that uniform approaches are applied. Specification tailoring (page 343) is now a common feature of development contracting and this is an important aspect in the negotiation phase, requiring experience and knowledge of the situation of other contracts being operated or negotiated.

The supplier's reliability plan, prepared in response to the purchaser's requirement, should also be reviewed by the central organization, particularly for major contracts.

The contractor's reporting tasks are often specified in the statement of work (SOW). These usually include:

1. The reliability plan.
2. Design analysis reports and updates (prediction, FMECA, FTA, etc.).
3. Test reports.

Reporting should be limited to what would be useful for monitoring performance. For example, a 50-page FMECA report, tabulating every failure mode in a system, is unlikely to be useful to the purchaser. Therefore the statement of work should specify the content, format and size of reports. The detailed analyses leading to the reports should be available for specific queries or for audit.

The purchaser should observe the supplier's design reviews. Some large organizations assign staff to suppliers' premises, to monitor development and to advise on problems such as interpretation of specifications. This can be very useful on major projects such as aircraft, ships and plant, particularly if the assigned staff are subsequently involved in operation and maintenance of the system.

There are many purchasers of equipment who do not specify complete systems or let total development contracts. Also, many such purchasers do not have their own reliability standards. Nevertheless, they can usually influence the reliability and availability of equipment they buy. We will use an example to illustrate how a typical purchaser might do this.

Example 14.3

A medium-sized food-processing plant is being planned by a small group of entrepreneurs. Among other things, the plant will consist of:

1. Two large continuous-feed ovens, which are catalogue items but have some modifications added by the supplier, to the purchaser's specification. These are the most expensive items in the plant. There is only one potential supplier.
2. A conveyor feed system.
3. Several standard machines (flakers, packaging machines, etc.).
4. A process control system, operated by a central computer, for which both the hardware and software will be provided by a specialist supplier to the purchaser's specification.

The major installations except item 4 will be designed and fitted by a specialist contractor; the process control system integration will be handled by the purchasers. The plant must comply with the statutory safety standards, and the group is keen that both safety and plant availability are maximized. What should they do to ensure this?

The first step is to ensure that every supplier has, as far as can be ascertained, a good reputation for reliability and service. The purchasers should survey the range of equipment available, and if possible obtain information on reliability and service from other users. Equipment and supplier selection should be based to a large extent on these factors.

For the standard machines, the warranties provided should be studied. Since plant availability is important, the purchasers should attempt to negotiate service agreements which will guarantee up-time, e.g. for guaranteed repair or replacement within 24 hours. If this is not practicable, they should consider, in conjuction with the supplier, what spares they should hold.

Since the ovens are critical items and are being modified, the purchasers should ensure that the supplier's normal warranty applies, and service support should be guaranteed as for the standard items. They should consider negotiating an extended warranty for these items.

The process control system, being a totally new development (except for the computer), should be very carefully specified, with particular attention given to reliability, safety and maintainability, as described below. Key features of the specification and contract should be:

1. Definition of safety-critical failure effects.
2. Definition of operational failure effects.
3. Validation of correct operation when installed.
4. Guaranteed support for hardware and software, covering all repairs and corrections found to be necessary.
5. Clear, comprehensive documentation (test, operating and maintenance instructions, program listings, program notes).

For this development work, the purchasers should consider invoking appropriate standards in the contract, such as BS 5760. For example, FMECA and FTA could be very valuable for this system, and the software development should be properly controlled and documented. The supplier should be required to show how those aspects of the specification and contract will be addressed, to ensure that the requirements are fully understood. A suitable consultant engineer might be employed to specify and manage this effort.

The installation contract should also cover reliability, safety and maintainability, and service.

During commissioning, all operating modes should be tested. Safety aspects should be particularly covered, by simulating as far as possible all safety-critical failure modes.

The purchasers should formulate a maintenance plan, based upon the guidelines given in Chapter 9. A consultant engineer might be employed for this work also.

Finally, the purchasers should insure themselves against the risks. They should use the record of careful risk control during development to negotiate favourable terms with their insurers.

SPECIFYING RELIABILITY

Designs are usually based upon specifications, which define the problems the designer must overcome. Where the specification is deficient, it is possible that the unspecified parameter or feature will be traded against when compromises are needed. This is accepted without question for features which are deterministic, such as weight, size, and input and output parameter values.

However, it is more difficult to specify non-deterministic features, such as appearance or reliability. It is relatively easy to assess a design for conformance with deterministic features, or even for appearance. The inherent reliability of a design is not as easily apparent, since it depends upon external factors such as load cycles and the environment in which it will be used.

Deterministic performance characteristics can be accurately tested and assessed with 'breadboard' or prototype models. Reliability cannot be assessed adequately from such tests. If a prototype model passes specified performance tests, and reliability has not been specified, it is likely that production will be authorized, without any reliability testing being performed. The effects of this deficiency are apparent in many domestic, commercial and military products, which enter production with irritating design deficiencies which should have been eliminated by even a cursory design review or a limited reliability test programme.

In order to ensure that reliability does not suffer in this way, the requirement must be specified. Before describing how to specify reliability adequately, we will cover some of the ways *not* to do it:

1. Do not write vague requirements, such as 'as reliable as possible', 'high reliability is to be a feature of the design', or 'the target reliability is to be 99 per cent'. Such statements do not provide assurance against reliability being compromised.
2. Do not write unrealistic requirements. 'Will not fail under the specified operating conditions' is a realistic requirement in many cases. However, an unrealistically high reliability requirement for, say, a complex electronic equipment will not be accepted as a credible design parameter, and is likely therefore to be ignored.

The reliability specification *must* contain:

1. A definition of failure related to the product's function. The definition should cover all failure modes relevant to the function.
2. A full description of the environments in which the product will be stored, transported, operated and maintained.
3. A statement of the reliability requirement, and/or a statement of failure modes and effects which are particularly critical and which must therefore have a very low (or zero) probability of occurrence. The reliability specification may be a separate document, or it may be covered in the range of other specifications (design, test, maintenance, etc.). Reference 3 in the Bibliography covers the preparation of reliability specifications in detail.

Definition of failure

Care must be taken in defining failure to ensure that the failure criteria are unambiguous. Failure should always be related to a measurable parameter or to a clear indication. A seized bearing indicates itself clearly, but a leaking seal might or might not constitute a failure, depending upon the leak rate, or whether or not the leak can be rectified by a simple adjustment. An electronic equipment may have modes of failure which do not affect function in normal operation, but which may do so under other conditions. For example, the failure of a diode used to block transient voltage spikes may not be apparent during functional test, and will probably not affect normal function. Defects such as changes in appearance or minor degradation that do not affect function are not usually relevant to reliability. However, sometimes a perceived degradation is an indication that failure will occur and therefore such incidents can be classified as failures.

Inevitably there will be subjective variations in assessing failure, particularly when data are not obtained from controlled tests. For example, failure data from repairs carried out under warranty might differ from data on the same equipment after the end of the warranty period, and both will differ from data derived from a controlled reliability demonstration. The failure criteria in reliability specifications cannot eliminate such variations, but clear unambiguous definitions can go a long way to reducing the uncertainty of relating failure data to the specification and in helping the designer to understand the reliability requirement.

Environmental specifications

The environmental specification must cover all aspects of the many loads and other effects that can influence the product's strength or probability of failure. Without a clear definition of the conditions which the product will face, the designer will not be briefed on what he is designing against. Of course, aspects of the environmental specification might sometimes be taken for granted and the designer might be expected to cater for these conditions without an explicit instruction. It is generally preferable, though, to prepare a complete environmental specification for a new product, since the discipline of considering and analysing the likely usage conditions is a worthwhile exercise if it focuses attention on any aspect which might otherwise be overlooked in the design. For most design groups only a limited number of standard environmental specifications is necessary. For example, the environmental requirements and methods of test for military equipment are covered in specifications such as US MIL-STD-810 and UK Defence Standard 07-55.

The environments to be covered must include handling, transport, storage, normal use, foreseeable misuse, maintenance and any special conditions. For example, the type of test equipment likely to used, the skill level of users and test technicians, and the conditions under which testing might be performed should be stated if these factors might affect the observed reliability.

Stating the reliability requirement

The reliability requirement should be stated in a way which can be verified, and which makes sense relative to the use of the product. The simplest requirement to state is that no failure will occur under the stated conditions. The requirement should not include statements on the s-confidence levels of the measured reliability. The requirement relates to the population; s-confidence levels apply to the results of tests or other limited data. s-Confidence limits may be used for pass/fail decision-making and test-planning, but they should not be included in the requirement. The designer is concerned with reliability, but not with planning the reliability tests.

Levels of reliability less than unity can be stated as a success ratio, or as a life. For 'one-shot' items the success ratio is the only relevant criterion. For example, a missile solid propellant gas generator might be required to have a success ratio of 99.5 per cent. after a stated period in specified storage conditions. The definition of success in terms of pressure rise, burning time and pressure variations in a specified environment would have to be stated. This specification could then form the basis for batch acceptance testing, with the test plan selected to provide the required s-confidence.

Reliability specifications based on life parameters must be framed in relation to the appropriate life distributions. Two common parameters used are MTBF, when a constant failure rate is assumed, and B-life, related to Weibull life distributions. MTBFs should not be specified if a constant failure rate assumption cannot be justified. This assumption can usually be made for complex, repairable systems (see Chapter 5), e.g. complex domestic equipment and military electronic systems. Otherwise a B-life (e.g. B_{10}, the life by which not more than 10 per cent. of the population will be expected to have failed) should be specified.

Specified life parameters must clearly state the life characteristic. For example, the life of a switch, a sequence valve or a radio-cassette recorder cannot be usefully stated merely as a number of hours. The life must be related to the duty cycle (in these cases switch reversals and frequency, sequencing operations and frequency, and anticipated operating cycles on receive, record, playback and switch on/off). The life parameter may be stated as some time-dependent function, e.g. miles travelled, switching cycles, load reversals, or it may be stated as a time, with a stipulated operating cycle. US MIL-STD-781 is often used as a basis for specifying time-dependent reliability tests for constant failure rate items. MIL-STD-781 tests are covered in detail in Chapter 12.

CONTRACTING FOR RELIABILITY ACHIEVEMENT

Users of equipment which can have high unreliability costs have for some time imposed contractual conditions relative to reliability. Of course, every product warranty is a type of reliability contract. However, contracts which stipulate specific incentives or penalties related to reliability achievement have been developed, mainly by the military, but also by other major equipment users such as airlines and public utilities.

The most common form reliability contract is one which ties an incentive or penalty to a reliability demonstration. The demonstration may either be a formal test (e.g. MIL-STD-781) or may be based upon the user's experience. In either case, careful definition of what constitutes a relevant failure is necessary, and a procedure for failure classification must be agreed. If the contract is based only on incentive payments, it can be agreed that the customer will classify failures and determine the award, since no penalty is

Figure 14.3 Reliability incentive structure

involved. A well-known form of reliability incentive contract is that used for spacecraft, whereby the customer pays an incentive fee for successful operation for up to, say, 2 years. Straight incentive payments have advantages over incentive/ penalty arrangements. It is important to create a positive motivation, rather than a framework which can result in argument or litigation, and incentives are preferable in this respect. Also, an incentive is easier to negotiate, as it is likely to be accepted as offered. Incentive payments can be structured so that, whilst they represent a relatively small percentage of the customer's saving due to increased reliability, they provide a substantial increase in profit to the supplier. The receipt of an incentive fee has significant indirect advantages, as a morale booster and as a point worth quoting in future bid situations. A typical award fee structure is shown in Fig. 14.3.

When planning incentive contracts it is necessary to ensure that other performance aspects are sufficiently well specified and, if appropriate, also covered by financial provisions such as incentives or guarantees, so that the supplier is not motivated to aim for the reliability incentive at the expense of other features. Incentive contracting requires careful planning so that the supplier's motivation is aligned with the customer's requirements. The parameter values selected must provide a realistic challenge and the fee must be high enough to make extra effort worth while.

Reliability improvement warranty

A type of reliability incentive contract which has recently attracted a lot of attention is reliability improvement warranty (RIW). It has been employed by the US Department of Defense and others, such as airlines. An RIW contract requires that the supplier carries out all repairs and provides all spares needed, for a fixed period, typically 4 years, for a once-off fee. The supplier's motivation is then to maximize his profit by ensuring that the repair arising rate is minimized. The customer benefits by not having to be involved in monitoring the reliability programme and not having to administer the supply and repair work. Guidelines have been developed for RIW contracting, based upon experience. The main points which need to be considered are:

1. RIW contracts should only be applied to products where there is not a high development risk, and where the utilization will be reasonably stable and centralized, so that both parties to the contract can agree on the likely reliability achievement. The product should be manufactured and used in quantity, to ensure even work flow.
2. The contract fee must provide a good chance for the supplier to make a high profit, and yet contain a reasonable risk element. There is no advantage to either side in an RIW contract which bankrupts the supplier.
3. Real difficulties can arise in administering RIW contracts in conjunction with other, conventional, repair policies. For example, hardware subject to RIW needs special handling and marking, to ensure that units are not opened or repaired by anyone other than the RIW contractor. Personnel involved must be trained, and procedures must be written to cover the operation of the RIW contract. For example, arrangements have to be made for dealing with units which are reported defective, but which are shown to have no failure on investigation.
4. The contractor must be given freedom to modify the equipment to improve its reliability (which he would do at his own expense). However, the customer may wish to have some control over modifications, as they may affect interchangeability or performance. The contract must define the arrangements for notification, approval and incorporation of changes.
5. The contract should stipulate the way in which the equipment will be used and maintained, if there is possibility that these factors can affect reliability.
6. The equipment should be put into service immediately when delivered, and failures must be reported promptly.

RIW contracts sometimes include an MTBF option, which provides for a further incentive payment related to reliability achievement.

RIW contracts can be very rewarding to both parties, but there are pitfalls to be avoided. Careful planning and management, coupled with goodwill and willingness to compromise, are essential to their success.

ORGANIZATION FOR RELIABILITY

Because several different activities contribute to the reliability of a product it is difficult to be categorical about the optimum organization to ensure effective management of reliability. Reliability is affected by design, development, production quality control, control of subcontractors and maintenance. These activities need to be coordinated, and the resources applied to them must be related to the requirements of the product. The requirements may be determined by a market assessment, by warranty cost considerations or by the customer. The amount of customer involvement in the reliability effort varies. The military and other public organizations often stipulate the activities required in detail, and demand access to design data, test records and other information, particularly when the procurement agency funds the development. At the other extreme, domestic customers are not involved in any way directly with the development and production programme. Different activities will have greater or lesser importance depending on whether the product involves innovative or complex design, or is simple and based upon considerable experience. The reliability effort also varies as the project moves through the development, production and in-use phases, so that the design department will be very much involved to begin with, but later the emphasis will shift to the production, quality control and maintenance functions. However, the design must take account of production, test and maintenance, so these downstream activities must be considered by the specification writers and designers.

The reliability organization should be responsive to all of the factors described above, and should control the functions that contribute to the product's reliability. To be effective it must be executive, not merely advisory. On the other hand, it must not usurp the specialist design and production functions. The reliability team does not design, produce or maintain. However, it should be able to give or withhold approval, and to decide how the product should be tested to assure reliability.

Reliability management must be integrated with other project management functions, to ensure that reliability is given the appropriate attention and resources in relation to all the other project requirements and constraints.

Two main forms of reliability organization have evolved. These are described below.

Quality assurance based organization

The quality assurance (QA) based organization places responsibility for reliability with QA management, which then controls the 'quality' of design, maintenance, etc., as well as of production. This organizational form is based upon the definition of *quality* as *the totality of features which bear on a product's ability to satisfy the requirement*. This is the formal European (including UK) definition of quality. Consequently, in Europe the QA department or project QA manager is often responsible for all aspects of product reliability. Figure 14.4 shows a typical organization. The reliability engineering team interfaces mainly with the engineering departments, while quality control is mainly concerned with production. However, there is close coordination of reliability engineering and quality control, and shared functions: e.g. a common failure data collection and analysis system can be operated, covering development, production and in-use. The QA department then provides the feedback loop from in-use experience to future design and production.

The QA manager is responsible for a much wider range of engineering

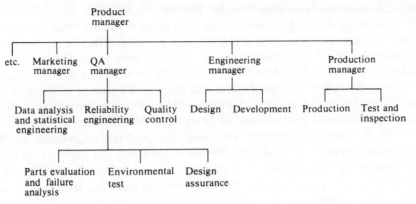

Figure 14.4 QA based reliability organization

activity than was implied in the traditional test and inspection approach to QA. The reliability engineering team demands the higher talents in statistical engineering, design analysis, stress test planning and failure investigation, so the reliability manager should be the second in line to the QA manager. The sections reporting to the reliability manager perform the functions described in earlier chapters.

Engineering based organization

In the engineering based organization, reliability is made the responsibility of the engineering manager. The QA (or quality control) manager is responsible only for controlling production quality and may report direct to the product manager or to the production manager. Figure 14.5 shows a typical organization. This type of organization is more common in the United States.

Comparison of types of organization

The QA based organization for reliability allows easier integration of some tasks that are common to design, development and production. The ability to

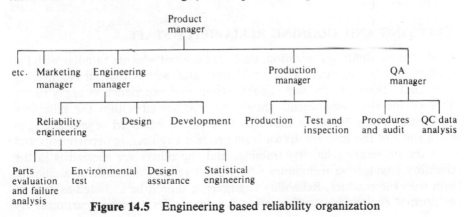

Figure 14.5 Engineering based reliability organization

operate a common failure data system has been mentioned. In addition, the statistical methods used to analyse development test data and production failure data are the same, as is much of the test equipment and test methods. For example, the environmental test equipment used to perform reliability qualification tests in development will probably be the same as that used for production reliability acceptance tests and burn-in. Engineers with experience or qualifications in QA are often familiar with reliability engineering methods, as their professional associations on both sides of the Atlantic include reliability in their areas of interest. However, for products or systems where a considerable amount of innovative design is required, the engineering based organization has advantages, since more of the reliability effort will have to be directed towards design assurance, such as stress analysis, design review and development testing.

The main question to be addressed in deciding upon which type of organization should be adopted is whether the split of responsibility for reliability activities inherent in the engineering based organization can be justified in relation to the amount of reliability effort considered necessary during design and development. In fact the type of organization adopted is much less important than the need to ensure integrated management of the reliability programme. So long as the engineers performing reliability activities are properly absorbed into the project team, and report to the same manager as the engineers responsible for other perfomance aspects, functional departmental loyalties are of secondary importance. To ensure an integrated team approach reliability engineers should be attached to and work alongside the design and other staff directly involved with the project. These engineers should have access to the departmental supporting services, such as data analysis and component evaluation, but their prime responsibility should be to the project. Since reliability is so much more dependent upon variation than most other performance features, reliability effort is more likely to be considered inessential and abstruse if it is not integrated into the project. The effort will then be less effective, and will lose credibility in the eyes of the 'sharp-end' project staff.

SELECTING AND TRAINING RELIABILITY STAFF

Within the reliability organization, staff are required who are familiar with the product (its design, manufacture and test) and with statistical engineering techniques. Therefore the same qualifications and experience as apply to the other engineering departments should be represented within the reliability organization. The objective shoud be to create a balanced organization, in which some of the staff are drawn from product engineering departments and given the necessary reliability training, and the others are specialists in the reliability engineering techniques who should receive training to familiarize them with the product. Reliability engineering should be included as part of the normal engineering staff rotation for career development purposes. By having a balanced department, and engineers in other departments with experience of reliability engineering, the reliability effort will have credibility and will make the most effective contribution.

Reliability engineers need not necessarily be specialists in particular disciplines, such as electronic circuit design or metallurgy. Rather, a more widely based experience and sufficient knowledge to understand the specialists' problems is appropriate. The reliability engineer's task is not to solve design or production problems but to demonstrate the reasons for failure and the margins of safety. He must, therefore, be a communicator, competent to participate with the engineering specialists in the team and able to demonstrate the value and relevance of the reliability methods applied. Failure analysis may occasionally demand highly specialized knowledge, but in such cases the specialist would be contributing to the reliability effort. Of course high quality work from the specialist engineering groups makes the reliability task easier.

Since reliability and quality control have much in common, quality control work often provides suitable experience from which to draw, provided that the QC experience has been deeper than the traditional test and inspection approach, with no design or development involvement. For those in the reliability organization providing data analysis and statistical engineering support, specialist reliability training is relatively more important than product familiarity.

The qualities required of the reliability engineering staff obviously are equally relevant for the head of the reliability function. Since reliability engineering should involve interfaces with several other functions, including such non-engineering areas as marketing and forecasting, this position should not be viewed as the end of the line for moderately competent engineers, but rather as one in which potential top management staff can develop general talents and further insight into the overall business, as well as providing further reliability awareness at higher levels in due course.

Since reliability engineering is often suited to general as opposed to specialist engineers, it has often been made the final resting place for 'failed' engineers. Also, because there is a strong statistical element, statisticians have been given reliability engineering titles and responsibilities. A combination of second-rate engineers, known as such in the other departments, and mathematical statisticians masquerading as engineers, is a recipe for an ineffective reliability organization which will lack credibility. Since reliability is a relatively new branch of engineering there is often a shortage of suitable candidates, but it is usually wiser to operate undermanned than to recruit unsuitable staff, since the reputation of the reliability organization within the firm is essential to its success. It may be that an engineer who does not excel in a specialist area can, with the right training in reliability methods, make a useful contribution, but the reliability organization must not be seen to attract more of this type of staff than do other departments. Any statisticians within the organization must be seen to provide a specialist supporting function, and no more. In fact only the largest reliability or quality assurance departments should be able to justify full-time employment of a statistician, and then only if he or she is familiar with the analytical methods used in the work.

Reliability training

Some colleges and universities include reliability engineering courses, often combined with quality assurance, in their engineering curricula. However, the subject is not covered in many engineering courses, and when it is included it is normally provided as an optional subject. Most engineers have little or no formal training in reliability methods, or even in the statistical approach to engineering and science. This fact occasionally makes it difficult for design and development engineers and engineering management to appreciate the need for distinct reliability effort. Therefore it is often useful to provide training in the basic reliability concepts and the underlying cost justification for management and for engineering staff, to generate awareness and appreciation. More detailed courses should be provided for engineers taking up reliability responsibilities.

Reliability training should be tailored to the needs of the product and of the business situation. For example, a company manufacturing domestic electrical equipment would need to train its staff in the reliability management aspects of product liability and warranty costs, and a defence equipment supplier would need to ensure that the staff were familiar with military standards and procurement policies. For this reason many large corporations and organizations operate internal reliability training courses, tailored to their needs. Where this cannot be justified, training can be provided by sending staff to some of the many short courses run by universities, colleges and other organizations such as the Reliability Analysis Center (USA) or the National Centre for Systems Reliability (UK). Postgraduate full-time courses also exist, providing deeper and wider coverage to Masters degree standard. The major national professional institutions for quality assurance personnel also include reliability management and engineering in their training and qualification schemes.

USE OF EXTERNAL SERVICES

The retention of staff and facilities for analysis and test and the maintenance of procedures and training can only be cost-effective when the products involve fairly intensive and continuous development. In advanced product areas such as defence and aerospace, electronic instrumentation, control and communcations, vehicles, and for large manufacturers of less advanced products such as domestic equipment and less complex industrial equipment, a dedicated reliability engineering organization is necessary, even if it is not a contractual requirement. Smaller companies with less involvement in risk-type development may have as great a need for reliability engineering expertise, but not on a continuous basis. External reliability engineering services can fulfil the requirements of smaller companies by providing the specialist support and facilities when needed. Reliability engineering consultants and specialist test establishments can often be useful to larger companies also, in support of internal staff and facilities. Since they are engaged full time across a number of different types of project they should be considered whenever new problems arise.

A summary of organizations which could be consulted is given in Appendix 8.

THE RELIABILITY MANUAL

Just as most medium to large design and development organizations have internal manuals covering design practices, organizational structure, QA procedures, etc., so reliability management and engineering should be covered. Depending on the type of product and the organization adopted, reliability can be adequately covered by appropriate sections in the engineering and QA manuals, or a separate reliability manual may be necessary. In-house reliability procedures should not attempt to teach basic principles in detail, but rather should refer to appropriate standards and literature, which should of course then be made available. Since most reliability programme activities, particularly as required for defence contracting, are described in military standards and handbooks as well as in other national and industry standards, these should always be referred to and followed when appropriate. The bibliographies at the end of each chapter of this book list the major references. The in-house documents should cover, as a minimum, the following subjects:

1. Corporate policy for reliability.
2. Organization for reliability.
3. Reliability procedures in design (e.g. design analysis, parts derating policy, parts, materials and process selection, approval and review, critical items listing, design review).
4. Reliability test procedures.
5. Reliability data collection, analysis and action system, including data from test, warranty, etc. (See Appendix 9).

Table 14.1 Reliability manual: responsibilities.

		Department responsible		
Task	Reference	Prime	Resources	Support
Stress analysis (electronic)	MIL-HDBK-217	Project design	Reliability	Reliability
Reliability test	MIL-STD-781D	Reliability	Environmental test	Project design
Reliability data	MIL-STD-781D	Reliability	Reliability	QA

The written procedures must state, in every case, who carries responsibility for action and who is responsible for providing the resources and capability. They must also state who provides supporting services. A section from the reliability manual may appear as shown in Table 14.1.

THE PROJECT RELIABILITY PLAN

It is normal for customer-funded development projects to include a requirement for a reliability plan to be produced. MIL-STD-785 and Defence Standard 00-40, for example, stipulate this, and define the contents of the plans.

The reliability plan should include:

1. A brief statement of the reliability requirement.
2. The supplier's organization for reliability.
3. The reliability activities that will be performed (design analysis, test, reports).
4. The timing of all major activities, in relation to the project development milestones.
5. Reliability management of subcontractors.
6. The standards, specifications and internal procedures (e.g. the reliability manual) which will be used, as well as cross-references to other plans such as for safety, maintainability and quality assurance.

When a reliability plan is submitted as part of a response to a customer request for proposals (RFP) in a competitive bid situation, it is important that the plan reflects complete awareness and understanding of the requirements and competence in compliance. A response to a competitive RFP should never query the requirements, since the responses will be judged by the people who prepared the RFP. If it is considered sensible to propose any alternative approaches, these should be included in a separate annex to the main proposal, and separately costed.

RFP responses should not include references to internal procedures, but should be complete in themselves. They should be brief, and should not repeat material in the RFP.

A reliability plan prepared as part of a project development (e.g. in response to a statement of work), after a contract has been accepted, is more comprehensive than an RFP response, since it gives more detail of activities, timescales and reporting. The project reliability plan may include references to internal procedures, if these are made available to the customer. Since the project reliability plan usually forms part of the contract once accepted by the customer, it is important that every aspect is covered clearly and explicitly.

A well-prepared reliability plan is useful for instilling confidence in the supplier's competence to undertake the tasks, and for providing a sound reliability management plan for the project to follow.

Specification tailoring

Specification tailoring is a term used to describe the process of suggesting alternatives to the customer's specification. 'Tailoring' is often invited in RFPs and in development contracts. A typical example occurs when a customer specifies a system and requires a formal reliability demonstration. If a potential supplier can supply a system for which adequate in-service reliability records exist, the specification could be tailored by proposing that these data be used in place of the reliability demonstration. This could save the customer the considerable expense of the demonstration. Other examples might arise out of trade-off studies, which might show, for instance, that a reduced performance parameter could lead to cost savings or reliability improvement.

MANAGING PRODUCTION QUALITY CONTROL

The production department should have ultimate responsibility for the manufacturing quality of the product. It is often said that quality cannot be 'inspected in' or 'tested in' to a product. The QC department is responsible for assessing the quality of production but not for operations which determine quality. QC thus has the same relationship to production as reliability engineering has to design.

In much modern production, inspection and test operations have become integrated with production operations. For example, in operator control of NC machine tools the machine operator might carry out workpiece gauging and machine calibration. Also, the costs of inspection and test can be considered to be production costs, particularly when it is not practicable to separate the functions or when, as in electronics production, the test policy can have a great impact on production costs. For these reasons, there is a trend towards routine inspection and test work being made the responsibility of production, with QC providing support services such as patrol and sample inspection, and possibly final inspection, as well as training, calibration, etc. Determination of inspection and test policy methods and staffing should then be primarily a production responsibility, with QC in a supporting role, providing advice and ensuring that quality standards will be met.

This approach to modern QC results in much smaller QC departments than under the older system whereby production produced and passed the products to QC for inspection and test at each stage. It also obviously reduces the total cost of production (production cost plus inspection and test costs). Motivation for quality is enhanced and QC staff are better placed to contribute positively, rather than acting primarily in a policing role. The quality circles movement has also heavily influenced this trend; quality circles could not operate effectively under the old approach.

The QC department is responsible for:

1. Setting production quality standards.

2. Monitoring production quality performance and costs.
3. Quality audit of production.
4. QC training.
5. Specialist facilities and services.

These will be discussed in turn.

Setting production quality standards

The QC manager must decide the production quality standards to be met. These might have been set by the customer, as is increasingly the case in commercial as well as in defence equipment manufacture, in which case the QC manager is the interface with the customer on production quality matters, leaving production people to concentrate on producing. Quality standards apply to the finished product, to production processes and to bought-in materials and components. Therefore the QC manager should decide, or approve, the final inspection and test to ensure conformance. He or she should also determine such details as quality levels of components (e.g. screening requirements for semiconductor devices), quality control of suppliers and calibration requirements for test and measuring equipment.

Monitoring production quality performance and costs

The QC manager must be satisfied that the quality objectives are being attained or that action is being taken to ensure this. These include quality cost objectives, as described earlier. Therefore the resources must be available to perform this task, as far as practicable independently of the production organization. QC staff should therefore oversee and monitor functions such as defect reporting and final conformance inspection and test. The QC department should prepare or approve quality performance and cost reports, and should monitor and assist with problem-solving. The Pareto analysis method described in Chapter 13 is particularly appropriate for this task.

Quality audit

Quality audit is an independent appraisal of all of the operations, processes and management activities that can affect the quality of a product. The objective is to ensure that procedures are effective, that they are understood and that they are being followed. The idea of independent quality audit originated in the United States, with the objective of reducing the level of government inspection at factories. Instead of products being manufactured, then being submitted for a final customer inspection, with customer inspection of intermediate processes, the control of quality was passed to the manufacturer, with the government retaining the authority to audit the effectiveness of the suppliers' QC system. MIL-STD-9858 describes the policy, and all military contracts for equipment manufacture call up this standard. Quality audit has become the common approach for most large defence and non-defence customers. In the United Kingdom, Defence Standard 05-21 is the controlling document, and AQAP-1 (Allied Quality Assurance Procedure) describes the policy for NATO contracts. Similar non-defence standards exist, e.g. BS 5750.

Quality audit, like financial audit, requires both internal and external audit. Internal audit is a continuing function whereby independent QA staff review the operations and controls, and report on discrepancies. External audit is imposed by the customer, on a regular schedule, typically annually or bi-annually. Companies which pass the audit are 'approved suppliers', a status which can confer advantages in selling to other customers, particularly in the defence equipment field. To a large extent approval to one national quality audit standard is accepted by customers from other countries.

Quality audit includes review of all production, test and inspection operations, as well as associated procedures and documentation. An important aspect is the assurance that personnel know and understand their role in the quality system, including relevant procedures and responsibilities. Areas covered by quality audit include:

Organization
Procedures
Problem-reporting and corrective action
Calibration of measuring and test equipment
Quality cost monitoring
Material and process traceability

Reliability aspects can be effectively included in quality audit, since there is so much common ground, particularly in relation to failure reporting and corrective action, and this is done in some quality audit systems.

An important feature of quality audit is that equipment manufacturers in turn audit their suppliers. In practice this has led to companies which supply many customers being overwhelmed by external auditors, and subcontractor audit can be a controversial aspect of the system.

The quality manager's responsibility for audit includes all internal audit, for ensuring that the company is successful in customer audits and for quality auditing of suppliers. Preparation for external audit and being subjected to it can be a very important task, and much effort is involved. The quality department should be skilled in undertaking this responsibility with minimum disruption to normal design and production work. This demands thorough knowledge of the appropriate standards and the ways in which they are applied. Training is an important feature to ensure that personnel will respond correctly during audit.

Quality training

The quality manager is responsible for all quality training. This is particularly important in training for operator control and quality circles, since all production people must understand and apply basic quality concepts such as simple SQC and data analysis.

Specialist facilities and services

The QC department provides facilities such as calibration services and records, component and material assessment, and defect data collection and analysis.

The assessment facilities used for testing components and materials so that their use can be approved are also the best services to use for failure investigation, since this makes the optimum use of expensive resources such as spectroscopic analysis equipment and scanning electron microscopes, and the associated specialist staff.

The defect data collection and analysis service must cover all failures detected during production test and inspection.

TOTAL QUALITY ASSURANCE

The terms 'total quality assurance' and 'total quality control' are often used to describe a system whereby all the activities that contribute to product quality, not just production quality control, are appraised and controlled by the QA manager. In this context quality is defined as the totality of features which determine a product's acceptability, and as such includes appearance, performance, reliability, support, etc.

Under this concept the QA manager has very wide authority for setting and monitoring quality standards, in this wide sense, throughout all functions of the organization. The QA manager then reports directly to the chief executive. It remains essential for line functions such as design, test and production to retain responsibility for their contributions to quality and reliability. However, the QA manager is responsible for ensuring that the total approach is co-ordinated, through the setting of standards, training and performance monitoring.

The total QA approach to reliability and quality can be very effective, particularly when applied to correct a situation in which quality is perceived as being lower than is required, but the reasons cannot be pinned down.

However, there are problems in the total QA approach. It is not easy to find people who can effectively fulfil the total QA management role. The task demands rather exceptional talents of persuasiveness and ability. It is easy for the QA manager and the organization to become dissociated from corporate realities, and the authority of the QA manager might be questioned by line departments and project managers.

The obvious solution to this is for the chief executive to undertake the responsibility. This has the supreme advantage of showing that quality and reliability are of top level concern. Functions such as design reliability and production quality control can then be integrated with design and production, and coordination of standards, training, etc., can be achieved through a chief executive's QA committee.

Only the chief executive can ensure total integration of the quality and reliability functions with the management of specifying, designing, producing and supporting the product. The increasing integration of design and production, and the pace and competitiveness of modern markets for technology-based products, demand that a fully integrated approach be used. This is to be found in many of the modern high technology companies that have grown up in the last 20 years, and in those older companies that have perceived quality and reliability as being matters too important to be left to chance or to lower levels of management. Their success has been largely due to this recognition, and to the commitment and involvement of the most senior executives.

CONCLUSIONS: GREED, FEAR AND FREEDOM

The last 20 years or so have been marked by a revolution in reliability and quality which has quietly affected the lives of all of us. Most people take the effects for granted and are unaware of the forces involved and the methods used to provide the reliability we now so often take for granted. Of course, people are hardly ever satisfied with less than perfect reliability, and we all experience failures of products. However, a few facts should set the perspective:

1. Modern electronic equipment, despite the incredible complexity of individual parts, hardly ever fails. Twenty years ago equivalent systems built with thermionic valves or discrete electronic transistors, if feasible at all, would have been extremely unreliable.
2. Automobile manufacturers used to grow fat on the sale of spare parts for their unreliable products. Now cars are much more reliable, and reliability is a selling point in advertising.
3. Spacecraft travel to the outer planets, functioning for years with no maintenance.
4. Nuclear power stations are so safe that no one has ever been killed by a radiation leak, as far as is known.

So much for the facts. What have been the driving forces behind such performance?

There have been two forces — one to show what could be done, the other a new awareness that people want reliability and can design and produce reliable products if properly motivated. There has been a change in the balance between greed and fear in relation to reliability.

The American space programme has provided the most dramatic demonstration of what could be achieved. Here the balance between greed and fear was very heavily weighted on the side of fear. If a mission failed, particularly if there was loss of life, the future funding for NASA projects could have been drastically curtailed. Therefore everything had to work. NASA developed new techniques for reliability analysis, testing of new designs and quality control that ensured this success. We know of the two accidents, one which cost the lives of three astronauts in the capsule fire and the other the Apollo 13 failure which nearly ended in disaster; but consider the magnitude of the success in reliability terms. For example, in all the Apollo launches, over 200 rocket engines worked with only 1 failure, and these were new, complex, highly stressed systems unproven in previous applications.

The awareness that ordinary people want reliable products, and that reliability is a saleable property, has been most highly developed in Japan. The Japanese have the knack, apparently not prevalent in other countries, of deciding at a high level what needs to be done and then executing the decision right down to where the action is. They decided on quality and reliability as national priorities, and we can all see the results.

The awareness of a need for high reliability (the fear side of the balance) has also been a powerful motivator in the other fields mentioned. The new electronics gave us the opportunity to manufacture far more complex systems for everyday use, but these had to be reliable. Competition, reputation and the costs of warranties have made manufacturers aware as never before. Once national pride ('British is Best') provided motivation for quality, but sadly we do not see much of that now, and anyhow it is not enough for the new technology.

Public safety consciousness has been a powerful spur to quality and reliability in some areas, such as nuclear power and air travel, where government agencies set up and monitor reliability and safety standards. Product liability legislation is a new factor which is striking fear, if not dread, into manufacturers' hearts.

Competition has provided a sharp awareness of the need for more reliable products in many consumer and industrial areas. Again, the Japanese have generally led the way, though some Western manufacturers have also been in the forefront. Now we are all trying to catch up, in quality, reliability and value, and the consumer benefits, as of course do the leading suppliers.

There has also been a drive for high reliability in military products, and in fact most of the formal methods for design reliability improvement started in the military area in the United States, so that the US military standards on reliability analysis and methods are now used in many non-military areas. The military interest in reliability stemmed from the need to reduce the enormous cost of maintaining very complex systems. Whilst the commercial world adopted many of the techniques developed by the military, there is now a swing the other way, with military buyers insisting on commercial-type warranties.

The awareness that people can produce very reliable products has evolved from the debunking of the theories of scientific management. Scientific management was born in the 1930s, and taught that people will produce economically if given good conditions and if work was carefully planned. People did produce more economically, but were separated from involvement in the product. Quality of work was a separate 'scientific' function of measurement and control. Scientific management could result in higher quality, but a worker was not supposed to think, take decisions or solve problems. Scientific management led to the production line, and it dehumanized labour by forgetting that people must think and solve problems to obtain fulfillment at work.

Peter Drucker buried scientific management with publication of his book *The Practice of Management,* published in 1955. Despite this, the theories of the 1930s are still taught in Western countries. Production Drucker-style involves making the worker assess and solve problems. The solution to problems is usually known at the point where the problems originate. Scientific management made it the job of management to find solutions. Drucker makes the worker interested in finding and either implementing solutions or in telling management how to solve them. This is, of course, only a small part of Drucker's teaching, but it is particularly appropriate to our subject. For example, the quality circles movement is obviously Drucker-inspired.

Quality and freedom

It is notable that no undemocratic state has been able to make any significant contribution to the reliability and quality revolution. Quality represents the essence of freedom — freedom to make decisions at work and as a consumer. Centralized bureaucratic state systems do not allow this freedom. They are stuck with scientific management, if they can even get that far. The products needed today must be complex to be competitive, whether they are toys, domestic equipment or weaponry. Modern technology has given us the ability to mass-produce complex products at low cost, but complexity is the enemy of reliability. The new techniques for controlling the reliability of design and quality of production enable us to produce complex yet reliable products, but the techniques are dependent for their success on the motivation that comes only with personal freedom.

BIBLIOGRAPHY

1. US MIL-STD-785. *Reliability Programs for Systems and Equipments — Development and Production.* Available from the National Technical Information Service, Springfield, Virginia.
2. UK Defence Standard 00—40: *The Management of Reliability and Maintainability.* HMSO.
3. British Standard, BS 5760: *Reliability of Systems, Equipments and Components.* British Standards Institution, London.
4. A. V. Feigenbaum, *Total Quality Control,* 3rd edn. McGraw-Hill, New York (1983).
5. P. F. Drucker, *The Practice of Management.* Heinemann (1955).
6. P. B. Crosby, *Quality is Free.* McGraw-Hill (1979).

Appendix 1. The standard cumulative normal distribution function

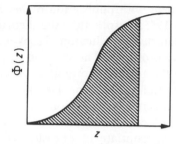

$$\Phi(z) = \frac{1}{(2\pi)^{1/2}} \int_{-\infty}^{z} \exp\left(\frac{-x^2}{2}\right) dx$$

for $0.00 \leqslant z \leqslant 4.00$

$$1 - \Phi(z) = \Phi(-z)$$

z	0.00	0.01	0.02	0.03	0.04	0.05	0.06	0.07	0.08	0.09
0.0	0.5000	0.5040	0.5080	0.5120	0:5160	0.5199	0.5239	0.5279	0.5319	0.5359
0.1	0.5398	0.5438	0.5478	0.5517	0.5557	0.5596	0.5636	0.5675	0.5714	0.5753
0.2	0.5793	0.5832	0.5871	0.5910	0.5948	0.5987	0.6026	0.6064	0.6103	0.6141
0.3	0.6179	0.6217	0.6255	0.6293	0.6331	0.6368	0.6406	0.6443	0.6480	0.6517
0.4	0.6554	0.6591	0.6628	0.6664	0.6700	0.6736	0.6772	0.6808	0.6844	0.6879
0.5	0.6915	0.6985	0.6985	0.7019	0.7054	0.7088	0.7123	0.7157	0.7190	0.7224
0.6	0.7257	0.7291	0.7324	0.7357	0.7389	0.7422	0.7454	0.7486	0.7517	0.7549
0.7	0.7580	0.7611	0.7642	0.7673	0.7703	0.7734	0.7764	0.7794	0.7823	0.7852
0.8	0.7881	0.7910	0.7939	0.7967	0.7995	0.8023	0.8051	0.8078	0.8106	0.8133
0.9	0.8159	0.8186	0.8212	0.8238	0.8264	0.8289	0.8315	0.8340	0.8365	0.8389
1.0	0.8413	0.8438	0.8461	0.8485	0.8508	0.8531	0.8554	0.8577	0.8599	0.8621
1.1	0.8643	0.8665	0.8686	0.8708	0.8729	0.8749	0.8770	0.8790	0.8810	0.8830
1.2	0.8849	0.8869	0.8888	0.8907	0.8925	0.8944	0.8962	0.8980	0.8997	0.9015
1.3	0.9032	0.9049	0.9066	0.9082	0.9099	0.9115	0.9131	0.9147	0.9162	0.9177
1.4	0.9192	0.9207	0.9222	0.9236	0.9251	0.9265	0.9279	0.9292	0.9306	0.9319
1.5	0.9332	0.9345	0.9357	0.9370	0.9382	0.9394	0.9406	0.9418	0.9430	0.9440
1.6	0.9452	0.9463	0.9474	0.9485	0.9495	0.9505	0.9515	0.9525	0.9535	0.9545
1.7	0.9554	0.9564	0.9573	0.9582	0.9591	0.9599	0.9608	0.9616	0.9625	0.9633
1.8	0.9641	0.9649	0.9656	0.9664	0.9671	0.9678	0.9686	0.9693	0.9700	0.9706
1.9	0.9713	0.9719	0.9726	0.9732	0.9738	0.9744	0.9750	0.9756	0.9762	0.9767
2.0	0.9773	0.9778	0.9783	0.9788	0.9793	0.9798	0.9803	0.9808	0.9812	0.9817
2.1	0.9821	0.9826	0.9830	0.9834	0.9838	0.9842	0.9846	0.9850	0.9854	0.9857
2.2	0.9861	0.9865	0.9868	0.9871	0.9875	0.9878	0.9881	0.9884	0.9887	0.9890
2.3	0.9893	0.9896	0.9898	0.9^2010	0.9^2061	0.9^2035	0.9^2086	0.9^2111	0.9^2134	0.9^2158
2.4	0.9^2180	0.9^2202	0.9^2224	0.9^2245	0.9^2266	0.9^2286	0.9^2305	0.9^2324	0.9^2343	0.9^2361
2.5	0.9^2379	0.9^2396	0.9^2413	0.9^2430	0.9^2446	0.9^2461	0.9^2477	0.9^2492	0.9^2506	0.9^2520
2.6	0.9^2534	0.9^2547	0.9^2560	0.9^2573	0.9^2586	0.9^2598	0.9^2609	0.9^2621	0.9^2632	0.9^2643
2.7	0.9^2653	0.9^2664	0.9^2674	0.9^2683	0.9^2693	0.9^2702	0.9^2711	0.9^2720	0.9^2728	0.9^2737
2.8	0.9^2745	0.9^2752	0.9^2760	0.9^2767	0.9^2774	0.9^2781	0.9^2788	0.9^2795	0.9^2801	0.9^2807
2.9	0.9^2813	0.9^2819	0.9^2825	0.9^2831	0.9^2836	0.9^2841	0.9^2846	0.9^2851	0.9^2856	0.9^2861

z	0.00	0.01	0.02	0.03	0.04	0.05	0.06	0.07	0.08	0.09
3.0	0.9^2865	0.9^2869	0.9^2874	0.9^2878	0.9^2882	0.9^2886	0.9^2889	0.9^2893	0.9^2897	0.9^2900
3.1	0.9^3032	0.9^3065	0.9^3096	0.9^3126	0.9^3155	0.9^3184	0.9^3211	0.9^3238	0.9^3264	0.9^3289
3.2	0.9^3313	0.9^3336	0.9^3359	0.9^3381	0.9^3402	0.9^3423	0.9^3443	0.6^3462	0.9^3481	0.9^3499
3.3	0.9^3517	0.9^3534	0.9^3550	0.9^3566	0.9^3581	0.9^3596	0.9^3610	0.9^3624	0.9^3638	0.9^3651
3.4	0.9^3663	0.9^3675	0.9^3687	0.9^3698	0.9^3709	0.9^3720	0.9^3730	0.9^3740	0.9^3749	0.9^3759
3.5	0.9^3767	0.9^3776	0.9^3784	0.9^3792	0.9^3800	0.9^3807	0.9^3815	0.9^3822	0.9^3822	0.9^3835
3.6	0.9^3841	0.9^3847	0.9^3853	0.9^3858	0.9^3864	0.9^3869	0.9^3874	0.9^3879	0.9^3883	0.9^3888
3.7	0.9^3892	0.9^3896	0.9^4004	0.9^4043	0.9^4116	0.9^4116	0.9^4150	0.9^4184	0.9^4216	0.9^4257
3.8	0.9^4277	0.9^4305	0.9^4333	0.9^4359	0.9^4385	0.9^4409	0.9^4433	0.9^4456	0.9^4478	0.9^4499
3.9	0.9^4519	0.9^4539	0.9^4557	0.9^4575	0.9^4593	0.9^4609	0.9^4625	0.9^4641	0.9^4655	0.9^4670

Appendix 2. Values of $y = \exp(-x)$

x	e^{-z}	x	e^{-z}	x	e^{-z}	x	e^{-z}	x	e^{-z}	x	e^{-z}
0.00	1.0000	0.50	0.6065	1.00	0.3679	1.50	0.2231	2.00	0.1353	2.50	0.0821
0.01	0.9900	0.51	0.6005	1.01	0.3642	1.51	0.2209	2.01	0.1340	2.51	0.0813
0.02	0.9802	0.52	0.5945	1.02	0.3606	1.52	0.2187	2.02	0.1327	2.52	0.0805
0.03	0.9704	0.53	0.5886	1.03	0.3570	1.53	0.2165	2.03	0.1313	2.53	0.0797
0.04	0.9608	0.54	0.5827	1.04	0.3535	1.54	0.2144	2.04	0.1300	2.54	0.0789
0.05	0.9512	0.55	0.5769	1.05	0.3499	1.55	0.2122	2.05	0.1287	2.55	0.0781
0.06	0.9418	0.56	0.5712	1.06	0.3465	1.56	0.2101	2.06	0.1275	2.56	0.0773
0.07	0.9324	0.57	0.5655	1.07	0.3430	1.57	0.2080	2.07	0.1262	2.57	0.0765
0.08	0.9231	0.58	0.5599	1.08	0.3396	1.58	0.2060	2.08	0.1249	2.58	0.0758
0.09	0.9139	0.59	0.5543	1.09	0.3362	1.59	0.2039	2.09	0.1237	2.59	0.0750
0.10	0.9048	0.60	0.5488	1.10	0.3329	1.60	0.2019	2.10	0.1225	2.60	0.0743
0.11	0.8958	0.61	0.5434	1.11	0.3296	1.61	0.1999	2.11	0.1212	2.61	0.0735
0.12	0.8869	0.62	0.5379	1.12	0.3263	1.62	0.1979	2.12	0.1200	2.62	0.0728
0.13	0.8781	0.63	0.5326	1.13	0.3230	1.63	0.1959	2.13	0.1188	2.63	0.0721
0.14	0.8694	0.64	0.5273	1.14	0.3198	1.64	0.1940	2.14	0.1177	2.64	0.0714
0.15	0.8607	0.65	0.5220	1.15	0.3166	1.65	0.1920	2.15	0.1165	2.65	0.0707
0.16	0.8521	0.66	0.5169	1.16	0.3135	1.66	0.1901	2.16	0.1153	2.66	0.0699
0.17	0.8437	0.67	0.5117	1.17	0.3104	1.67	0.1882	2.17	0.1142	2.67	0.0693
0.18	0.8353	0.68	0.5066	1.18	0.3073	1.68	0.1864	2.18	0.1130	2.68	0.0686
0.19	0.8270	0.69	0.5016	1.19	0.3042	1.69	0.1845	2.19	0.1119	2.69	0.0679
0.20	0.8187	0.70	0.4966	1.20	0.3012	1.70	0.1827	2.20	0.1108	2.70	0.0672
0.21	0.8106	0.71	0.4916	1.21	0.2982	1.71	0.1809	2.21	0.1097	2.71	0.0665
0.22	0.8025	0.72	0.4868	1.22	0.2952	1.72	0.1791	2.22	0.1086	2.72	0.0659
0.23	0.7945	0.73	0.4819	1.23	0.2923	1.73	0.1773	2.23	0.1075	2.73	0.0652
0.24	0.7866	0.74	0.4771	1.24	0.2894	1.74	0.1755	2.24	0.1065	2.74	0.0646
0.25	0.7788	0.75	0.4724	1.25	0.2865	1.75	0.1738	2.25	0.1054	2.75	0.0639
0.26	0.7711	0.76	0.4677	1.26	0.2837	1.76	0.1720	2.26	0.1044	2.76	0.0633
0.27	0.7634	0.77	0.4630	1.27	0.2808	1.77	0.1703	2.27	0.1033	2.77	0.0627
0.28	0.7558	0.78	0.4584	1.28	0.2780	1.78	0.1686	2.28	0.1023	2.78	0.0620
0.29	0.7483	0.79	0.4538	1.29	0.2753	1.79	0.1670	2.29	0.1013	2.79	0.0614
0.30	0.7408	0.80	0.4493	1.30	0.2725	1.80	0.1653	2.30	0.1003	2.80	0.0608
0.31	0.7334	0.81	0.4449	1.31	0.2698	1.81	0.1637	2.31	0.0993	2.81	0.0602
0.32	0.7261	0.82	0.4404	1.32	0.2671	1.82	0.1620	2.32	0.0983	2.82	0.0596
0.33	0.7189	0.83	0.4360	1.33	0.2645	1.83	0.1604	2.33	0.0973	2.83	0.0590
0.34	0.7118	0.84	0.4317	1.34	0.2618	1.84	0.1588	2.34	0.0963	2.84	0.0584
0.35	0.7047	0.85	0.4274	1.35	0.2592	1.85	0.1572	2.35	0.0954	2.85	0.0578
0.36	0.6977	0.86	0.4322	1.36	0.2567	1.86	0.1557	2.36	0.0944	2.86	0.0573
0.37	0.6907	0.87	0.4190	1.37	0.2541	1.87	0.1541	2.37	0.0935	2.87	0.0567
0.38	0.6839	0.88	0.4148	1.38	0.2516	1.88	0.1526	2.38	0.0926	2.88	0.0561
0.39	0.6771	0.89	0.4107	1.39	0.2491	1.89	0.1511	2.39	0.0916	2.89	0.0556
0.40	0.6703	0.90	0.4066	1.40	0.2466	1.90	0.1496	2.40	0.0907	2.90	0.0550
0.41	0.6637	0.91	0.4025	1.41	0.2441	1.91	0.1481	2.41	0.0898	2.91	0.0545
0.42	0.6570	0.92	0.3985	1.42	0.2417	1.92	0.1466	2.42	0.0889	2.92	0.0539
0.43	0.6505	0.93	0.3946	1.43	0.2393	1.93	0.1451	2.43	0.0880	2.93	0.0534
0.44	0.6440	0.94	0.3906	1.44	0.2369	1.94	0.1437	2.44	0.0872	2.94	0.0529
0.45	0.6376	0.95	0.3867	1.45	0.2346	1.95	0.1423	2.45	0.0863	2.95	0.0523
0.46	0.6313	0.96	0.3829	1.46	0.2322	1.96	0.1409	2.46	0.0854	2.96	0.0518
0.47	0.6250	0.97	0.3791	1.47	0.2299	1.97	0.1395	2.47	0.0846	2.97	0.0513
0.48	0.6188	0.98	0.3753	1.48	0.2277	1.98	0.1381	2.48	0.0837	2.98	0.0508
0.49	0.6126	0.99	0.3716	1.49	0.2254	1.99	0.1367	2.49	0.0829	2.99	0.0503
0.50	0.6065	1.00	0.3679	1.50	0.2231	2.00	0.1353	2.50	0.0821	3.00	0.0498

For $x < 0.01$, use the approximation $y \approx 1 - x$. If greater accuracy is needed, use $y \approx 1 - x + \dfrac{x^2}{2}$.

Appendix 3. Percentiles of the χ^2 distribution (continued overleaf)

Degrees of freedom ν	α 0.005	0.010	0.025	0.05	0.10	0.20	0.30	0.40
1	0.0^4393	0.0^3157	0.0^3982	0.0^2393	0.0158	0.0642	0.148	0.275
2	0.0100	0.0201	0.0506	0.103	0.211	0.446	0.713	1.02
3	0.0717	0.115	0.216	0.352	0.584	1.00	1.42	1.87
4	0.207	0.297	0.484	0.711	1.06	1.65	2.19	2.75
5	0.412	0.554	0.831	1.15	1.61	2.34	3.00	3.66
6	0.676	0.872	1.24	1.64	2.20	3.07	3.83	4.57
7	0.989	1.24	1.69	2.17	2.83	3.82	4.67	5.49
8	1.34	1.65	2.18	2.73	3.49	4.59	5.53	6.42
9	1.73	2.09	2.70	3.33	4.17	5.38	6.39	7.36
10	2.16	2.56	3.25	3.94	4.87	6.18	7.27	8.30
11	2.60	3.05	3.82	4.57	5.58	6.99	8.15	9.24
12	3.07	3.57	4.40	5.23	6.30	7.81	9.03	10.2
13	3.57	4.11	5.01	5.89	7.04	8.63	9.93	11.1
14	4.07	4.66	5.63	6.57	7.79	9.47	10.8	12.1
15	4.60	5.23	6.26	7.26	8.55	10.3	11.7	13.0
16	5.14	5.81	6.91	7.96	9.31	11.2	12.6	14.0
17	5.70	6.41	7.56	8.67	10.1	12.0	13.5	14.9
18	6.26	7.01	8.23	9.39	10.9	12.9	14.4	15.9
19	6.84	7.63	8.91	10.1	11.7	13.7	15.4	16.9
20	7.43	8.26	9.59	10.9	12.4	14.6	16.3	17.8
21	8.03	8.90	10.3	11.6	13.2	15.4	17.2	18.8
22	8.64	9.54	11.0	12.3	14.0	16.3	18.1	19.7
23	9.26	10.2	11.7	13.1	14.8	17.2	19.0	20.7
24	9.89	10.9	12.4	13.8	15.7	18.1	19.9	21.7
25	10.5	11.5	13.1	14.6	16.5	18.9	20.9	22.6
26	11.2	12.2	13.8	15.4	17.3	19.8	21.8	23.6
27	11.8	12.9	14.6	16.2	18.1	20.7	22.7	24.5
28	12.5	13.6	15.3	16.9	18.9	21.6	23.6	25.5
29	13.1	14.3	16.0	17.7	19.8	22.5	24.6	26.5
30	13.8	15.0	16.8	18.5	20.6	23.4	25.5	27.4
35	17.2	18.5	20.6	22.5	24.8	27.8	30.2	32.3
40	20.7	22.2	24.4	26.5	29.1	32.3	34.9	37.1
45	24.3	25.9	28.4	30.6	33.4	36.9	39.6	42.0
50	28.0	29.7	32.4	34.8	37.7	41.4	44.3	46.9
75	47.2	49.5	52.9	56.1	59.8	64.5	68.1	71.3
100	67.3	70.1	74.2	77.9	82.4	87.9	92.1	95.8

0.50	0.60	0.70	0.80	0.90	0.95	0.975	0.990	α 0.995	v
0.455	0.708	1.07	1.64	2.71	3.84	5.02	6.63	7.88	1
1.39	1.83	2.41	3.22	4.61	5.99	7.38	9.21	10.6	2
2.37	2.95	3.67	4.64	6.25	7.81	9.35	11.3	12.8	3
3.36	4.04	4.88	5.99	7.78	9.49	11.1	13.3	14.9	4
4.35	5.13	6.06	7.29	9.24	11.1	12.8	15.1	16.7	5
5.35	6.21	7.23	8.56	10.6	12.6	14.4	16.8	18.5	6
6.35	7.28	8.38	9.80	12.0	14.1	16.0	18.5	20.3	7
7.34	8.35	9.52	11.0	13.4	15.5	17.5	20.1	22.0	8
8.34	9.41	10.7	12.2	14.7	16.9	19.0	21.7	23.6	9
9.34	10.5	11.8	13.4	16.0	18.3	20.5	23.2	25.2	10
10.3	11.5	12.9	14.6	17.3	19.7	21.9	24.7	26.8	11
11.3	12.6	14.0	15.8	18.5	21.0	23.3	26.2	28.3	12
12.3	13.6	15.1	17.0	19.8	22.4	24.7	27.7	29.8	13
13.3	14.7	16.2	18.2	21.1	23.7	26.1	29.1	31.3	14
14.3	15.7	17.3	19.3	22.3	25.0	27.5	30.6	32.8	15
15.3	16.8	18.4	20.5	23.5	26.3	28.8	32.0	34.3	16
16.3	17.8	19.5	21.6	24.8	27.6	30.2	33.4	35.7	17
17.3	18.9	20.6	22.8	26.0	28.9	31.5	34.8	37.2	18
18.3	19.9	21.7	23.9	27.2	30.1	32.9	36.2	38.6	19
19.3	21.0	22.8	25.0	28.4	31.4	34.2	37.6	40.0	20
20.3	22.0	23.9	26.2	29.6	32.7	35.5	38.9	41.4	21
21.3	23.0	24.9	27.3	30.8	33.9	36.8	40.3	42.8	22
22.3	24.1	26.0	28.4	32.0	35.2	38.1	41.6	44.2	23
23.3	25.1	27.1	29.6	33.2	36.4	39.4	43.0	45.6	24
24.3	26.1	28.2	30.7	34.4	37.7	40.6	44.3	46.9	25
25.3	27.2	29.2	31.8	35.6	38.9	41.9	45.6	48.3	26
26.3	28.2	30.3	32.9	36.7	40.1	43.2	47.0	49.6	27
27.3	29.2	31.4	34.0	37.9	41.3	44.5	48.3	51.0	28
28.3	30.3	32.5	35.1	39.1	42.6	45.7	49.6	52.3	29
29.3	31.3	33.5	36.3	40.3	43.8	47.0	50.9	53.7	30
34.3	36.5	38.9	41.8	46.1	49.8	53.2	57.3	60.3	35
39.3	41.6	44.2	47.3	51.8	55.8	59.3	63.7	66.8	40
44.3	46.8	49.5	52.7	57.5	61.7	65.4	70.0	73.2	45
49.3	51.9	54.7	58.2	63.2	67.5	71.4	76.2	79.5	50
74.3	77.5	80.9	85.1	91.1	96.2	100.8	106.4	110.3	75
99.3	102.9	106.9	111.7	118.5	124.3	129.6	135.6	140.2	100

Appendix 4. Values of the F-distribution

(A. Hald, *Statistical Tables and Formulas,* copyright © 1952 John Wiley & Sons Inc. Reprinted by permission of John Wiley & Sons Inc.)

Tabulation of the values of $F_{0.10; v_1; v_2}$ versus v_1 and v_2

		1	2	3	4	5	6	7	8	9	10	15	20	30	50	100	200	500	∞
	1	39.9	49.5	53.6	55.8	57.2	58.2	58.9	59.4	59.9	60.2	61.2	61.7	62.3	62.7	63.0	63.2	63.3	63.3
	2	8.53	9.00	9.16	9.24	9.29	9.33	9.35	9.37	9.38	9.39	9.42	9.44	9.46	9.47	9.48	9.49	9.49	9.49
	3	5.54	5.46	5.39	5.34	5.31	5.28	5.27	5.25	5.24	5.23	5.20	5.18	5.17	5.15	5.14	5.14	5.14	5.13
	4	4.54	4.32	4.19	4.11	4.05	4.01	3.98	3.95	3.94	3.92	3.87	3.84	3.82	3.80	3.78	3.77	3.76	3.76
	5	4.06	3.78	3.62	3.52	3.45	3.40	3.37	3.34	3.32	3.30	3.24	3.21	3.17	3.15	3.13	3.12	3.11	3.10
	6	3.78	3.46	3.29	3.18	3.11	3.05	3.01	2.98	2.96	2.94	2.87	2.84	2.80	2.77	2.75	2.73	2.73	2.72
	7	3.59	3.26	3.07	2.96	2.88	2.83	2.78	2.75	2.72	2.70	2.63	2.59	2.56	2.52	2.50	2.48	2.48	2.47
	8	3.46	3.11	2.92	2.81	2.73	2.67	2.62	2.59	2.56	2.54	2.46	2.42	2.38	2.35	2.32	2.31	2.30	2.29
	9	3.36	3.01	2.81	2.69	2.61	2.55	2.51	2.47	2.44	2.42	2.34	2.30	2.25	2.22	2.19	2.17	2.17	2.16
	10	3.28	2.92	2.73	2.61	2.52	2.46	2.41	2.38	2.35	2.32	2.24	2.20	2.16	2.12	2.09	2.07	2.06	2.06
	11	3.23	2.86	2.66	2.54	2.45	2.39	2.34	2.30	2.27	2.25	2.17	2.12	2.08	2.04	2.00	1.99	1.98	1.97
	12	3.18	2.81	2.61	2.48	2.39	2.33	2.28	2.24	2.21	2.19	2.10	2.06	2.01	1.97	1.94	1.92	1.91	1.90
	13	3.14	2.76	2.56	2.43	2.35	2.28	2.23	2.20	2.16	2.14	2.05	2.01	1.96	1.92	1.88	1.86	1.85	1.85
	14	3.10	2.73	2.52	2.39	2.31	2.24	2.19	2.15	2.12	2.10	2.01	1.96	1.91	1.87	1.83	1.82	1.80	1.80
	15	3.07	2.70	2.49	2.36	2.27	2.21	2.16	2.12	2.09	2.06	1.97	1.92	1.87	1.83	1.79	1.77	1.76	1.76
	16	3.05	2.67	2.46	2.33	2.24	2.18	2.13	2.09	2.06	2.03	1.94	1.89	1.84	1.79	1.76	1.74	1.73	1.72
	17	3.03	2.64	2.44	2.31	2.22	2.15	2.10	2.06	2.03	2.00	1.91	1.86	1.81	1.76	1.73	1.71	1.69	1.69
	18	3.01	2.62	2.42	2.29	2.20	2.13	2.08	2.04	2.00	1.98	1.89	1.84	1.78	1.74	1.70	1.68	1.67	1.66
	19	2.99	2.61	2.40	2.27	2.18	2.11	2.06	2.02	1.98	1.96	1.86	1.81	1.76	1.71	1.67	1.65	1.64	1.63
	20	2.97	2.59	2.38	2.25	2.16	2.09	2.04	2.00	1.96	1.94	1.84	1.79	1.74	1.69	1.65	1.63	1.62	1.61
	22	2.95	2.56	2.35	2.22	2.13	2.06	2.01	1.97	1.93	1.90	1.81	1.76	1.70	1.65	1.61	1.59	1.58	1.57
	24	2.93	2.54	2.33	2.19	2.10	2.04	1.98	1.94	1.91	1.88	1.78	1.73	1.67	1.62	1.58	1.56	1.54	1.53
	26	2.91	2.52	2.31	2.17	2.08	2.01	1.96	1.92	1.88	1.86	1.76	1.71	1.65	1.59	1.55	1.53	1.51	1.50
	28	2.89	2.50	2.29	2.16	2.06	2.00	1.94	1.90	1.87	1.84	1.74	1.69	1.63	1.57	1.53	1.50	1.49	1.48
	30	2.88	2.49	2.28	2.14	2.05	1.98	1.93	1.88	1.85	1.82	1.72	1.67	1.61	1.55	1.51	1.48	1.47	1.46
	40	2.84	2.44	2.23	2.09	2.00	1.93	1.87	1.83	1.79	1.76	1.66	1.61	1.54	1.48	1.43	1.41	1.39	1.38
	50	2.81	2.41	2.20	2.06	1.97	1.90	1.84	1.80	1.76	1.73	1.63	1.57	1.50	1.44	1.39	1.36	1.34	1.33
	60	2.79	2.39	2.18	2.04	1.95	1.87	1.82	1.77	1.74	1.71	1.60	1.54	1.48	1.41	1.36	1.33	1.31	1.29
	80	2.77	2.37	2.15	2.02	1.92	1.85	1.79	1.75	1.71	1.68	1.57	1.51	1.44	1.38	1.32	1.28	1.26	1.24
	100	2.76	2.36	2.14	2.00	1.91	1.83	1.78	1.73	1.70	1.66	1.56	1.49	1.42	1.35	1.29	1.26	1.23	1.21
	200	2.73	2.33	2.11	1.97	1.88	1.80	1.75	1.70	1.66	1.63	1.52	1.46	1.38	1.31	1.24	1.20	1.17	1.14
	500	2.72	2.31	2.10	1.96	1.86	1.79	1.73	1.68	1.64	1.61	1.50	1.44	1.36	1.28	1.21	1.16	1.12	1.09
	∞	2.71	2.30	2.08	1.94	1.85	1.77	1.72	1.67	1.63	1.60	1.49	1.42	1.34	1.26	1.18	1.13	1.08	1.00

Degrees of freedom for the numerator (v_1)

Degrees of freedom for the denominator (v_2)

Tabulation of the values of $F_{0.05; \nu_1; \nu_2}$ versus ν_1 and ν_2

			Degrees of freedom for the numerator (ν_1)																
		1	2	3	4	5	6	7	8	9	10	11	12	13	14	15	16	17	18
	1	161	200	216	225	230	234	237	239	241	242	243	244	245	245	246	246	247	247
	2	18.5	19.0	19.2	19.2	19.3	19.3	19.4	19.4	19.4	19.4	19.4	19.4	19.4	19.4	19.4	19.4	19.4	19.4
	3	10.1	9.55	9.28	9.12	9.01	8.94	8.89	8.85	8.81	8.79	8.76	8.74	8.73	8.71	8.70	8.69	8.68	8.67
	4	7.71	6.94	6.59	6.39	6.26	6.16	6.09	6.04	6.00	5.96	5.94	5.91	5.82	5.87	5.86	5.84	5.83	5.82
	5	6.61	5.79	5.41	5.19	5.05	4.95	4.88	4.82	4.77	4.74	4.70	4.08	4.66	4.64	4.62	4.60	4.59	4.58
	6	5.99	5.14	4.76	4.53	4.39	4.28	4.21	4.15	4.10	4.06	4.03	4.00	3.98	3.96	3.94	3.92	3.91	3.90
	7	5.59	4.74	4.35	4.12	3.97	3.87	3.79	3.73	3.68	3.64	3.60	3.57	3.55	3.53	3.51	3.49	3.48	3.47
	8	5.32	4.46	4.07	3.84	3.69	3.58	3.50	3.44	3.39	3.35	3.31	3.28	3.26	3.24	3.22	3.20	3.19	3.17
	9	5.12	4.26	3.86	3.63	3.48	3.37	3.29	3.23	3.18	3.14	3.10	3.07	3.05	3.03	3.01	2.99	2.97	2.96
	10	4.96	4.10	3.71	3.48	3.33	3.22	3.14	3.07	3.02	2.98	2.94	2.91	2.89	2.86	2.85	2.83	2.81	2.80
	11	4.84	2.98	3.50	3.36	3.20	3.01	2.95	2.90	2.85	2.82	2.82	2.79	2.76	2.74	2.72	2.70	2.69	2.67
	12	4.75	3.89	3.49	3.26	3.11	3.00	2.91	2.85	2.80	2.75	2.72	2.69	2.66	2.64	2.62	2.60	2.58	2.57
	13	4.67	3.81	3.41	3.18	3.03	2.92	2.83	2.77	2.71	2.67	2.63	2.60	2.58	2.55	2.53	2.51	2.50	2.48
	14	4.60	3.74	3.34	3.11	2.96	2.85	2.76	2.70	2.65	2.60	2.57	2.53	2.51	2.48	2.46	2.44	2.43	2.41
	15	4.54	3.68	3.29	3.06	2.90	2.79	2.71	2.64	2.59	2.54	2.51	2.48	2.45	2.42	2.40	2.38	2.37	2.35
	16	4.49	3.63	3.24	3.01	2.85	2.74	2.66	2.59	2.54	2.49	2.46	2.42	2.40	2.37	2.35	2.33	2.32	2.30
	17	4.45	3.59	3.20	2.96	2.81	2.70	2.61	2.55	2.49	2.45	2.41	2.38	2.36	2.33	2.31	2.29	2.27	2.26
	18	4.41	3.55	3.16	2.93	2.77	2.66	2.58	2.51	2.46	2.41	2.37	2.34	2.31	2.29	2.27	2.25	2.23	2.22
	19	4.38	3.52	3.13	2.90	2.74	2.63	2.54	2.48	2.42	2.38	2.34	2.31	2.28	2.26	2.23	2.21	2.20	2.18
	20	4.35	3.49	3.10	2.87	2.71	2.60	2.51	2.45	2.39	2.35	2.31	2.28	2.25	2.22	2.20	2.18	2.17	2.15
	21	4.32	3.47	3.07	2.82	2.68	2.57	2.49	2.42	2.37	2.32	2.28	2.25	2.22	2.20	2.18	2.16	2.14	2.12
	22	4.30	3.44	3.05	2.84	2.66	2.55	2.46	2.40	2.34	2.30	2.26	2.23	2.20	2.17	2.15	2.13	2.11	2.10
	23	4.28	3.42	3.03	2.80	2.64	2.53	2.44	2.37	2.32	2.27	2.23	2.20	2.18	2.15	2.13	2.11	2.09	2.07
	24	4.26	3.40	3.01	2.78	2.62	2.51	2.42	2.36	2.30	2.25	2.21	2.18	2.15	2.13	2.11	2.09	2.07	2.05
	25	4.24	3.39	2.99	2.76	2.60	2.49	2.40	2.34	2.28	2.24	2.20	2.16	2.14	2.11	2.09	2.07	2.05	2.04
	26	4.23	3.37	2.98	2.74	2.59	2.47	2.39	2.32	2.27	2.22	2.18	2.15	2.12	2.09	2.07	2.05	2.03	2.02
	27	4.21	3.35	2.96	2.73	2.57	2.46	2.37	2.31	2.25	2.20	2.17	2.13	2.10	2.08	2.06	2.04	2.02	2.00
	28	4.20	3.34	2.95	2.71	2.56	2.45	2.36	2.29	2.24	2.19	2.15	2.12	2.09	2.06	2.04	2.02	2.00	1.99
	29	4.18	3.33	2.93	2.70	2.55	2.43	2.35	2.28	2.22	2.18	2.14	2.10	2.08	2.05	2.03	2.01	1.99	1.97
	30	4.17	3.32	2.92	2.69	2.53	2.42	2.33	2.27	2.21	2.16	2.13	2.09	2.06	2.04	2.01	1.99	1.98	1.96
	32	4.15	3.29	2.90	2.67	2.51	2.40	2.31	2.24	2.19	2.14	2.10	2.07	2.04	2.01	1.99	1.97	1.95	1.94
	34	4.13	3.28	2.88	2.65	2.49	2.38	2.29	2.23	2.17	2.12	2.08	2.05	2.02	1.99	1.97	1.95	1.93	1.92
	36	4.11	3.26	2.87	2.63	2.48	2.36	2.28	2.21	2.15	2.11	2.07	2.03	2.00	1.98	1.95	1.93	1.92	1.90
	38	4.10	3.24	2.85	2.62	2.46	2.35	2.26	2.19	2.14	2.09	2.05	2.02	1.99	1.96	1.94	1.92	1.90	1.88
	40	4.08	3.23	2.84	2.61	2.45	2.34	2.25	2.18	2.12	2.08	2.04	2.00	1.97	1.95	1.92	1.90	1.89	1.87

Degrees of freedom for the denominator (ν_2)

Degrees of freedom for the numerator (v_1)																	
19	20	22	24	26	28	30	35	40	45	50	60	80	100	200	500	∞	
248	248	249	249	249	250	250	251	251	251	252	252	252	253	254	254	254	1
19.4	19.4	19.5	19.5	19.5	19.5	19.5	19.5	19.5	19.5	19.5	19.5	19.5	19.5	19.5	19.5	19.5	2
8.67	8.66	8.65	8.64	8.63	8.62	8.62	8.60	8.59	8.59	8.58	8.57	8.56	8.55	8.54	8.53	8.53	3
5.81	5.80	5.79	5.77	5.76	5.75	5.75	5.73	5.72	5.71	5.70	5.69	5.67	5.66	5.65	5.64	5.63	4
4.57	4.56	4.54	4.53	4.52	4.50	4.50	4.48	4.46	4.45	4.44	4.43	4.41	4.41	4.30	4.37	4.37	5
3.88	3.87	3.88	3.84	3.83	3.82	3.81	3.79	3.77	3.76	3.75	3.74	3.72	3.71	3.09	3.63	3.67	6
3.46	3.44	3.43	3.41	3.40	3.39	3.38	3.36	3.34	3.33	3.32	3.30	3.20	3.29	2.25	3.24	3.23	7
3.16	3.15	3.13	3.12	3.10	3.09	3.08	3.06	3.04	3.03	3.02	3.01	2.99	2.97	2.95	2.94	2.93	8
2.95	2.94	2.92	2.90	2.89	2.87	2.86	2.84	2.83	2.81	2.80	2.79	2.77	2.76	2.23	2.72	2.71	9
2.78	2.77	2.75	2.74	2.72	2.71	2.70	2.68	2.66	2.65	2.64	2.62	2.60	2.59	2.55	2.55	2.54	10
2.66	2.65	2.63	2.61	2.59	2.58	2.57	2.55	2.53	2.52	2.51	2.49	2.47	2.46	2.43	2.42	2.40	11
2.56	2.54	2.52	2.51	2.49	2.48	2.47	2.44	2.43	2.41	2.40	2.38	2.36	2.35	2.32	2.31	2.30	12
2.47	2.46	2.44	2.42	2.41	2.39	2.38	2.36	2.34	2.33	2.31	2.30	2.27	2.26	2.23	2.22	2.21	13
2.40	2.39	2.37	2.35	2.33	2.32	2.31	2.28	2.27	2.26	2.24	2.22	2.20	2.19	2.16	2.14	2.13	14
2.34	2.33	2.31	2.29	2.27	2.26	2.25	2.22	2.20	2.19	2.18	2.16	2.14	2.12	2.10	2.08	2.07	15
2.29	2.28	2.25	2.24	2.22	2.21	2.19	2.17	2.15	2.14	2.12	2.11	2.08	2.07	2.04	2.02	2.01	16
2.24	2.23	2.21	2.19	2.17	2.16	2.15	2.12	2.10	2.09	2.08	2.06	2.03	2.02	1.09	1.97	1.96	17
2.20	2.19	2.17	2.15	2.13	2.12	2.11	2.08	2.06	2.05	2.04	2.02	1.99	1.98	1.95	1.93	1.92	18
2.17	2.16	2.13	2.11	2.10	2.08	2.07	2.05	2.03	2.01	2.00	1.98	1.96	1.94	1.91	1.89	1.88	19
2.14	2.12	2.10	2.08	2.07	2.05	2.04	2.01	1.99	1.98	1.97	1.95	1.92	1.91	1.88	1.86	1.84	20
2.11	2.10	2.07	2.05	2.04	2.02	2.01	1.98	1.96	1.95	1.94	1.92	1.89	1.88	1.84	1.82	1.81	21
2.08	2.07	2.03	2.03	2.01	2.00	1.98	1.96	1.94	1.92	1.91	1.89	1.86	1.85	1.82	1.80	1.78	22
2.06	2.05	2.02	2.00	1.99	1.97	1.96	1.93	1.91	1.90	1.88	1.85	1.84	1.82	1.79	1.77	1.76	23
2.04	2.03	2.00	1.98	1.97	1.95	1.94	1.91	1.89	1.88	1.86	1.84	1.82	1.80	1.77	1.75	1.73	24
2.02	2.01	1.98	1.96	1.95	1.93	1.92	1.89	1.87	1.85	1.84	1.82	1.80	1.78	1.75	1.73	1.71	25
2.00	1.99	1.97	1.95	1.93	1.91	1.90	1.87	1.85	1.84	1.82	1.80	1.78	1.76	1.73	1.71	1.69	26
1.99	1.97	1.95	1.93	1.91	1.90	1.88	1.86	1.84	1.82	1.81	1.79	1.76	1.74	1.71	1.69	1.67	27
1.97	1.96	1.93	1.91	1.90	1.88	1.87	1.84	1.82	1.80	1.79	1.77	1.74	1.73	1.60	1.67	1.65	28
1.96	1.94	1.92	1.90	1.88	1.87	1.85	1.83	1.81	1.79	1.77	1.75	1.73	1.71	1.67	1.65	1.64	29
1.95	1.93	1.91	1.89	1.87	1.85	1.84	1.81	1.79	1.77	1.76	1.74	1.71	1.70	1.68	1.64	1.62	30
1.92	1.91	1.88	1.86	1.85	1.83	1.82	1.79	1.77	1.75	1.74	1.71	1.69	1.67	1.63	1.61	1.59	32
1.90	1.89	1.86	1.84	1.82	1.80	1.80	1.77	1.75	1.73	1.71	1.69	1.66	1.65	1.61	1.59	1.57	34
1.88	1.87	1.85	1.82	1.81	1.79	1.78	1.75	1.73	1.71	1.69	1.67	1.64	1.62	1.59	1.56	1.55	36
1.87	1.85	1.83	1.81	1.79	1.77	1.76	1.73	1.71	1.69	1.68	1.65	1.62	1.61	1.57	1.54	1.53	38
1.85	1.84	1.81	1.79	1.77	1.76	1.74	1.72	1.69	1.67	1.66	1.64	1.61	1.50	1.55	1.53	1.51	40

Degrees of freedom for the denominator (v_2)

		Degrees of freedom for the numerator (v_1)																
		1	2	3	4	5	6	7	8	9	10	11	12	13	14	15	16	17	18
Degrees of freedom for the denominator (v_2)	42	4.07	3.22	2.83	2.59	2.44	2.32	2.24	2.16	2.11	2.06	2.03	1.99	1.96	1.93	1.91	1.89	1.87	1.86
	44	4.06	3.21	2.82	2.58	2.43	2.31	2.23	2.16	2.10	2.05	2.01	1.98	1.95	1.92	1.90	1.88	1.86	1.84
	46	4.05	3.20	2.81	2.57	2.42	2.30	2.22	2.15	2.09	2.04	2.00	1.97	1.94	1.91	1.89	1.87	1.85	1.83
	48	4.04	3.19	2.80	2.57	2.41	2.29	2.21	2.14	2.08	2.03	1.99	1.96	1.93	1.90	1.88	1.86	1.84	1.82
	50	4.03	3.18	2.79	2.56	2.40	2.29	2.20	2.13	2.07	2.03	1.99	1.95	1.92	1.89	1.87	1.85	1.83	1.81
	55	4.02	3.16	2.77	2.54	2.38	2.27	2.18	2.11	2.06	2.01	1.97	1.93	1.90	1.88	1.85	1.83	1.81	1.79
	60	4.00	3.15	2.76	2.53	2.37	2.25	2.17	2.10	2.04	1.99	1.95	1.92	1.89	1.86	1.84	1.82	1.80	1.78
	65	3.99	3.14	2.75	2.51	2.36	2.24	2.15	2.08	2.03	1.98	1.94	1.90	1.87	1.85	1.82	1.80	1.78	1.76
	70	3.98	3.13	2.74	2.50	2.35	2.23	2.14	2.07	2.02	1.97	1.93	1.89	1.86	1.84	1.81	1.79	1.77	1.75
	80	3.96	3.11	2.73	2.49	2.33	2.21	2.13	2.06	2.00	1.95	1.91	1.88	1.84	1.82	1.79	1.77	1.75	1.73
	90	3.95	3.10	2.71	2.47	2.32	2.20	2.11	2.04	1.99	1.94	1.90	1.86	1.83	1.80	1.78	1.76	1.74	1.72
	100	3.94	3.09	2.70	2.46	2.31	2.19	2.10	2.03	1.97	1.93	1.89	1.85	1.82	1.79	1.77	1.75	1.73	1.71
	125	3.92	3.07	2.68	2.44	2.29	2.17	2.08	2.01	1.96	1.91	1.87	1.83	1.80	1.77	1.76	1.72	1.70	1.69
	150	3.90	3.08	2.66	2.43	2.27	2.16	2.07	2.00	1.94	1.89	1.85	1.82	1.79	1.76	1.73	1.71	1.69	1.67
	200	3.89	3.04	2.65	2.42	2.26	2.14	2.06	1.98	1.93	1.88	1.84	1.80	1.77	1.74	1.72	1.69	1.67	1.65
	300	3.87	3.03	2.63	2.40	2.24	2.13	2.04	1.97	1.91	1.86	1.82	1.78	1.75	1.72	1.70	1.68	1.66	1.64
	500	3.86	3.01	2.62	2.39	2.23	2.12	2.03	1.96	1.90	1.85	1.81	1.77	1.74	1.71	1.69	1.66	1.64	1.62
	1000	3.85	3.00	2.61	2.38	2.22	2.11	2.02	1.95	1.89	1.84	1.80	1.76	1.73	1.70	1.68	1.65	1.63	1.61
	∞	3.84	3.00	2.60	2.37	2.21	2.10	2.01	1.94	1.88	1.83	1.79	1.75	1.72	1.69	1.67	1.64	1.62	1.60

Degrees of freedom for the numerator (ν_1)

19	20	22	24	26	28	30	35	40	45	50	60	80	100	200	500	∞	ν_2
1.84	1.83	1.80	1.78	1.76	1.74	1.73	1.70	1.68	1.66	1.65	1.62	1.59	1.57	1.53	1.51	1.49	42
1.83	1.81	1.79	1.77	1.75	1.73	1.72	1.69	1.67	1.65	1.63	1.61	1.58	1.56	1.52	1.49	1.48	44
1.82	1.80	1.78	1.76	1.74	1.72	1.71	1.68	1.65	1.64	1.62	1.60	1.57	1.55	1.51	1.48	1.46	46
1.81	1.79	1.77	1.75	1.73	1.71	1.70	1.67	1.64	1.62	1.61	1.59	1.56	1.54	1.49	1.47	1.45	48
1.80	1.78	1.76	1.74	1.72	1.70	1.69	1.66	1.63	1.61	1.60	1.58	1.54	1.52	1.48	1.46	1.44	50
1.78	1.76	1.74	1.72	1.70	1.68	1.67	1.64	1.61	1.59	1.58	1.55	1.52	1.50	1.46	1.43	1.41	55
1.76	1.75	1.72	1.70	1.68	1.66	1.65	1.62	1.59	1.57	1.56	1.53	1.50	1.48	1.44	1.41	1.39	60
1.75	1.73	1.71	1.69	1.67	1.65	1.63	1.60	1.58	1.56	1.54	1.52	1.49	1.48	1.42	1.39	1.37	65
1.74	1.72	1.70	1.67	1.65	1.64	1.62	1.59	1.57	1.55	1.53	1.50	1.47	1.45	1.40	1.37	1.35	70
1.72	1.70	1.68	1.65	1.63	1.62	1.60	1.57	1.54	1.52	1.51	1.48	1.45	1.43	1.38	1.35	1.32	80
1.70	1.69	1.66	1.64	1.62	1.60	1.59	1.55	1.53	1.51	1.49	1.46	1.43	1.41	1.36	1.32	1.30	90
1.69	1.68	1.65	1.63	1.61	1.59	1.57	1.54	1.52	1.49	1.48	1.45	1.41	1.39	1.34	1.31	1.28	100
1.67	1.65	1.63	1.60	1.58	1.57	1.55	1.52	1.49	1.47	1.45	1.42	1.39	1.36	1.31	1.27	1.25	125
1.66	1.64	1.61	1.59	1.57	1.55	1.53	1.50	1.48	1.45	1.44	1.41	1.37	1.34	1.29	1.23	1.22	150
1.64	1.62	1.60	1.57	1.55	1.53	1.52	1.48	1.46	1.43	1.41	1.39	1.35	1.32	1.26	1.22	1.19	200
1.62	1.61	1.58	1.55	1.53	1.51	1.50	1.46	1.43	1.41	1.39	1.38	1.32	1.30	1.23	1.19	1.15	300
1.61	1.50	1.56	1.54	1.52	1.50	1.48	1.45	1.42	1.40	1.38	1.34	1.30	1.28	1.21	1.16	1.11	500
1.60	1.58	1.55	1.53	1.51	1.49	1.47	1.44	1.41	1.38	1.36	1.33	1.29	1.26	1.19	1.13	1.08	1000
1.59	1.57	1.54	1.52	1.50	1.48	1.46	1.42	1.39	1.37	1.35	1.32	1.27	1.24	1.17	1.11	1.00	∞

Degrees of freedom for the denominator (ν_2)

360

Tabulation of the values of $F_{0.025; \nu_1; \nu_2}$ versus ν_1 and ν_2

						Degrees of freedom for the numerator (ν_1)												
	1	2	3	4	5	6	7	8	9	10	11	12	13	14	15	16	17	18
1	648	800	864	900	922	937	948	957	963	969	975	977	980	983	985	987	989	990
2	38.5	39.0	39.2	29.2	39.3	39.3	39.4	39.4	39.4	39.4	39.4	39.4	39.4	39.4	39.4	39.4	39.4	39.4
3	17.4	16.0	15.4	15.1	14.9	14.7	14.6	14.5	14.5	14.4	14.4	14.3	14.3	14.3	14.3	14.2	14.2	14.2
4	12.2	10.6	9.98	9.60	9.36	9.20	9.07	8.98	8.90	8.84	8.79	8.75	8.79	8.69	8.66	8.64	8.62	8.60
5	10.0	8.43	7.76	7.39	7.15	6.98	6.85	6.76	6.68	6.62	6.57	6.52	6.49	6.46	6.43	6.41	6.39	6.37
6	8.81	7.28	6.00	6.23	5.90	5.82	5.70	5.60	5.52	5.48	5.41	5.37	5.33	5.30	5.27	5.25	5.23	5.21
7	8.07	6.54	5.89	5.52	5.29	5.12	4.99	4.90	4.82	4.76	4.71	4.67	4.63	4.60	4.57	4.54	4.52	4.50
8	7.57	6.06	5.42	5.05	4.62	4.65	4.53	4.43	4.66	4.30	4.24	4.20	4.16	4.13	4.10	4.06	4.05	4.03
9	7.21	5.71	5.08	4.72	4.48	4.32	4.20	4.10	4.03	3.96	3.91	3.87	3.83	3.80	3.77	3.74	3.72	3.70
10	6.94	5.46	4.83	4.47	4.24	4.07	3.95	3.85	3.78	3.72	3.86	3.67	3.89	3.55	3.52	3.50	3.47	3.45
11	6.72	5.26	4.63	4.28	4.04	3.88	3.76	3.66	3.59	3.53	3.47	3.43	3.39	3.36	3.33	3.30	3.28	3.26
12	6.55	5.10	4.47	4.12	3.89	3.73	3.61	3.51	3.44	3.37	3.32	3.28	3.24	3.21	3.18	3.15	3.13	3.11
13	6.41	4 97	4.35	4.00	3.77	3.70	3.48	3.39	3.31	3.24	3.20	3.14	3.12	3.08	3.05	3.03	3.00	2.98
14	6.30	4.86	4.24	3.89	3.66	3.50	3.38	3.29	3.21	3.15	3.09	3.05	3.01	2.98	2.95	2.92	2.90	2.88
15	6.20	4.76	4.15	3.80	3.58	3.41	3.29	3.20	3.12	3.06	3.01	2.96	2.92	2.89	2.86	2.84	2.81	2.79
16	6.12	4.69	4.08	3.73	3.50	3.34	3.22	3.12	3.05	2.99	2.93	2.89	2.85	2.82	2.79	2.76	2.74	2.72
17	6.04	4.62	4.01	3.66	3.44	3.28	3.16	3.06	2.98	2.92	2.87	2.82	2.79	2.75	2.72	2.70	2.67	2.65
18	5.98	4.56	3.95	3.61	3.38	3.22	3.10	3.01	2.93	2.87	2.81	2.77	2.73	2.70	2.67	2.64	2.62	2.60
19	5.92	4.51	3.90	3.56	3.33	3.17	3.05	2.96	2.88	2.82	2.76	2.72	2.68	2.65	2.62	2.59	2.57	2.55
20	5.87	4.48	3.86	3.51	3.29	3.13	3.01	2.91	2.84	2.77	2.72	2.68	2.64	2.60	2.57	2.55	2.52	2.50
21	5.83	4.42	3.82	3.48	3.25	3.09	2.97	2.87	2.80	2.73	2.68	2.64	2.60	2.56	2.53	2.51	2.48	2.46
22	5.79	4.38	3.78	3.44	3.22	3.05	2.93	2.84	2.76	2.70	2.65	2.60	2.56	2.53	2.50	2.47	2.45	2.43
23	5.75	4.35	3.75	3.41	3.18	3.02	2.90	2.81	2.73	2.67	2.62	2.57	2.53	2.50	2.47	2.46	2.42	2.39
24	5.72	4.32	3.72	3.38	3.15	2.99	2.87	2.78	2.70	2.64	2.59	2.54	2.50	2.47	2.44	2.41	2.39	2.36
25	5.69	4.29	3.69	3.35	3.13	2.97	2.85	2.75	2.68	2.61	2.56	2.51	2.48	2.44	2.41	2.38	2.36	2.34
26	5.66	4.27	3.67	3.33	3.10	2.94	2.82	2.73	2.65	2.59	2.54	2.49	2.45	2.42	2.39	2.36	2.34	2.31
27	5.63	4.24	3.65	3.31	3.08	2.92	2.80	2.71	2.63	2.57	2.51	2.47	2.43	2.39	2.36	2.34	2.31	2.29
28	5.61	4.22	3.63	3.29	3.06	2.90	2.78	2.69	2.61	2.55	2.49	2.45	2.41	2.37	2.34	2.32	2.29	2.27
29	5.59	4.20	3.61	3.27	3.04	2.88	2.76	2.67	2.59	2.53	2.48	2.43	2.39	2.36	2.32	2.30	2.27	2.25
30	5.57	4.18	3.59	3.25	3.03	2.87	2.75	2.65	2.57	2.51	2.46	2.41	2.37	2.34	2.31	2.28	2.26	2.23
32	5.53	4.15	3.56	3.22	3.00	2.84	2.72	2.62	2.54	2.48	2.43	2.38	2.34	2.31	2.28	2.25	2.22	2.20
34	5.50	4.12	3.53	3.19	2.97	2.81	2.69	2.59	2.52	2.45	2.40	2.35	2.31	2.28	2.25	2.22	2.19	2.17
36	5.47	4.09	3.51	3.17	2.94	2.79	2.56	2.57	2.49	2.43	2.37	2.33	2.29	2.25	2.22	2.20	2.17	2.15
38	5.45	4.07	3.48	3.15	2.92	2.76	2.64	2.55	2.47	2.41	2.35	2.31	2.27	2.23	2.20	2.17	2.15	2.13
40	5.42	4.05	3.46	3.13	2.90	2.74	2.62	2.53	2.45	2.39	2.33	2.29	2.25	2.21	2.18	2.15	2.13	2.11

Degrees of freedom for the denominator (ν_2)

Degrees of freedom for the numerator (ν_1)

19	20	22	24	25	28	30	35	40	45	50	60	80	100	200	500	∞	
992	923	995	997	999	1000	1001	1004	1008	1007	1008	1010	1012	1013	1016	1017	1018	1
23.4	30.4	29.5	39.5	39.5	39.5	39.5	39.5	39.5	39.5	39.5	39.5	39.5	39.5	39.5	39.5	39.5	2
14.2	14.2	14.1	14.1	14.1	14.1	14.1	14.1	10.0	14.0	14.0	14.0	14.0	14.0	13.9	13.9	13.9	3
8.53	8.56	8.53	8.51	8.49	8.48	8.48	8.44	8.41	8.39	8.38	8.36	8.33	8.32	8.29	8.27	8.26	4
6.35	6.33	6.30	6.28	6.26	6.24	6.23	6.20	6.18	6.16	6.14	6.12	6.10	6.06	6.05	6.03	6.02	5
5.19	5.17	5.14	5.12	5.10	5.08	5.07	5.04	5.01	4.99	4.96	4.96	4.93	4.92	4.88	4.86	4.85	6
4.48	4.47	4.44	4.42	4.92	4.38	4.38	4.33	4.31	4.29	4.28	4.25	4.23	4.21	4.18	4.16	4.14	7
4.02	4.00	3.97	3.95	3.93	3.91	3.90	3.86	3.84	3.82	3.81	3.78	3.76	3.74	3.70	3.78	3.67	8
3.68	3.67	3.64	3.61	3.59	3.58	3.56	3.53	3.51	3.49	3.47	3.45	3.42	3.40	3.37	3.35	3.33	9
3.44	3.42	3.39	3.37	3.34	3.33	3.31	3.28	3.26	3.24	3.22	3.20	3.17	3.15	3.12	3.09	3.08	10
3.24	3.23	3.20	3.17	3.15	3.13	3.12	3.09	3.06	3.04	3.03	3.00	2.97	2.96	2.92	2.90	2.88	11
3.09	3.07	3.04	3.02	3.00	2.98	2.96	2.93	2.91	2.89	2.87	2.85	2.82	2.80	2.76	2.74	2.72	12
2.96	2.95	2.92	2.89	2.87	2.85	2.84	2.80	2.78	2.76	2.74	2.72	2.69	2.67	2.63	2.61	2.60	13
2.86	2.84	2.81	2.79	2.77	2.75	2.73	2.70	2.67	2.65	2.64	2.61	2.58	2.56	2.53	2.50	2.49	14
2.77	2.76	2.73	2.70	2.68	2.66	2.64	2.61	2.58	2.56	2.55	2.52	2.49	2.47	2.44	2.41	2.40	15
2.70	2.68	2.65	2.63	2.60	2.58	2.57	2.53	2.51	2.49	2.47	2.45	2.42	2.40	2.36	2.33	2.32	16
2.63	2.62	2.59	2.56	2.54	2.52	2.50	2.47	2.44	2.42	2.41	2.38	2.35	2.33	2.29	2.28	2.25	17
2.58	2.56	2.53	2.50	2.48	2.46	2.44	2.41	2.38	2.36	2.35	2.32	2.29	2.27	2.23	2.20	2.19	18
2.53	2.51	2.48	2.45	2.43	2.41	2.39	2.36	2.33	2.31	2.30	2.27	2.24	2.22	2.18	2.15	2.13	19
2.43	2.46	2.43	2.41	2.39	2.37	2.35	2.31	2.29	2.27	2.25	2.22	2.19	2.17	2.13	2.10	2.09	20
2.44	2.42	2.39	2.37	2.34	2.33	2.31	2.27	2.25	2.23	2.21	2.18	2.15	2.13	2.09	2.06	2.04	21
2.41	2.39	2.36	2.33	2.31	2.29	2.27	2.24	2.21	2.19	2.17	2.14	2.11	2.09	2.05	2.02	2.00	22
2.37	2.36	2.33	2.30	2.28	2.26	2.24	2.20	2.18	2.15	2.14	2.11	2.08	2.06	2.01	1.99	1.97	23
2.35	2.33	2.30	2.27	2.25	2.23	2.21	2.17	2.15	2.12	2.11	2.08	2.05	2.02	1.98	1.95	1.94	24
2.32	2.30	2.27	2.24	2.22	2.20	2.18	2.13	2.12	2.10	2.08	2.05	2.02	2.00	1.95	1.92	1.91	25
2.29	2.28	2.24	2.22	2.19	2.17	2.16	2.12	2.09	2.07	2.05	2.03	1.99	1.97	1.92	1.90	1.88	26
2.27	2.25	2.22	2.19	2.17	2.15	2.13	2.10	2.07	2.05	2.03	2.00	1.97	1.94	1.90	1.87	1.85	27
2.25	2.23	2.20	2.17	2.15	2.13	2.11	2.08	2.05	2.03	2.01	1.98	1.94	1.92	1.88	1.85	1.83	28
2.23	2.21	2.18	2.15	2.13	2.11	2.09	2.06	2.03	2.01	1.99	1.96	1.92	1.90	1.86	1.83	1.81	29
2.21	2.20	2.16	2.14	2.11	2.08	2.07	2.04	2.01	1.99	1.97	1.94	1.90	1.88	1.84	1.81	1.79	30
2.18	2.16	2.13	2.10	2.08	2.06	2.04	2.00	1.98	1.95	1.93	1.91	1.87	1.85	1.80	1.77	1.75	32
2.15	2.13	2.10	2.07	2.05	2.03	2.01	1.97	1.95	1.92	1.90	1.88	1.84	1.82	1.77	1.74	1.72	34
2.12	2.11	2.08	2.05	2.03	2.00	1.99	1.95	1.92	1.90	1.88	1.85	1.81	1.79	1.74	1.71	1.69	36
2.11	2.09	2.05	2.03	2.00	1.98	1.96	1.93	1.90	1.87	1.85	1.82	1.79	1.76	1.71	1.68	1.66	38
2.08	2.07	2.03	2.01	1.98	1.96	1.94	1.90	1.88	1.85	1.83	1.80	1.76	1.74	1.69	1.66	1.64	40

Degrees of freedom for the denominator (ν_2)

		Degrees of freedom for the numerator (ν_1)																
		1	2	3	4	5	6	7	8	9	10	11	12	13	14	15	16	17	18
Degrees of freedom for the denominator (ν_2)	42	5.40	4.03	3.45	3.11	2.89	2.73	2.61	2.51	2.44	2.37	2.32	2.27	2.23	2.20	2.16	2.14	2.11	2.09
	44	5.39	4.02	3.43	3.09	2.87	2.71	2.59	2.50	2.42	2.36	2.30	2.26	2.21	2.18	2.15	2.12	2.10	2.07
	46	5.37	4.00	3.42	3.08	2.88	2.70	2.58	2.48	2.41	2.34	2.29	2.24	2.20	2.17	2.13	2.11	2.08	2.06
	48	5.35	3.99	3.40	3.07	2.84	2.69	2.57	2.47	2.39	2.33	2.27	2.23	2.19	2.15	2.12	2.09	2.07	2.05
	50	5.34	3.98	3.39	3.06	2.83	2.67	2.55	2.46	2.38	2.32	2.26	2.22	2.18	2.14	2.11	2.06	2.06	2.03
	55	5.31	3.95	3.36	3.03	2.81	2.65	2.53	2.43	2.36	2.29	2.24	2.19	2.15	2.11	2.08	2.05	2.03	2.01
	60	5.29	3.93	3.34	3.01	2.79	2.63	2.51	2.41	2.33	2.27	2.22	2.17	2.13	2.09	2.06	2.03	2.01	1.98
	65	5.27	3.91	3.32	2.99	2.77	2.61	2.49	2.39	2.32	2.25	2.20	2.15	2.11	2.07	2.04	2.01	1.99	1.97
	70	5.25	3.89	3.31	2.98	2.75	2.60	2.48	2.38	2.30	2.24	2.18	2.14	2.10	2.06	2.03	2.00	1.97	1.95
	80	5.22	3.86	3.28	2.95	2.73	2.57	2.45	2.36	2.28	2.21	2.16	2.11	2.07	2.03	2.00	1.97	1.95	1.93
	90	5.20	3.84	3.27	2.93	2.71	2.55	2.43	2.34	2.26	2.19	2.14	2.09	2.05	2.02	1.98	1.95	1.93	1.91
	100	5.18	3.83	3.25	2.92	2.70	2.54	2.42	2.32	2.24	2.18	2.12	2.08	2.04	2.00	1.97	1.94	1.91	1.89
	125	5.15	3.80	3.22	2.89	2.67	2.51	2.39	2.30	2.22	2.14	2.10	2.05	2.01	1.97	1.94	1.91	1.89	1.86
	150	5.13	3.78	3.20	2.87	2.65	2.49	2.37	2.28	2.20	2.13	2.08	2.03	1.99	1.95	1.92	1.89	1.87	1.84
	200	5.10	3.76	3.18	2.85	2.63	2.47	2.35	2.26	2.18	2.11	2.06	2.01	1.97	1.93	1.90	1.87	1.84	1.82
	300	5.08	3.74	3.16	2.83	2.61	2.45	2.33	2.23	2.16	2.09	2.04	1.99	1.95	1.91	1.88	1.85	1.82	1.80
	500	5.05	3.72	3.14	2.81	2.59	2.43	2.31	2.22	2.14	2.07	2.02	1.97	1.93	1.89	1.86	1.83	1.80	1.78
	1000	5.04	3.70	3.13	2.80	2.58	2.42	2.30	2.20	2.13	2.06	2.01	1.96	1.92	1.88	1.85	1.82	1.79	1.77
	∞	5.02	3.69	3.12	2.79	2.57	2.41	2.29	2.19	2.11	2.05	1.99	1.94	1.90	1.87	1.83	1.80	1.78	1.75

						Degrees of freedom for the numerator (v_1)												
19	*20*	*22*	*24*	*25*	*28*	*30*	*35*	*40*	*45*	*50*	*60*	*80*	*100*	*200*	*500*	∞		
2.07	2.05	2.02	1.99	1.96	1.94	1.92	1.89	1.86	1.83	1.81	1.78	1.74	1.72	1.67	1.64	1.62	42	
2.05	2.03	2.00	1.97	1.95	1.93	1.91	1.87	1.84	1.82	1.80	1.77	1.73	1.70	1.65	1.62	1.60	44	
2.04	2.02	1.99	1.96	1.93	1.91	1.89	1.85	1.82	1.80	1.78	1.75	1.71	1.69	1.63	1.60	1.58	46	
2.03	2.01	1.97	1.94	1.92	1.90	1.88	1.84	1.81	1.79	1.77	1.73	1.69	1.67	1.62	1.58	1.56	48	
2.01	1.99	1.96	1.93	1.91	1.88	1.87	1.83	1.80	1.77	1.75	1.72	1.68	1.66	1.59	1.57	1.55	50	
1.99	1.97	1.93	1.90	1.88	1.86	1.84	1.80	1.77	1.74	1.72	1.69	1.65	1.62	1.57	1.54	1.51	55	
1.96	1.94	1.91	1.88	1.86	1.83	1.82	1.78	1.74	1.72	1.70	1.67	1.62	1.60	1.54	1.51	1.48	60	
1.95	1.93	1.89	1.86	1.84	1.82	1.80	1.76	1.72	1.70	1.68	1.65	1.60	1.58	1.52	1.48	1.46	65	
1.93	1.91	1.88	1.85	1.82	1.80	1.78	1.74	1.71	1.68	1.66	1.63	1.58	1.56	1.50	1.46	1.44	70	
1.90	1.88	1.85	1.82	1.79	1.77	1.75	1.71	1.68	1.65	1.63	1.60	1.55	1.53	1.47	1.43	1.40	80	
1.88	1.86	1.83	1.80	1.77	1.75	1.73	1.69	1.66	1.63	1.61	1.58	1.53	1.50	1.44	1.40	1.37	90	
1.87	1.85	1.81	1.78	1.76	1.74	1.71	1.67	1.64	1.61	1.59	1.56	1.51	1.48	1.42	1.38	1.35	100	
1.84	1.82	1.79	1.75	1.73	1.71	1.68	1.64	1.61	1.58	1.56	1.52	1.48	1.45	1.38	1.34	1.30	125	
1.82	1.80	1.77	1.74	1.71	1.69	1.67	1.62	1.58	1.56	1.54	1.50	1.45	1.42	1.35	1.31	1.27	150	
1.80	1.78	1.74	1.71	1.68	1.66	1.64	1.60	1.56	1.53	1.51	1.47	1.42	1.39	1.32	1.27	1.23	200	
1.77	1.75	1.72	1.69	1.66	1.64	1.62	1.57	1.54	1.51	1.48	1.45	1.39	1.36	1.28	1.23	1.18	200	
1.78	1.74	1.70	1.67	1.64	1.62	1.60	1.55	1.51	1.49	1.46	1.42	1.37	1.34	1.25	1.19	1.14	500	
1.74	1.72	1.69	1.65	1.63	1.60	1.58	1.54	1.50	1.47	1.44	1.41	1.35	1.32	1.23	1.16	1.09	1000	
1.73	1.71	1.67	1.64	1.61	1.59	1.57	1.52	1.48	1.45	1.43	1.39	1.23	1.30	1.21	1.13	1.00	∞	

Degrees of freedom for the denominator (v_2)

Tabulation of the values of $F_{0.01;\,\nu_1;\,\nu_2}$ versus ν_1 and ν_2

	Degrees of freedom for the numerator (ν_1)																	
	1	*2*	*3*	*4*	*5*	*6*	*7*	*8*	*9*	*10*	*11*	*12*	*13*	*14*	*15*	*16*	*17*	*18*
	Multiply the numbers of the first row ($\nu_2 = 1$) by 10																	
1	*405*	*500*	*540*	*563*	*576*	*596*	*598*	*598*	*602*	*606*	*608*	*611*	*613*	*614*	*616*	*617*	*618*	*619*
2	93.5	99.0	99.3	99.3	99.3	99.3	99.4	99.4	99.4	99.4	99.4	99.4	99.4	99.4	99.4	99.4	99.4	99.4
3	34.1	30.8	20.5	28.7	28.2	27.9	27.7	27.5	27.3	27.2	27.1	27.1	27.0	26.9	26.9	25.8	26.8	26.8
4	21.2	18.0	16.7	16.0	15.5	15.2	15.0	14.8	14.7	14.5	14.4	14.4	14.3	14.2	14.2	14.2	14.1	14.1
5	16.8	13.2	12.1	11.4	11.0	10.7	10.5	10.3	10.2	10.1	9.06	9.89	9.82	9.77	9.72	9.68	9.64	9.61
6	13.7	10.9	9.78	9.15	8.75	8.47	8.28	8.10	7.98	7.87	7.79	7.72	7.66	7.60	7.56	7.52	7.48	7.45
7	12.2	9.55	8.45	7.85	7.46	7.19	6.99	6.94	6.72	6.62	6.54	6.47	6.41	6.36	5.31	6.27	6.24	6.21
8	11.3	8.65	7.89	7.01	6.63	6.37	6.18	6.03	5.91	5.81	5.73	5.67	5.61	5.56	5.52	5.48	5.44	5.41
9	10.6	8.02	6.99	6.42	6.06	5.80	5.61	5.47	5.35	5.26	5.13	5.11	5.05	5.00	4.96	4.92	4.89	4.86
10	10.0	7.56	6.55	5.99	5.64	5.39	5.20	5.06	4.94	4.85	4.77	4.71	4.65	4.60	4.56	4.52	4.49	4.46
11	9.65	7.21	6.22	5.67	5.32	5.07	4.89	4.74	4.63	4.54	4.46	4.40	4.34	4.39	4.25	4.21	4.18	4.15
12	9.33	6.93	5.95	5.41	5.06	4.82	4.64	4.50	4.30	4.30	4.22	4.16	4.10	4.05	4.01	3.97	3.94	3.91
13	9.07	6.70	5.74	5.21	4.86	4.62	4.44	4.30	4.19	4.10	4.02	3.96	3.91	3.86	3.82	3.78	3.75	3.72
14	8.86	6.51	5.58	5.04	4.70	4.46	4.28	4.14	4.03	3.94	3.88	3.80	3.75	3.70	3.66	3.62	3.50	3.56
15	8.68	6.26	5.42	4.89	4.56	4.32	4.14	4.00	3.89	3.80	3.72	3.67	3.61-3.56	3.52	3.49	3.45	3.42
16	8.53	6.22	5.29	4.77	4.44	4.20	4.03	3.89	3.78	3.69	3.62	3.55	3.50	3.45	3.41	3.37	3.34	3.31
17	8.60	6.11	5.18	4.67	4.34	4.10	3.93	3.79	3.68	3.59	3.52	3.46	3.40	3.35	3.31	3.27	3.24	3.21
18	8.20	6.01	5.09	4.58	4.25	4.01	3.84	3.71	3.60	3.51	3.43	3.37	3.32	3.27	3.22	3.19	3.16	3.13
19	8.18	5.93	5.01	4.50	4.17	3.94	3.77	3.68	3.52	3.43	3.36	3.30	3.24	3.19	3.15	3.12	3.08	3.05
20	8.10	5.85	4.94	4.43	4.10	3.87	3.70	3.56	3.46	3.37	3.29	3.23	3.18	3.13	3.09	3.05	3.02	2.99
21	8.02	5.78	4.87	4.37	4.04	3.81	3.64	3.51	3.40	3.31	3.24	3.17	3.12	3.07	3.03	2.99	2.96	2.93
22	7.95	5.72	4.82	4.31	3.99	3.76	3.59	3.45	3.35	3.26	3.18	3.12	3.07	3.02	2.98	2.94	2.91	2.88
23	7.86	5.66	4.76	4.26	3.94	3.71	3.54	3.41	3.30	3.21	3.14	3.07	3.02	2.97	2.93	2.89	2.86	2.83
24	7.82	5.61	4.72	4.22	3.90	3.67	3.50	3.36	3.26	3.17	3.09	3.03	2.98	2.93	2.89	2.85	2.83	2.79
25	7.77	5.57	4.68	4.18	3.86	3.63	3.46	3.32	3.22	3.13	3.06	2.99	2.94	2.89	2.85	2.81	2.78	2.75
26	7.72	5.53	4.64	4.14	3.82	3.59	3.42	3.29	3.18	3.09	3.02	2.96	2.90	2.86	2.82	2.78	2.74	2.72
27	7.66	5.49	4.60	4.11	3.78	3.56	3.39	3.26	3.15	3.06	2.99	2.93	2.87	2.82	2.78	2.75	2.71	2.68
28	7.64	5.45	4.57	4.07	3.75	3.53	3.36	3.23	3.12	3.03	2.96	2.90	2.84	2.79	2.75	2.72	2.68	2.65
29	7.60	5.42	4.54	4.04	3.73	3.50	3.33	3.20	3.09	3.00	2.93	2.87	2.81	2.77	2.73	2.69	2.66	2.63
30	7.56	5.39	4.51	4.03	3.70	3.47	3.30	3.17	3.07	2.98	2.91	2.84	2.79	2.74	2.70	2.66	2.63	2.60
32	7.50	5.34	4.46	3.97	3.65	3.43	3.26	3.13	3.02	2.93	2.86	2.80	2.74	2.70	2.66	2.62	2.58	2.55
34	7.44	5.29	4.42	3.93	3.61	3.39	3.23	3.09	2.96	2.89	2.82	2.76	2.70	2.66	2.62	2.58	2.55	2.51
36	7.40	5.25	4.36	3.89	3.57	3.35	3.18	3.05	2.95	2.86	2.79	2.72	2.67	2.62	2.58	2.54	2.51	2.48
38	7.35	5.21	4.34	3.86	3.54	3.32	3.15	3.02	2.92	2.83	2.75	2.69	2.64	2.59	2.55	2.51	2.48	2.45
40	7.31	5.18	4.31	3.83	3.51	3.29	3.12	2.99	2.80	2.80	2.73	2.66	2.61	2.56	2.52	2.48	2.45	2.42

Degrees of freedom for the denominator (ν_2)

Degrees of freedom for the numerator (ν_1)

Multiply the numbers of the first row ($\nu_2 = 1$) by 10

19	20	23	24	26	28	30	35	40	45	50	60	90	100	200	500	∞	Degrees of freedom for the denominator (ν_2)
020	621	623	623	624	625	626	628	629	630	630	631	633	633	635	636	637	1
99.4	99.4	99.5	99.5	99.5	99.5	99.5	99.5	99.5	99.5	99.5	99.5	99.5	99.5	99.5	99.5	99.5	2
26.7	26.7	26.6	26.6	26.6	26.5	26.5	26.5	26.4	26.4	26.4	26.3	26.3	26.2	26.2	26.1	26.1	3
14.0	14.0	14.0	13.9	13.9	13.9	13.8	13.8	13.7	13.7	13.7	13.7	13.6	13.6	13.5	13.5	13.5	4
9.58	9.55	9.51	9.47	9.43	9.40	9.38	9.33	9.39	9.36	9.24	9.20	9.16	9.13	9.08	9.04	9.02	5
7.42	7.40	7.35	7.31	7.28	7.25	7.23	7.18	7.14	7.11	7.09	7.06	7.01	6.99	6.93	6.90	6.88	6
6.18	6.16	6.11	6.07	6.04	6.02	5.99	5.94	5.91	5.88	5.88	5.82	5.78	5.75	5.70	5.67	5.65	7
5.38	5.28	5.32	5.28	5.25	5.22	5.20	5.15	5.12	5.09	5.07	5.03	4.99	4.96	4.91	4.88	4.86	8
4.53	4.81	4.77	4.73	4.70	4.67	4.65	4.60	4.57	4.54	4.52	4.48	4.44	4.42	4.36	4.33	4.31	9
4.43	4.41	4.36	4.33	4.30	4.27	4.25	4.20	4.17	4.14	4.12	4.08	4.04	4.01	3.96	3.93	3.91	10
4.12	4.10	4.06	4.02	3.99	3.95	3.94	3.89	3.88	3.83	3.81	3.78	3.73	3.71	3.66	3.62	3.60	11
3.88	3.86	3.82	3.78	3.75	3.72	3.70	3.65	3.62	3.59	3.57	3.54	3.49	3.47	3.41	3.38	3.36	12
3.69	3.66	3.62	3.59	3.56	3.53	3.51	3.46	3.43	3.40	3.38	3.34	3.30	3.27	3.22	3.19	3.17	13
3.53	3.51	3.46	3.43	3.40	3.37	3.35	3.30	3.27	3.24	3.22	3.18	3.14	3.11	3.06	3.03	3.00	14
3.40	3.37	3.33	3.29	3.26	3.24	3.21	3.17	3.13	3.10	3.08	3.05	3.00	2.98	2.92	2.89	2.87	15
3.28	3.26	3.22	3.18	3.15	3.12	3.10	3.05	3.02	2.99	2.97	2.93	2.89	2.88	2.81	2.78	2.75	16
3.18	3.16	3.12	3.08	3.05	3.03	3.00	2.96	2.92	2.89	2.87	2.83	2.79	2.76	2.71	2.68	2.68	17
3.10	3.08	3.03	3.00	2.97	2.94	2.92	2.87	2.84	2.81	2.78	2.75	2.70	2.68	2.62	2.59	2.57	18
3.03	3.00	2.96	2.92	2.89	2.87	2.84	2.80	2.76	2.73	2.71	2.67	2.63	2.60	2.55	2.51	2.49	19
2.93	2.94	2.90	2.88	2.83	2.80	2.78	2.73	2.69	2.67	2.64	2.61	2.56	2.54	2.48	2.44	2.42	20
2.90	2.88	2.84	2.80	2.77	2.74	2.72	2.67	2.64	2.61	2.58	2.55	2.50	2.46	2.42	2.38	2.36	21
2.85	2.83	2.78	2.75	2.72	2.69	2.67	2.62	2.58	2.55	2.53	2.50	2.45	2.42	2.36	2.33	2.31	22
2.80	2.78	2.74	2.70	2.67	2.64	2.62	2.57	2.54	2.51	2.48	2.45	2.40	2.37	2.32	2.28	2.26	23
2.76	2.74	2.70	2.66	2.63	2.60	2.58	2.53	2.49	2.46	2.44	2.40	2.36	2.33	2.27	2.24	2.21	24
2.72	2.70	2.66	2.62	2.50	2.56	2.54	2.49	2.45	2.42	2.40	2.36	2.32	2.29	2.23	2.19	2.17	25
2.69	2.66	2.62	2.58	2.55	2.53	2.50	2.45	2.42	2.39	2.36	2.33	2.28	2.25	2.19	2.16	2.13	26
2.66	2.63	2.59	2.55	2.52	2.49	2.47	2.42	2.38	2.35	2.33	2.29	2.25	2.22	2.16	2.12	2.10	27
2.63	2.60	2.56	2.52	2.49	2.46	2.44	2.39	2.35	2.32	2.30	2.26	2.22	2.19	2.13	2.09	2.06	28
2.60	2.57	2.53	2.49	2.46	2.44	2.41	2.36	2.33	2.30	2.27	2.23	2.19	2.16	2.10	2.06	2.03	29
2.57	2.55	2.51	2.47	2.44	2.41	2.39	2.34	2.30	2.27	2.24	2.21	2.16	2.13	2.07	2.03	2.01	30
2.53	2.50	2.46	2.42	2.39	2.36	2.34	2.29	2.25	2.22	2.20	2.16	2.11	2.08	2.02	1.98	1.96	32
2.49	2.46	2.42	2.38	2.35	2.32	2.30	2.25	2.21	2.18	2.16	2.12	2.07	2.04	1.98	1.94	1.91	34
2.45	2.43	2.38	2.35	2.32	2.29	2.26	2.21	2.17	2.14	2.12	2.08	2.03	2.00	1.94	1.90	1.87	36
2.42	2.40	2.35	2.32	2.28	2.26	2.23	2.18	2.14	2.11	2.09	2.05	2.00	1.97	1.90	1.86	1.84	38
2.39	2.37	2.33	2.39	2.26	2.23	2.20	2.15	2.11	2.08	2.06	2.02	1.97	1.94	1.87	1.83	1.80	40

		Degrees of freedom for the numerator (ν_1)																
		1	*2*	*3*	*4*	*5*	*6*	*7*	*8*	*9*	*10*	*11*	*12*	*13*	*14*	*15*	*16*	*17*	*18*

Multiply the numbers of the first row ($\nu_2 = 1$) by 10

Degrees of freedom for the denominator (ν_2)		*1*	*2*	*3*	*4*	*5*	*6*	*7*	*8*	*9*	*10*	*11*	*12*	*13*	*14*	*15*	*16*	*17*	*18*
	42	7.28	5.15	4.29	3.80	3.49	3.27	3.10	2.97	2.86	2.78	2.70	2.64	2.59	2.54	2.50	2.46	2.43	2.40
	44	7.25	5.12	4.26	3.78	3.47	3.24	3.08	2.95	2.84	2.75	2.68	2.62	2.56	2.52	2.47	2.44	2.40	2.37
	46	7.22	5.10	4.24	3.76	3.44	3.22	3.06	2.93	2.82	2.73	2.66	2.60	2.54	2.50	2.45	2.42	2.38	2.35
	48	7.19	5.08	4.22	3.74	3.43	3.20	3.04	2.91	2.80	2.72	2.64	2.58	2.53	2.48	2.44	2.40	2.37	2.33
	50	7.17	5.06	4.20	3.72	3.41	3.19	3.02	2.89	2.79	2.70	2.63	2.56	2.51	2.46	2.42	2.38	2.35	2.32
	55	7.12	5.01	4.16	3.68	3.37	3.15	2.98	2.85	2.75	2.66	2.59	2.53	2.47	2.42	2.38	2.34	2.31	2.28
	60	7.08	4.98	4.12	3.65	3.34	3.12	2.95	2.82	2.72	2.63	2.56	2.50	2.44	2.39	2.35	2.31	2.28	2.24
	65	7.04	4.95	4.10	3.62	3.31	3.09	2.93	2.80	2.69	2.61	2.53	2.47	2.42	2.37	2.33	2.29	2.26	2.23
	70	7.01	4.92	4.08	3.60	3.29	3.07	2.91	2.78	2.67	2.59	2.51	2.45	2.40	2.24	2.31	2.27	2.23	2.20
	80	6.98	4.88	4.04	3.56	3.26	3.04	2.87	2.74	2.64	2.55	2.48	2.42	2.36	2.21	2.27	2.23	2.20	2.17
	90	6.93	4.85	4.01	3.54	3.23	3.01	2.84	2.72	2.61	2.52	2.45	2.39	2.33	2.29	2.24	2.21	2.17	2.14
	100	6.90	4.83	3.96	3.51	3.21	2.99	2.82	2.69	2.59	2.50	2.43	2.37	2.31	2.26	2.22	2.19	2.15	2.12
	125	6.84	4.78	3.94	3.47	3.17	2.95	2.79	2.66	2.55	2.47	2.39	2.33	2.28	2.23	2.19	2.15	2.11	2.08
	150	6.81	4.75	3.92	3.45	3.14	2.92	2.76	2.63	2.53	2.44	2.37	2.31	2.25	2.20	2.16	2.12	2.09	2.06
	200	6.76	4.71	3.88	3.41	3.11	2.89	2.73	2.60	2.50	2.41	2.34	2.27	2.22	2.17	2.13	2.09	2.06	2.02
	300	6.72	4.68	3.85	3.38	3.08	2.86	2.70	2.57	2.47	2.36	2.31	2.24	2.19	2.14	2.10	2.06	2.03	1.99
	500	6.69	4.65	3.82	3.36	3.05	2.84	2.68	2.55	2.44	2.36	2.28	2.22	2.17	2.12	2.07	2.04	2.00	1.97
	1000	6.66	4.63	3.80	3.34	3.04	2.82	2.66	2.53	2.43	2.34	2.27	2.20	2.15	2.10	2.06	2.02	1.98	1.95
	∞	6.66	4.61	3.78	3.32	3.02	2.80	2.64	2.51	2.41	2.32	2.25	2.18	2.13	2.08	2.04	2.00	1.97	1.00

							Degrees of freedom for the numerator (ν_1)										
19	20	23	24	26	28	30	35	40	45	50	60	90	100	200	500	∞	
							Multiply the numbers of the first row ($\nu_2 = 1$) by 10										
2.37	2.34	2.30	2.26	2.23	2.20	2.18	2.13	2.09	2.08	2.03	1.99	1.94	1.91	1.85	1.80	1.78	42
2.35	2.32	2.28	2.24	2.21	2.18	2.15	2.10	2.06	2.03	2.01	1.97	1.92	1.89	1.82	1.78	1.75	44
2.33	2.30	2.26	2.22	2.19	2.16	2.13	2.08	2.04	2.01	1.99	1.95	1.90	1.86	1.80	1.75	1.73	46
2.31	2.28	2.24	2.20	2.17	2.14	2.12	2.06	2.02	1.99	1.97	1.93	1.88	1.84	1.78	1.73	1.70	48
2.29	2.27	2.22	2.18	2.15	2.12	2.10	2.05	2.01	1.97	1.95	1.91	1.86	1.82	1.76	1.71	1.68	50
2.25	2.23	2.18	2.15	2.11	2.08	2.06	2.01	1.97	1.92	1.91	1.87	1.81	1.78	1.71	1.67	1.64	55
2.22	2.20	2.15	2.12	2.08	2.05	2.03	1.98	1.94	1.90	1.88	1.84	1.78	1.75	1.68	1.63	1.60	60
2.20	2.17	2.13	2.09	2.06	2.03	2.00	1.95	1.91	1.88	1.85	1.81	1.75	1.72	1.65	1.60	1.57	65
2.18	2.15	2.11	2.07	2.03	2.01	1.98	1.95	1.89	1.85	1.83	1.78	1.73	1.70	1.62	1.57	1.54	70
2.14	2.12	2.07	2.03	2.00	1.97	1.94	1.89	1.85	1.81	1.79	1.75	1.69	1.66	1.58	1.53	1.49	80
2.11	2.09	2.04	2.00	1.97	1.94	1.92	1.86	1.82	1.79	1.76	1.72	1.66	1.62	1.54	1.49	1.46	90
2.09	2.07	2.02	1.98	1.94	1.92	1.89	1.84	1.80	1.76	1.73	1.69	1.63	1.60	1.52	1.47	1.43	100
2.05	2.03	1.98	1.94	1.91	1.88	1.85	1.80	1.79	1.72	1.69	1.65	1.59	1.55	1.47	1.41	1.37	125
2.03	2.00	1.96	1.92	1.88	1.85	1.83	1.77	1.73	1.69	1.66	1.63	1.56	1.52	1.43	1.35	1.33	150
2.00	1.97	1.93	1.89	1.84	1.82	1.79	1.74	1.69	1.66	1.63	1.58	1.52	1.48	1.39	1.33	1.28	200
1.97	1.94	1.89	1.85	1.82	1.79	1.76	1.71	1.68	1.62	1.58	1.55	1.48	1.44	1.35	1.28	1.22	300
1.94	1.92	1.87	1.83	1.79	1.76	1.74	1.68	1.63	1.60	1.55	1.53	1.45	1.41	1.31	1.23	1.16	500
1.92	1.90	1.85	1.81	1.77	1.74	1.72	1.66	1.61	1.57	1.54	1.50	1.43	1.38	1.28	1.19	1.11	1000
1.90	1.86	1.83	1.79	1.76	1.72	1.70	1.64	1.58	1.55	1.52	1.47	1.40	1.36	1.25	1.15	1.00	∞

Degrees of freedom for the denominator (ν_2)

Tabulation of the values of $F_{0.005;\, \nu_1;\, \nu_2}$ versus ν_1 and ν_2

ν_2	Degrees of freedom for the numerator (ν_1)																	
	1	2	3	4	5	6	7	8	9	10	11	12	13	14	15	16	17	18
	Multiply the numbers of the first row ($\nu_2 = 1$) by 100																	
1	162	200	216	225	231	234	237	239	241	242	243	244	245	246	246	247	247	248
2	198	199	199	199	199	199	199	199	199	199	199	199	199	199	199	199	199	199
3	55.6	49.8	47.5	46.2	45.4	44.3	44.4	44.1	43.9	43.7	43.5	43.4	43.3	43.2	43.1	43.0	42.9	42.9
4	21.2	26.3	24.3	23.2	22.5	22.0	21.6	21.4	21.1	21.0	20.8	20.7	20.6	20.5	20.4	20.4	20.3	20.2
5	22.8	18.2	16.5	15.6	14.3	14.5	14.2	14.0	13.8	13.6	13.5	13.4	13.3	12.2	12.1	12.1	13.0	13.0
6	18.6	14.5	12.9	13.0	11.5	11.1	10.8	10.6	10.4	10.2	10.1	10.0	9.95	9.88	9.81	9.76	9.71	9.08
7	16.2	12.4	10.9	10.0	9.52	9.16	8.89	8.68	8.51	8.38	8.27	8.18	8.10	8.03	7.97	7.93	7.87	7.83
8	14.7	11.0	9.60	8.81	8.30	7.95	7.69	7.50	7.34	7.21	7.10	7.01	6.94	6.37	6.34	6.76	6.72	6.68
9	13.6	10.1	8.72	7.96	7.47	7.13	6.88	6.69	6.54	6.42	6.31	6.23	6.15	6.09	6.03	5.98	5.94	5.90
10	12.8	9.43	8.06	7.34	6.87	6.54	6.30	6.12	5.97	5.85	5.75	5.66	5.50	5.53	5.47	5.42	5.38	5.34
11	12.2	8.91	7.60	6.88	6.42	6.10	5.86	5.68	5.54	5.42	5.32	5.24	5.16	5.10	5.05	5.00	4.96	4.92
12	11.8	8.51	7.23	6.52	6.07	5.76	5.52	5.35	5.20	5.09	4.99	4.91	4.84	4.77	4.72	4.67	4.63	4.50
13	11.4	8.19	6.93	6.23	5.79	5.48	5.25	5.08	4.94	4.82	4.72	4.54	4.57	4.54	4.46	4.41	4.37	4.23
14	11.1	7.92	6.68	6.00	5.56	5.26	5.03	4.86	4.72	4.60	4.51	4.43	4.36	4.30	4.25	4.20	4.18	4.12
15	10.8	7.70	6.48	5.80	5.37	5.07	4.85	4.67	4.54	4.42	4.33	4.25	4.18	4.12	4.07	4.02	3.93	3.95
16	10.6	7.51	6.30	5.64	5.21	4.91	4.69	4.52	4.48	4.27	4.18	4.10	4.03	3.97	3.92	3.87	3.83	3.80
17	10.4	7.35	6.16	5.50	5.07	4.78	4.56	4.39	4.25	4.14	4.05	3.97	3.90	3.84	3.79	3.75	3.71	3.67
18	10.2	7.21	6.03	5.37	4.98	4.66	4.44	4.28	4.14	4.03	3.94	3.86	3.79	3.73	3.68	3.64	3.60	3.56
19	10.1	7.09	5.92	5.27	4.85	4.56	4.34	4.18	4.04	3.93	3.84	3.76	3.70	3.64	3.59	3.54	3.50	3.46
20	9.94	6.99	5.82	5.17	4.76	4.47	4.28	4.09	3.96	3.85	3.76	3.68	3.61	3.55	3.50	3.48	3.42	3.28
21	9.82	6.89	5.73	5.09	4.39	4.63	4.18	4.01	3.88	3.77	3.68	3.60	3.54	3.48	3.43	3.38	3.34	3.31
22	9.73	6.81	5.65	5.02	4.61	4.32	4.11	3.94	3.81	3.70	3.61	3.54	3.47	3.41	3.36	3.31	3.27	3.24
23	9.63	6.72	5.58	4.95	4.54	4.26	4.05	3.85	3.75	3.64	3.55	3.47	3.41	3.35	3.30	3.25	3.21	3.18
24	9.55	6.66	5.52	4.89	4.49	4.20	3.99	3.83	3.69	3.59	3.50	3.42	3.35	3.30	3.25	3.20	3.16	3.12
25	9.48	6.60	5.46	4.84	4.43	4.15	3.94	3.78	3.64	3.54	3.45	3.37	3.30	3.25	3.20	3.15	3.11	3.08
26	9.41	6.54	5.41	4.79	4.36	4.10	3.89	3.73	3.60	3.49	3.40	3.33	3.26	3.20	3.15	3.17	3.07	3.03
27	9.34	6.49	5.36	4.74	3.44	4.06	3.85	3.69	3.56	3.45	3.36	3.28	3.22	3.16	3.11	3.07	3.03	2.99
28	9.28	6.44	5.32	4.70	4.30	4.02	3.81	3.65	3.52	3.41	3.32	3.25	3.18	3.12	3.07	2.08	2.99	2.95
29	9.23	6.40	5.26	4.66	4.26	3.98	3.77	3.61	3.48	3.38	3.29	3.21	3.15	3.09	3.04	2.09	2.95	2.92
30	9.18	6.35	5.24	4.62	4.23	3.95	3.74	3.58	3.45	3.34	3.25	3.18	3.11	3.06	3.01	2.98	2.92	2.89
32	9.09	6.28	5.17	4.56	4.17	3.89	3.68	3.52	3.39	3.29	3.20	3.12	3.06	3.00	2.95	2.90	2.86	2.83
34	9.01	6.22	5.11	4.50	4.11	3.84	3.63	3.47	3.34	3.24	3.15	3.07	3.01	2.95	2.90	2.85	2.81	2.78
36	8.94	6.15	5.06	4.46	4.06	3.79	3.58	3.42	3.30	3.19	3.10	3.03	2.96	2.90	2.85	2.81	2.77	2.73
38	8.88	6.11	5.02	4.41	4.02	3.75	3.54	3.39	3.25	3.15	3.06	2.99	2.92	2.87	2.82	2.77	2.73	2.70
40	8.83	6.07	4.98	4.37	3.99	3.71	3.51	3.35	3.22	3.12	3.03	2.95	2.90	2.83	2.78	2.74	2.70	2.66

Degrees of freedom for the denominator (ν_2)

19	20	22	24	26	28	30	35	40	45	50	60	80	100	200	500	∞	ν₂

Degrees of freedom for the numerator (ν_1)

Columns: 19 20 | 22 24 26 28 30 | 35 40 45 50 60 | 80 100 200 500 ∞ | (denominator ν_2)

Multiply the numbers of the first row $(\nu_2 = 1)$ by 100

19	20	22	24	26	28	30	35	40	45	50	60	80	100	200	500	∞	ν_2
248	248	249	249	250	250	250	251	251	252	252	253	253	253	254	254	255	1
199	199	199	199	199	199	199	199	199	199	199	199	199	199	199	200	200	2
42.8	42.8	42.7	42.6	42.6	42.5	42.5	42.4	42.3	42.3	42.2	42.1	42.1	42.0	41.9	41.9	41.8	3
20.2	20.2	20.1	20.0	20.0	19.9	19.9	19.8	19.8	19.7	19.7	19.6	19.5	19.5	19.4	19.4	19.3	4
12.9	12.9	12.8	12.8	12.7	12.6	12.7	12.6	12.5	12.5	12.5	12.4	12.3	12.3	12.2	12.2	12.1	5
9.62	9.59	9.53	9.47	9.43	9.39	9.36	9.29	9.24	9.20	9.17	9.12	9.06	9.03	8.95	8.91	8.83	6
7.79	7.75	7.69	7.64	7.60	7.57	7.53	7.47	7.42	7.38	7.25	7.21	7.25	7.22	7.15	7.10	7.08	7
6.64	6.61	6.55	6.50	6.46	6.43	6.40	6.33	6.29	6.25	6.22	6.18	6.12	6.09	6.02	5.98	5.95	8
5.86	5.83	5.78	5.73	5.69	5.65	5.62	5.56	5.52	5.48	5.45	5.41	5.36	5.32	5.26	5.21	5.19	9
5.30	5.27	5.22	5.17	5.13	5.10	5.07	5.01	4.97	4.93	4.90	4.98	4.80	4.77	4.71	4.67	4.64	10
4.89	4.86	4.80	4.76	4.72	4.68	4.65	4.60	4.55	4.52	4.49	4.44	4.39	4.36	4.20	4.25	4.23	11
4.56	4.53	4.48	4.43	4.39	4.36	4.33	4.27	4.23	4.19	4.17	4.12	4.07	4.04	3.97	3.93	3.90	12
4.30	4.27	4.22	4.17	4.13	4.10	4.07	4.01	3.97	3.94	3.91	3.87	3.81	3.78	3.71	3.67	3.65	13
4.09	4.08	4.01	3.96	3.92	3.89	3.86	3.80	3.76	3.73	3.70	3.66	3.60	3.57	3.50	3.46	3.44	14
3.91	3.88	3.83	3.79	3.75	3.72	3.69	3.63	3.58	3.55	3.52	3.43	3.43	3.30	3.33	3.29	3.28	15
3.76	3.73	3.68	3.64	3.60	3.57	3.54	3.48	3.44	3.40	3.37	3.33	3.28	3.25	3.18	3.14	3.11	16
3.64	3.61	3.56	3.51	3.47	3.44	3.41	3.35	3.31	3.28	3.25	3.21	3.15	3.12	3.05	3.01	2.98	17
3.53	3.50	3.45	3.40	3.36	3.33	3.33	3.25	3.20	3.17	3.14	3.10	3.04	3.01	2.94	2.90	2.87	18
3.43	3.40	3.35	3.31	3.27	3.24	3.21	3.15	3.11	3.07	3.04	3.00	2.95	2.91	2.85	2.80	2.78	19
3.35	3.32	3.27	3.22	3.18	3.15	3.12	3.07	3.02	2.99	2.96	2.92	2.86	2.83	2.76	2.72	2.69	20
3.27	3.24	3.19	3.15	3.11	3.08	3.05	2.99	2.95	2.91	2.88	2.84	2.78	2.75	2.68	2.64	2.61	21
3.20	3.18	3.12	3.08	3.04	3.01	2.98	2.92	2.88	2.84	2.82	2.77	2.72	2.69	2.62	2.57	2.55	22
3.15	3.12	3.06	3.02	2.98	2.95	2.92	2.86	2.82	2.78	2.76	2.71	2.66	2.62	2.56	2.51	2.48	23
3.09	3.06	3.01	2.97	2.93	2.90	2.87	2.81	2.77	2.73	2.70	2.68	2.60	2.57	2.50	2.46	2.43	24
3.04	3.01	2.96	2.92	2.88	2.85	2.82	2.76	2.72	2.68	2.65	2.61	2.55	2.52	2.45	2.41	2.38	25
3.00	2.97	2.92	2.87	2.83	2.80	2.77	2.72	2.67	2.64	2.61	2.56	2.51	2.47	2.40	2.36	2.33	26
2.96	2.93	2.88	2.83	2.79	2.76	2.73	2.67	2.63	2.58	2.57	2.52	2.47	2.43	2.36	2.32	2.29	27
2.92	2.89	2.84	2.79	2.76	2.72	2.69	2.64	2.59	2.56	2.53	2.48	2.43	2.39	2.32	2.28	2.25	28
2.88	2.86	2.80	2.76	2.72	2.69	2.66	2.60	2.56	2.52	2.49	2.45	2.39	2.36	2.28	2.24	2.21	29
2.85	2.82	2.77	2.73	2.69	2.66	2.63	2.57	2.52	2.49	2.46	2.42	2.36	2.32	2.25	2.21	2.18	30
2.80	2.77	2.71	2.67	2.63	2.60	2.57	2.51	2.47	2.43	2.40	2.36	2.30	2.26	2.19	2.15	2.11	32
2.75	2.72	2.66	2.62	2.58	2.55	2.52	2.46	2.42	2.38	2.35	2.30	2.25	2.21	2.14	2.09	2.06	34
2.70	2.67	2.62	2.58	2.54	2.50	2.48	2.42	2.37	2.33	2.30	2.26	2.20	2.17	2.09	2.04	2.01	36
2.66	2.63	2.58	2.54	2.50	2.47	2.44	2.38	2.33	2.29	2.27	2.22	2.16	2.12	2.05	2.00	1.97	38
2.63	2.60	2.55	2.50	2.46	2.43	2.40	2.34	2.30	2.28	2.23	2.18	2.12	2.09	2.01	1.96	1.93	40

Degrees of freedom for the denominator (ν_2)

						Degrees of freedom for the numerator (v_1)												
	1	2	3	4	5	6	7	8	9	10	11	12	13	14	15	16	17	18
					Multiply the numbers of the first row ($v_2 = 1$) by 100													
42	8.78	6.03	4.94	4.34	3.95	3.68	3.48	3.32	3.19	3.09	3.00	2.92	2.86	2.80	2.75	2.71	2.67	2.63
44	8.74	5.99	4.91	4.31	3.92	3.65	3.45	3.29	3.16	3.06	2.97	2.89	2.83	2.77	2.72	2.68	2.64	2.60
46	8.70	5.96	4.88	4.28	3.90	3.62	3.42	3.26	3.14	3.03	2.94	2.87	2.80	2.75	2.70	2.65	2.61	2.58
48	8.66	5.93	4.85	4.25	3.87	3.60	3.40	3.24	3.11	3.01	2.92	2.85	2.78	2.72	2.67	2.63	2.59	2.55
50	8.63	5.90	4.83	4.23	3.85	3.58	3.38	3.22	3.09	2.99	2.90	2.82	2.76	2.70	2.65	2.61	2.57	2.53
55	8.55	5.84	4.77	4.18	3.80	3.53	3.33	3.17	3.05	2.94	2.85	2.78	2.71	2.66	2.61	2.56	2.52	2.49
60	8.49	5.80	4.73	4.14	3.76	3.49	3.29	3.12	3.01	2.90	2.82	2.74	2.68	2.62	2.57	2.53	2.49	2.45
65	8.44	5.75	4.68	4.11	3.73	3.45	3.26	3.10	2.98	2.87	2.79	2.71	2.63	2.59	2.54	2.49	2.45	2.42
70	8.40	5.72	4.65	4.08	3.70	3.43	3.23	3.08	2.95	2.85	2.76	2.68	2.62	2.56	2.51	2.47	2.43	2.39
80	8.33	5.67	4.61	4.03	3.65	3.39	3.19	3.03	2.91	2.80	2.72	2.64	2.58	2.52	2.47	2.43	2.39	2.35
90	8.28	5.62	4.57	3.99	3.62	3.35	3.15	3.00	2.87	2.77	2.68	2.61	2.54	2.49	2.44	2.39	2.35	2.32
100	8.24	5.59	4.54	3.96	3.59	3.33	3.13	2.97	2.85	2.74	2.66	2.58	2.52	2.46	2.41	2.37	2.33	2.29
125	8.17	5.53	4.49	3.91	3.54	3.28	3.08	2.93	2.80	2.70	2.61	2.54	2.47	2.42	2.37	2.32	2.28	2.24
150	8.12	5.49	4.43	3.88	3.51	3.25	3.05	2.89	2.77	2.67	2.58	2.51	2.44	2.38	2.33	2.29	2.25	2.21
200	8.06	5.44	4.41	3.84	3.47	3.21	3.01	2.85	2.73	2.63	2.54	2.47	2.40	2.35	2.30	2.26	2.21	2.18
300	8.00	5.39	4.37	3.80	3.43	3.17	2.97	2.81	2.69	2.59	2.51	2.43	2.37	2.31	2.26	2.21	2.17	2.14
500	7.95	5.36	4.33	3.76	3.40	3.14	2.94	2.79	2.66	2.56	2.48	2.40	2.34	2.28	2.23	2.19	2.14	2.11
1000	7.92	5.33	4.31	3.74	3.37	3.11	2.92	2.77	2.64	2.54	2.45	2.38	2.32	2.28	2.21	2.16	2.12	2.09
∞	7.88	5.30	4.28	3.72	3.35	2.09	2.90	2.74	2.62	2.52	2.43	2.36	2.29	2.24	2.19	2.14	2.10	2.08

Degrees of freedom for the denominator (v_2)

Degrees of freedom for the numerator (v_1)

19	20		22	24	26	28	30		35	40	45	50	60		80	100	200	500	∞		
						Multiply the numbers of the first row ($v_2 = 1$) by 100															
2.60	2.57		2.52	2.47	2.43	2.40	2.37		2.31	2.26	2.23	2.20	2.15		2.09	2.06	1.98	1.93	1.90		42
2.57	2.54		2.49	2.44	2.40	2.37	2.34		2.28	2.24	2.20	2.17	2.12		2.06	2.03	1.95	1.90	1.87		44
2.54	2.51		2.46	2.42	2.38	2.34	2.32		2.26	2.21	2.17	2.14	2.10		2.04	2.00	1.92	1.87	1.84		46
2.52	2.49		2.44	2.39	2.36	2.32	2.29		2.23	2.19	2.15	2.12	2.07		2.01	1.97	1.89	1.84	1.81		48
2.50	2.47		2.42	2.37	2.33	2.30	2.27		2.21	2.16	2.13	2.10	2.05		1.99	1.95	1.87	1.82	1.79		50
2.45	2.42		2.37	2.33	2.29	2.26	2.23		2.16	2.12	2.08	2.05	2.00		1.94	1.90	1.82	1.77	1.73		55
2.42	2.39		2.33	2.29	2.24	2.22	2.19		2.13	2.08	2.04	2.01	1.98		1.90	1.86	1.78	1.73	1.69		60
2.39	2.36		2.30	2.26	2.22	2.19	2.16		2.09	2.05	2.01	1.98	1.93		1.87	1.83	1.74	1.69	1.65		65
2.36	2.33		2.28	2.23	2.19	2.16	2.13		2.07	2.02	1.98	1.93	1.90		1.84	1.80	1.71	1.65	1.62		70
2.32	2.29		2.23	2.19	2.15	2.11	2.08		2.02	1.97	1.93	1.90	1.85		1.79	1.75	1.66	1.60	1.56		80
2.28	2.25		2.20	2.15	2.12	2.08	2.05		1.99	1.94	1.90	1.87	1.82		1.75	1.71	1.62	1.56	1.52		90
2.26	2.23		2.17	2.13	2.09	2.05	2.02		1.96	1.91	1.87	1.84	1.79		1.72	1.68	1.59	1.53	1.49		100
2.21	2.18		2.13	2.08	2.04	2.01	1.98		1.91	1.86	1.82	1.79	1.74		1.67	1.63	1.53	1.47	1.42		125
2.18	2.15		2.10	2.05	2.01	1.98	1.94		1.88	1.83	1.79	1.76	1.70		1.63	1.59	1.49	1.42	1.37		150
2.14	2.11		2.06	2.01	1.97	1.94	1.91		1.84	1.79	1.75	1.71	1.66		1.58	1.54	1.44	1.37	1.31		200
2.10	2.07		2.02	1.97	1.93	1.90	1.87		1.80	1.75	1.71	1.67	1.61		1.54	1.50	1.39	1.31	1.25		300
2.07	2.04		1.99	1.94	1.90	1.87	1.84		1.77	1.72	1.67	1.64	1.58		1.51	1.46	1.35	1.26	1.18		500
2.05	2.02		1.97	1.92	1.88	1.84	1.81		1.75	1.69	1.65	1.61	1.56		1.48	1.43	1.31	1.22	1.12		1000
2.03	2.00		1.95	1.90	1.86	1.82	1.79		1.72	1.67	1.63	1.59	1.53		1.45	1.40	1.28	1.17	1.00		∞

Degrees of freedom for the denominator (v_2)

Tabulation of the values of $F_{0.001; \nu_1; \nu_2}$ versus ν_1 and ν_2

		Degrees of freedom for the numerator (ν_1)																	
		1	2	3	4	5	6	7	8	9	10	15	20	30	50	100	200	500	∞
		Multiply the numbers of the first row ($\nu_2 = 1$) by 1000																	
	1	405	500	540	562	576	586	593	598	602	606	616	621	626	630	633	635	636	637
	2	998	999	999	999	999	999	999	999	999	999	999	999	999	999	999	999	999	999
	3	168	148	141	137	135	133	132	131	130	129	127	126	125	125	124	124	124	124
	4	74.1	61.2	56.2	53.4	51.7	50.5	49.7	49.0	48.5	48.0	46.8	46.1	45.4	44.9	44.5	44.3	44.1	44.0
	5	47.0	36.6	33.2	31.1	29.8	28.8	28.2	27.6	27.2	26.9	25.9	25.4	24.9	24.4	24.1	23.9	23.8	23.8
	6	35.5	27.0	23.7	21.9	20.8	20.0	19.5	19.0	18.7	18.4	17.6	17.1	16.7	16.3	16.0	15.9	15.8	15.8
	7	29.2	21.7	18.8	17.2	16.2	15.5	15.0	14.6	14.3	14.1	13.3	12.9	12.5	12.2	11.9	11.8	11.7	11.7
	8	25.4	18.5	15.8	14.4	13.5	12.9	12.4	12.0	11.8	11.5	10.8	10.5	10.1	9.80	9.57	9.46	9.39	9.34
	9	22.9	16.4	13.9	12.6	11.7	11.1	10.7	10.4	10.1	9.89	9.24	8.90	8.55	8.26	8.04	7.93	7.86	7.81
	10	21.0	14.9	12.6	11.3	10.5	9.92	9.52	9.20	8.96	8.75	8.13	7.80	7.47	7.19	6.98	6.87	6.81	6.76
	11	19.7	13.8	11.6	10.4	9.58	9.05	8.66	8.35	8.12	7.92	7.32	7.01	6.68	6.41	6.21	6.10	6.04	6.00
	12	18.6	13.0	10.8	9.63	8.89	8.38	8.00	7.71	7.48	7.29	6.71	6.40	6.09	5.83	5.63	5.52	5.46	5.42
	13	17.8	12.3	10.2	9.07	8.35	7.86	7.49	7.21	6.98	6.80	6.23	5.93	5.62	5.37	5.17	5.07	5.01	4.97
	14	17.1	11.8	9.73	8.62	7.92	7.43	7.08	6.80	6.58	6.40	5.85	5.56	5.25	5.00	4.80	4.70	4.64	4.60
	15	16.6	11.3	9.34	8.25	7.57	7.09	6.74	6.47	6.26	6.08	5.53	5.25	4.95	4.70	4.51	4.41	4.35	4.31
	16	16.1	11.0	9.00	7.94	7.27	6.81	6.46	6.19	5.98	5.81	5.27	4.99	4.70	4.45	4.26	4.16	4.10	4.06
	17	15.7	10.7	8.73	7.68	7.02	6.56	6.22	5.96	5.75	5.58	5.05	4.78	4.48	4.24	4.05	3.95	3.89	3.85
	18	15.4	10.4	8.49	7.46	6.81	6.35	6.02	5.76	5.56	5.39	4.87	4.59	4.30	4.06	3.87	3.77	3.71	3.67
	19	15.1	10.2	8.28	7.26	6.61	6.18	5.84	5.59	5.39	5.22	4.70	4.43	4.14	3.90	3.71	3.61	3.55	3.51
	20	14.8	9.95	8.10	7.10	6.46	6.02	5.69	5.44	5.24	5.08	4.56	4.29	4.01	3.77	3.58	3.48	3.42	3.38
	22	14.4	9.61	7.80	6.81	6.19	5.76	5.44	5.19	4.99	4.83	4.32	4.06	3.77	3.53	3.34	3.25	3.19	3.15
	24	14.0	9.34	7.55	6.59	5.98	5.55	5.23	4.99	4.80	4.64	4.14	3.87	3.59	3.35	3.16	3.07	3.01	2.97
	26	13.7	9.12	7.36	6.41	5.80	5.38	5.07	4.83	4.64	4.48	3.99	3.72	3.45	3.20	3.01	2.92	2.86	2.82
	28	13.5	8.93	7.19	6.25	5.66	5.24	4.93	4.69	4.50	4.35	3.86	3.60	3.32	3.08	2.89	2.79	2.73	2.70
	30	13.3	8.77	7.05	6.12	5.53	5.12	4.82	4.58	4.39	4.24	3.75	3.49	3.22	2.98	2.79	2.69	2.63	2.59
	40	12.6	8.25	6.60	5.70	5.13	4.73	4.43	4.21	4.02	3.87	3.40	3.15	2.87	2.64	2.44	2.34	2.28	2.23
	50	12.2	7.95	6.34	5.46	4.90	4.51	4.22	4.00	3.82	3.67	3.20	2.95	2.68	2.44	2.24	2.14	2.07	2.03
	60	12.0	7.76	6.17	5.31	4.76	4.37	4.09	3.87	3.69	3.54	3.08	2.83	2.56	2.31	2.11	2.01	1.93	1.89
	80	11.7	7.54	5.97	5.13	4.58	4.21	3.92	3.70	3.53	3.39	2.93	2.68	2.40	2.16	1.95	1.84	1.77	1.72
	100	11.5	7.41	5.85	5.01	4.48	4.11	3.83	3.61	3.44	3.30	2.84	2.59	2.32	2.07	1.87	1.75	1.68	1.62
	200	11.2	7.15	5.64	4.81	4.29	3.92	3.65	3.43	3.26	3.12	2.67	2.42	2.15	1.90	1.68	1.55	1.46	1.39
	500	11.0	7.01	5.51	4.69	4.18	3.82	3.54	3.33	3.16	3.02	2.58	2.33	2.05	1.80	1.57	1.43	1.32	1.23
	∞	10.8	6.91	5.42	4.62	4.10	3.74	3.47	3.27	3.10	2.96	2.51	2.27	1.99	1.73	1.49	1.34	1.21	1.00

Degrees of freedom for the denominator (ν_2)

Appendix 5. Kolmogorov–Smirnov tables

Critical values, $d_\alpha(n)^a$, of the maximum absolute difference between sample $F_n(x)$ and population $F(x)$ cumulative distribution.

Number of trials, n	Level of significance, α			
	0.10	0.05	0.02	0.01
1	0.95000	0.97500	0.99000	0.99500
2	0.77639	0.84189	0.90000	0.92929
3	0.63604	0.70760	0.78456	0.82900
4	0.56522	0.62394	0.68887	0.73424
5	0.50945	0.56328	0.62718	0.66853
6	0.46799	0.51926	0.57741	0.61661
7	0.43607	0.48342	0.53844	0.57581
8	0.40962	0.45427	0.50654	0.54179
9	0.38746	0.43001	0.47960	0.51332
10	0.36866	0.40925	0.45662	0.48893
11	0.35242	0.39122	0.43670	0.46770
12	0.33815	0.37543	0.41918	0.44905
13	0.32549	0.36143	0.40362	0.43247
14	0.31417	0.34890	0.38970	0.41762
15	0.30397	0.33760	0.37713	0.40420
16	0.29472	0.32733	0.36571	0.39201
17	0.28627	0.31796	0.35528	0.38086
18	0.27851	0.30936	0.34569	0.37062
19	0.27136	0.30143	0.33685	0.36117
20	0.26473	0.29408	0.32866	0.35241
21	0.25858	0.28724	0.32104	0.34427
22	0.25283	0.28087	0.31394	0.33666
23	0.24746	0.27490	0.30728	0.32954
24	0.24242	0.26931	0.30104	0.32286
25	0.23768	0.26404	0.29516	0.31657
26	0.23320	0.25907	0.28962	0.31064
27	0.22898	0.25438	0.28438	0.30502
28	0.22497	0.24993	0.27942	0.29971
29	0.22117	0.24571	0.27471	0.29466
30	0.21756	0.24170	0.27023	0.28987
31	0.21412	0.23788	0.26596	0.28530
32	0.21085	0.23424	0.26189	0.28094
33	0.20771	0.23076	0.25801	0.27677
34	0.20472	0.22743	0.25429	0.27279
35	0.20185	0.22425	0.25073	0.26897

Critical values, $d_\alpha(n)$[a], of the maximum absolute difference between sample $F_n(x)$ and population $F(x)$ cumulative distribution.

Number of trials, n	Level of significance, α			
	0.10	0.05	0.02	0.01
36	0.19910	0.22119	0.24732	0.26532
37	0.19646	0.21826	0.24404	0.26180
38	0.19392	0.21544	0.24089	0.25843
39	0.19148	0.21273	0.24089	0.25518
40[b]	0.18913	0.21012	0.23494	0.25205

[a]Values of $d_\alpha(n)$ such that $p(\max)|F''(x) - F(x)| d_\alpha(n) = \alpha$.

[b]$N > 40 \approx \dfrac{1.22}{N^{1/2}}, \dfrac{1.36}{N^{1/2}}, \dfrac{1.51}{N^{1/2}}$ and $\dfrac{1.63}{N^{1/2}}$ for the four levels of significance.

Appendix 6. Rank tables (median, 5%, 95%)

MEDIAN RANKS

$j\backslash n$	1	2	3	4	5	6	7	8	9	10
					SAMPLE SIZE					
1	50.000	29.289	20.630	15.910	12.945	10.910	9.428	8.300	7.412	6.697
2		70.711	50.000	38.573	31.381	26.445	22.849	20.113	17.962	16.226
3			79.370	61.427	50.000	42.141	36.412	32.052	28.624	25.857
4				84.090	68.619	57.859	50.000	44.015	39.308	35.510
5					87.055	73.555	63.588	55.984	50.000	45.169
6						89.090	77.151	67.948	60.691	54.831
7							90.572	79.887	71.376	64.490
8								91.700	82.038	74.142
9									92.587	83.774
10										93.303

MEDIAN RANKS

$j\backslash n$	11	12	13	14	15	16	17	18	19	20
					SAMPLE SIZE					
1	6.107	5.613	5.192	4.830	4.516	4.240	3.995	3.778	3.582	3.406
2	14.796	13.598	12.579	11.702	10.940	10.270	9.678	9.151	8.677	8.251
3	23.578	21.669	20.045	18.647	17.432	16.365	15.422	14.581	13.827	13.147
4	32.380	29.758	27.528	25.608	23.939	22.474	21.178	20.024	18.988	18.055
5	41.189	37.853	35.016	32.575	30.452	28.589	26.940	25.471	24.154	22.967
6	50.000	45.951	42.508	39.544	36.967	34.705	32.704	30.921	29.322	27.880
7	58.811	54.049	50.000	46.515	43.483	40.823	38.469	36.371	34.491	32.795
8	67.620	62.147	57.492	53.485	50.000	46.941	44.234	41.823	39.660	37.710
9	76.421	70.242	64.984	60.456	56.517	53.059	50.000	47.274	44.830	42.626
10	85.204	78.331	72.472	67.425	63.033	59.177	55.766	52.726	50.000	47.542
11	93.893	86.402	79.955	74.392	69.548	65.295	61.531	58.177	55.170	52.458
12		94.387	87.421	81.353	76.061	71.411	67.296	63.629	60.340	57.374
13			94.808	88.298	82.568	77.525	73.060	69.079	65.509	62.289
14				95.169	89.060	83.635	78.821	74.529	70.678	67.205
15					95.484	89.730	84.578	79.976	75.846	72.119
16						95.760	90.322	85.419	81.011	77.033
17							96.005	90.849	86.173	81.945
18								96.222	91.322	86.853
19									96.418	91.749
20										96.594

MEDIAN RANKS

j\n	21	22	23	24	25	26	27	28	29	30
					SAMPLE SIZE					
1	3.247	3.101	2.969	2.847	2.734	2.631	2.534	2.445	2.362	2.284
2	7.864	7.512	7.191	6.895	6.623	6.372	6.139	5.922	5.720	5.532
3	12.531	11.970	11.458	10.987	10.553	10.153	9.781	9.436	9.114	8.814
4	17.209	16.439	15.734	15.088	14.492	13.942	13.432	12.958	12.517	12.104
5	21.890	20.911	20.015	19.192	18.435	17.735	17.086	16.483	15.922	15.397
6	26.574	25.384	24.297	23.299	22.379	21.529	20.742	20.010	19.328	18.691
7	31.258	29.859	28.580	27.406	26.324	25.325	24.398	23.537	22.735	21.986
8	35.943	34.334	32.863	31.513	30.269	29.120	28.055	27.065	26.143	25.281
9	40.629	38.810	37.147	35.621	34.215	32.916	31.712	30.593	29.550	28.576
10	45.314	43.286	41.431	39.729	38.161	36.712	35.370	34.121	32.958	31.872
11	50.000	47.762	45.716	43.837	42.107	40.509	39.027	37.650	36.367	35.168
12	54.686	52.238	50.000	47.946	46.054	44.305	42.685	41.178	39.775	38.464
13	59.371	56.714	54.284	52.054	50.000	48.102	46.342	44.707	43.183	41.760
14	64.057	61.190	58.568	56.162	53.946	51.898	50.000	48.236	46.592	45.056
15	68.742	65.665	62.853	60.271	57.892	55.695	53.658	51.764	50.000	48.352
16	73.426	70.141	67.137	64.379	61.839	59.491	57.315	55.293	53.408	51.648
17	78.109	74.616	71.420	68.487	65.785	63.287	60.973	58.821	56.817	54.944
18	82.791	79.089	75.703	72.594	69.730	67.084	64.630	62.350	60.225	58.240
19	87.469	83.561	79.985	76.701	73.676	70.880	68.288	65.878	63.633	61.536
20	92.136	88.030	84.266	80.808	77.621	74.675	71.945	69.407	67.041	64.852
21	96.753	92.488	88.542	84.912	81.565	78.471	75.602	72.935	70.450	68.128
22		96.898	92.809	89.013	85.507	82.265	79.258	76.463	73.857	71.424
23			97.031	93.105	89.447	86.058	82.914	79.990	77.265	74.719
24				97.153	93.377	89.847	86.568	83.517	80.672	78.014
25					97.265	93.628	90.219	87.042	84.078	81.309
26						97.369	93.861	90.564	87.483	84.603
27							97.465	94.078	90.865	87.896
28								97.555	94.280	91.186
29									97.638	94.468
30										97.716

MEDIAN RANKS

$j\backslash n$	SAMPLE SIZE									
	31	32	33	34	35	36	37	38	39	40
1	2.211	2.143	2.078	2.018	1.961	1.907	1.856	1.807	1.762	1.718
2	5.355	5.190	5.034	4.887	4.749	4.618	4.495	4.377	4.266	4.160
3	8.533	8.269	8.021	7.787	7.567	7.359	7.162	6.975	6.798	6.629
4	11.716	11.355	11.015	10.694	10.391	10.105	9.835	9.578	9.335	9.103
5	14.905	14.445	14.011	13.603	13.218	12.855	12.510	12.184	11.874	11.580
6	18.094	17.535	17.009	16.514	16.046	15.605	15.187	14.791	14.415	14.057
7	21.284	20.626	20.007	19.425	18.875	18.355	17.864	17.398	16.956	16.535
8	24.474	23.717	23.006	22.336	21.704	21.107	20.541	20.005	19.497	19.013
9	27.664	26.809	26.005	25.247	24.533	23.858	23.219	22.613	22.038	21.492
10	30.855	29.901	29.004	28.159	27.362	26.609	25.897	25.221	24.580	23.971
11	34.046	32.993	32.003	31.071	30.192	29.361	28.575	27.829	27.122	26.449
12	37.236	36.085	35.003	33.983	33.022	32.113	31.253	30.437	29.664	28.928
13	40.427	39.177	38.002	36.895	35.851	34.865	33.931	33.046	32.206	31.407
14	43.618	42.269	41.001	39.807	38.681	37.616	36.609	35.654	34.748	33.886
15	46.809	45.362	44.001	42.720	41.511	40.368	39.287	38.262	37.290	36.365
16	50.000	48.454	47.000	45.632	44.340	43.120	41.965	40.871	39.832	38.844
17	53.191	51.546	50.000	48.544	47.170	45.872	44.644	43.479	42.374	41.323
18	56.382	54.638	52.999	51.456	50.000	48.624	47.322	46.087	44.916	43.802
19	59.573	57.731	55.999	54.368	52.830	51.376	50.000	48.696	47.458	46.281
20	62.763	60.823	58.998	57.280	55.660	54.128	52.678	51.304	50.000	48.760
21	65.954	63.915	61.998	60.193	58.489	56.830	55.356	53.913	52.542	51.239
22	69.145	67.007	64.997	63.105	61.319	59.632	58.035	56.521	55.084	53.719
23	72.335	70.099	67.997	66.017	64.149	62.383	60.713	59.129	57.626	56.198
24	75.526	73.191	70.996	68.929	66.973	65.135	63.391	61.738	60.168	58.677
25	78.716	76.283	73.995	71.841	69.808	67.837	66.069	64.346	62.710	61.156
26	81.906	79.374	76.994	74.752	72.637	70.639	68.747	66.954	65.252	63.635
27	85.094	82.465	79.993	77.664	75.467	73.391	71.425	69.562	67.794	66.114
28	88.282	85.555	82.991	80.575	78.296	76.142	74.103	72.171	70.336	68.593
29	91.467	88.644	85.989	83.486	81.125	78.899	76.781	74.779	72.878	71.072
30	94.645	91.731	88.985	86.397	83.954	81.645	79.459	77.387	75.420	73.550
31	97.789	94.810	91.979	89.306	86.782	84.395	82.136	79.994	77.962	76.029
32		97.857	94.966	92.213	89.608	87.145	84.813	82.602	80.503	78.508
33			97.921	95.113	92.433	89.894	87.490	85.209	83.044	80.986
34				97.982	95.251	92.641	90.165	87.816	85.585	83.465
35					98.039	95.382	92.838	90.422	88.126	85.943
36						98.093	95.505	93.025	90.665	88.420
37							98.144	95.622	93.202	90.897
38								98.192	95.734	93.371
39									98.238	95.839
40										98.282

MEDIAN RANKS

| | | | | SAMPLE SIZE | | | | | |
$j\backslash n$	41	42	43	44	45	46	47	48	49	50
1	1.676	1.637	1.599	1.563	1.528	1.495	1.464	1.434	1.405	1.377
2	4.060	3.964	3.872	3.785	3.702	3.622	3.545	3.472	3.402	3.334
3	6.469	6.316	6.170	6.031	5.898	5.771	5.649	5.532	5.420	5.312
4	8.883	8.673	8.473	8.282	8.099	7.925	7.757	7.597	7.443	7.295
5	11.300	11.033	10.778	10.535	10.303	10.080	9.867	9.663	9.467	9.279
6	13.717	13.393	13.084	12.789	12.507	12.237	11.979	11.731	11.493	11.265
7	16.135	15.754	15.391	15.043	14.712	14.394	14.090	13.799	13.519	13.250
8	18.554	18.115	17.697	17.298	16.917	16.551	16.202	15.867	15.545	15.236
9	20.972	50.477	20.004	19.553	19.122	18.709	18.314	17.935	17.571	17.222
10	23.391	22.838	22.311	21.808	21.327	20.867	20.426	20.003	19.598	19.209
11	25.810	25.200	24.618	24.063	23.532	23.025	22.538	22.072	21.625	21.195
12	28.228	27.562	26.926	26.318	25.738	25.182	24.650	24.140	23.651	23.181
13	30.647	29.924	29.233	28.574	27.943	27.340	26.763	26.209	25.678	25.168
14	33.066	32.285	31.540	30.829	30.149	29.498	28.875	28.278	27.705	27.154
15	35.485	34.647	33.848	33.084	32.355	31.656	30.988	30.347	29.731	29.141
16	37.905	37.009	36.155	35.340	34.560	33.814	33.100	32.415	31.758	31.127
17	40.324	39.371	38.463	37.595	36.766	35.972	35.212	34.484	33.785	33.114
18	42.743	41.733	40.770	39.851	38.972	38.130	37.325	36.553	35.822	35.100
19	45.162	44.095	43.078	42.106	41.177	40.289	39.437	38.622	37.839	37.087
20	47.581	46.457	45.385	44.361	43.383	42.447	41.550	40.690	39.866	39.074
21	50.000	48.819	47.692	46.617	45.589	44.605	43.662	42.759	41.892	41.060
22	52.419	51.181	50.000	48.872	47.794	46.763	45.775	44.825	43.919	43.047
23	54.838	53.543	52.307	51.128	50.000	48.921	47.887	46.897	45.946	45.033
24	57.257	55.905	54.615	53.383	52.206	51.079	50.000	48.966	47.973	47.020
25	59.676	58.267	56.922	55.639	54.411	53.237	52.112	51.034	50.000	49.007
26	62.095	60.629	59.230	57.894	56.617	55.395	54.225	53.103	52.027	50.993
27	64.514	62.991	61.537	60.149	58.823	57.553	56.337	55.172	54.054	52.980
28	66.933	65.353	63.845	62.405	61.028	59.711	58.450	57.241	56.081	54.966
29	69.352	67.714	66.152	64.660	63.234	61.869	60.562	59.310	58.107	56.953
30	71.771	70.076	68.459	66.916	65.440	64.027	62.675	61.378	60.134	58.940
31	74.190	72.438	70.767	69.171	67.645	66.186	64.767	63.447	62.161	60.926
32	76.609	74.800	73.074	71.426	69.851	68.344	66.900	65.516	64.188	62.913
33	79.028	77.162	75.381	73.681	72.056	70.502	69.012	67.585	66.215	64.899
34	81.446	79.523	77.689	75.937	74.262	72.660	71.125	69.653	68.242	66.886
35	83.865	81.885	79.996	78.192	76.467	74.817	73.237	71.722	70.268	68.873
36	86.283	84.246	82.303	80.447	78.673	76.975	75.349	73.791	72.295	70.859
37	88.700	86.607	84.609	82.702	80.878	79.133	77.462	75.859	74.322	72.646
38	91.117	86.967	86.916	84.956	83.083	81.291	79.574	77.928	76.349	74.832
39	93.531	91.327	89.222	87.211	85.283	83.448	81.686	79.997	78.375	76.819
40	95.940	93.684	91.527	89.465	87.493	85.606	83.798	82.065	80.402	78.805
41	98.324	96.036	93.830	91.718	89.697	87.763	85.910	84.133	82.428	80.791
42		95.363	96.127	93.969	91.900	89.920	88.021	86.201	84.455	82.778
43			98.401	96.215	94.102	92.075	90.132	88.269	86.481	84.764
44				98.437	96.298	94.229	92.243	90.337	88.507	86.750
45					98.471	96.378	94.351	92.403	90.532	88.735
46						98.504	96.455	94.468	92.557	90.721
47							98.536	96.528	94.580	92.705
48								98.566	96.598	94.688
49									98.595	96.666
50										98.623

5 PER CENT. RANKS

$j \backslash n$	1	2	3	4	SAMPLE SIZE 5	6	7	8	9	10
1	5.000	2.532	1.695	1.274	1.021	0.851	0.730	0.639	0.568	0.512
2		22.361	13.535	9.761	7.644	6.285	5.337	4.639	4.102	3.677
3			36.840	24.860	18.925	15.316	12.876	11.111	9.775	8.726
4				47.237	34.259	27.134	22.532	19.290	16.875	15.003
5					54.928	41.820	34.126	28.924	25.137	22.244
6						60.696	47.930	40.031	34.494	30.354
7							65.184	52.932	45.036	39.338
8								68.766	57.086	49.310
9									71.687	60.584
10										74.113

5 PER CENT. RANKS

$j \backslash n$	11	12	13	14	SAMPLE SIZE 15	16	17	18	19	20
1	0.465	0.426	0.394	0.366	0.341	0.320	0.301	0.285	0.270	0.256
2	3.332	3.046	2.805	2.600	2.423	2.268	2.132	2.011	1.903	1.806
3	7.882	7.187	6.605	6.110	5.685	5.315	4.990	4.702	4.446	4.217
4	13.507	12.285	11.267	10.405	9.666	9.025	8.464	7.969	7.529	7.135
5	19.958	18.102	16.566	15.272	14.166	13.211	12.377	11.643	10.991	10.408
6	27.125	24.530	22.395	20.607	19.086	17.777	16.636	15.634	14.747	13.955
7	34.981	31.524	28.705	26.358	24.373	22.669	21.191	19.895	18.750	17.731
8	43.563	39.086	35.480	32.503	29.999	27.860	26.011	24.396	22.972	21.707
9	52.991	47.267	42.738	39.041	35.956	33.337	31.083	29.120	27.395	25.865
10	63.564	56.189	50.535	45.999	42.256	39.101	36.401	34.060	32.009	30.195
11	76.160	66.132	58.990	53.434	48.925	45.165	41.970	39.215	36.811	34.693
12		77.908	68.366	61.461	56.022	51.560	47.808	44.595	41.806	39.358
13			79.418	70.327	63.656	58.343	53.945	50.217	47.003	44.197
14				80.736	72.060	65.617	60.436	56.112	52.420	49.218
15					81.896	73.604	67.381	62.332	58.088	54.442
16						82.925	74.988	68.974	64.057	59.897
17							83.843	76.234	70.420	65.634
18								84.668	77.363	71.738
19									85.413	78.389
20										86.089

5 PER CENT. RANKS

$j\backslash n$	21	22	23	24	SAMPLE SIZE 25	26	27	28	29	30
1	0.244	0.233	0.223	0.213	0.205	0.197	0.190	0.183	0.177	0.171
2	1.719	1.640	1.567	1.501	1.440	1.384	1.332	1.284	1.239	1.198
3	4.010	3.822	3.651	3.495	3.352	3.220	3.098	2.985	2.879	2.781
4	6.781	6.460	6.167	5.901	5.656	5.431	5.223	5.031	4.852	4.685
5	9.884	9.411	8.981	8.588	8.229	7.899	7.594	7.311	7.049	6.806
6	13.245	12.603	12.021	11.491	11.006	10.560	10.148	9.768	9.415	9.087
7	16.818	15.994	15.248	14.569	13.947	13.377	12.852	12.367	11.917	11.499
8	20.575	19.556	18.634	17.796	17.030	16.328	15.682	15.085	14.532	14.018
9	24.499	23.272	22.164	21.157	20.238	19.396	18.622	17.908	17.246	16.633
10	28.580	27.131	25.824	24.639	23.559	22.570	21.662	20.824	20.050	19.331
11	32.811	31.126	29.609	28.236	26.985	25.842	24.793	23.827	22.934	22.106
12	37.190	35.254	33.515	31.942	30.513	29.508	28.012	26.911	25.894	24.953
13	41.720	39.516	37.539	35.756	34.139	32.664	31.314	30.072	28.927	27.867
14	46.406	43.913	41.684	39.678	37.862	36.209	34.697	33.309	32.030	30.846
15	51.261	48.454	45.954	43.711	41.684	39.842	38.161	36.620	35.200	33.889
16	56.302	53.151	50.356	47.858	45.607	43.566	41.707	40.004	38.439	36.995
17	61.559	58.020	54.902	52.127	49.636	47.384	45.336	43.464	41.746	40.163
18	67.079	63.091	59.610	56.531	53.779	51.300	49.052	47.002	45.123	43.394
19	72.945	68.409	64.507	61.086	58.048	55.323	52.861	50.621	48.573	46.691
20	79.327	74.053	69.636	65.819	62.459	59.465	56.770	54.327	52.099	50.056
21	86.705	80.188	75.075	70.773	67.039	63.740	60.790	58.127	55.706	53.493
22		87.269	80.980	76.020	71.828	68.176	64.936	62.033	59.403	57.007
23			87.788	81.711	76.896	72.810	69.237	66.060	63.200	60.605
24				88.265	82.388	77.711	73.726	70.231	67.113	64.299
25					88.707	83.017	78.470	74.583	71.168	68.103
26						89.117	83.603	79.179	75.386	72.038
27							89.498	84.149	79.844	76.140
28								89.853	84.661	80.467
29									90.185	85.140
30										90.497

5 PER CENT. RANKS

$j\backslash n$	SAMPLE SIZE									
	31	32	33	34	35	36	37	38	39	40
1	0.165	0.160	0.155	0.151	0.146	0.142	0.138	0.135	0.131	0.128
2	1.158	1.122	1.086	1.055	1.025	0.996	0.969	0.943	0.919	0.896
3	2.690	2.604	2.524	2.448	2.377	2.310	2.246	2.186	2.129	2.075
4	4.530	4.384	4.246	4.120	3.999	3.885	3.778	3.676	3.580	3.488
5	6.578	6.365	3.166	5.978	5.802	5.636	5.479	5.331	5.190	5.057
6	8.781	8.495	8.227	7.976	7.739	7.516	7.306	7.107	6.919	6.740
7	11.109	10.745	10.404	10.084	9.783	9.499	9.232	8.979	8.740	8.513
8	13.540	13.093	12.675	12.283	11.914	11.567	11.240	10.931	10.638	10.361
9	16.061	15.528	15.029	14.561	14.122	13.708	13.318	12.950	12.601	12.271
10	18.662	18.038	17.455	16.909	16.396	15.913	15.458	15.028	14.622	14.237
11	21.336	20.618	19.948	19.319	18.730	18.175	17.653	17.160	16.694	16.252
12	24.077	23.262	22.501	21.788	21.119	20.491	19.898	19.340	18.812	18.312
13	26.883	25.966	25.111	24.310	23.560	22.855	22.191	21.565	20.973	20.413
14	29.749	28.727	27.775	26.884	26.049	25.265	24.527	23.832	23.175	22.553
15	32.674	31.544	30.491	29.507	28.585	27.719	26.905	26.138	25.414	24.729
16	35.657	34.415	33.258	32.177	31.165	30.216	29.324	28.483	27.690	26.940
17	38.698	37.339	36.074	34.894	33.789	32.754	31.781	30.865	30.001	29.185
18	41.797	40.317	38.940	37.657	36.457	35.332	34.276	33.283	32.346	31.461
19	44.956	43.349	41.656	40.466	39.167	37.951	36.809	35.736	34.725	33.770
20	48.175	46.436	44.823	43.321	41.920	40.609	39.380	38.224	37.136	36.109
21	51.458	49.581	47.841	46.225	44.717	43.309	41.988	40.748	39.581	38.480
22	54.810	52.786	50.914	49.177	47.560	46.049	44.634	43.307	42.058	40.881
23	58.234	56.055	54.344	52.181	50.448	48.832	47.320	45.902	44.569	43.314
24	64.739	59.314	57.235	55.239	53.385	51.658	50.045	48.534	47.114	45.778
25	65.336	62.810	60.493	58.355	56.374	54.532	52.812	51.204	49.694	48.275
26	69.036	66.313	63.824	61.534	59.416	57.454	55.624	53.914	52.311	50.805
27	72.563	69.916	67.237	64.754	62.523	60.429	58.483	56.666	54.966	53.370
28	76.650	73.640	70.748	68.113	65.695	63.483	61.392	59.463	57.661	55.972
29	81.054	77.518	74.375	71.535	68.944	66.561	64.357	62.309	60.399	58.612
30	85.591	81.606	76.150	75.069	72.282	69.732	67.384	65.209	63.185	61.294
31	90.789	86.015	82.127	78.747	75.728	72.990	70.482	68.168	66.021	64.021
32		91.063	86.415	82.619	79.312	76.352	73.663	71.196	68.916	66.797
33			91.322	86.793	83.085	79.848	76.946	74.304	71.876	69.629
34				91.566	87.150	83.526	80.357	77.510	74.915	72.525
35					91.797	87.488	83.946	80.841	78.048	75.497
36						92.015	87.809	84.344	81.302	78.560
37							92.222	88.115	84.723	81.741
38								92.419	88.405	85.085
39									92.606	88.681
40										92.784

5 PER CENT. RANKS

				SAMPLE SIZE					
$j\backslash n$	41	42	43	44	45	46	47	48	49	50
1	0.125	0.122	0.119	0.116	0.114	0.111	0.109	0.107	0.105	0.102
2	0.874	0.853	0.833	0.814	0.795	0.778	0.761	0.745	0.730	0.715
3	2.024	1.975	1.928	1.884	1.842	1.801	1.762	1.725	1.689	1.655
4	3.402	3.319	3.240	3.165	3.093	3.025	2.959	2.897	2.836	2.779
5	4.930	4.810	4.695	4.586	4.481	4.382	4.286	4.195	4.108	4.024
6	6.570	6.409	6.256	6.109	5.969	5.836	5.708	5.586	5.469	5.357
7	8.298	8.093	7.898	7.713	7.536	7.366	7.205	7.050	6.902	6.760
8	10.097	9.847	9.609	9.382	9.166	8.959	8.762	8.573	8.392	8.218
9	11.958	11.660	11.377	11.107	10.850	10.605	10.370	10.146	9.931	9.725
10	13.872	13.525	13.195	12.881	12.582	12.296	12.023	11.762	11.512	11.272
11	15.833	15.436	15.058	14.698	14.355	14.028	13.715	13.416	13.130	12.856
12	17.838	17.389	16.961	16.554	16.166	15.796	15.443	15.105	14.782	14.472
13	19.883	19.379	18.901	18.445	18.012	17.598	17.203	16.825	16.464	16.117
14	21.964	21.406	20.875	20.370	19.889	19.430	18.993	18.574	18.174	17.790
15	24.081	23.466	22.881	22.326	21.796	21.292	20.810	20.350	19.910	19.488
16	26.230	25.557	24.918	24.311	23.732	23.180	22.654	22.151	21.671	21.210
17	28.412	27.679	26.984	26.323	25.694	25.095	24.523	23.977	23.455	22.955
18	30.624	29.831	29.078	28.363	27.683	27.034	26.416	25.825	25.261	24.721
19	32.867	32.011	31.200	30.429	29.696	28.997	28.331	27.696	27.088	26.507
20	35.138	34.219	33.348	32.520	31.733	30.984	30.269	29.588	28.936	28.313
21	37.440	36.455	35.522	34.636	33.794	32.993	32.229	31.500	30.804	30.138
22	39.770	38.719	37.722	36.777	35.879	35.025	34.210	33.434	32.692	31.980
23	42.129	41.009	39.949	38.943	37.987	37.078	36.212	35.387	34.599	33.845
24	44.518	43.328	42.201	41.133	40.118	39.154	38.235	37.360	36.524	35.726
25	46.937	45.674	44.480	43.347	42.273	41.251	40.279	39.353	38.469	37.625
26	49.388	48.050	46.785	45.587	44.451	43.371	42.344	41.366	10.432	39.541
27	51.869	50.454	49.117	47.852	46.652	45.513	44.430	43.398	42.415	41.476
28	54.385	52.889	51.478	50.143	48.878	47.678	46.537	45.451	44.416	43.428
29	56.935	55.356	53.868	52.461	51.129	49.866	48.666	47.524	46.436	45.399
30	59.522	57.857	56.288	54.807	53.406	52.078	50.817	49.618	48.477	47.388
31	62.149	60.393	58.741	57.183	55.710	54.315	52.991	51.734	50.537	49.396
32	64.820	62.968	61.228	59.590	58.042	56.578	55.190	53.871	52.617	51.423
33	67.539	56.585	63.753	62.029	60.404	58.868	57.413	56.032	54.720	53.470
34	70.311	68.248	66.318	64.505	62.798	61.187	59.662	58.217	56.844	55.538
35	73.146	70.963	68.927	67.020	65.227	63.537	61.940	60.427	58.991	57.627
36	76.053	73.738	71.587	69.578	67.694	65.921	64.247	62.664	61.164	59.738
37	79.049	76.584	74.306	72.185	70.203	68.341	66.587	64.931	63.362	61.874
38	82.160	79.517	77.093	74.849	72.759	70.805	68.963	67.228	65.589	64.034
39	85.429	82.561	79.964	77.580	75.370	73.309	71.378	69.561	67.846	66.222
40	88.945	85.759	82.944	80.392	78.046	75.870	73.838	71.932	70.136	68.440
41	92.954	89.196	86.073	83.310	80.802	78.494	76.350	74.347	72.465	70.691
42		93.116	89.437	86.374	83.661	81.196	78.924	76.812	74.836	72.978
43			93.270	89.666	86.662	83.998	81.573	79.337	77.256	75.306
44				93.418	89.887	86.939	84.321	81.936	79.734	77.683
45					93.560	90.098	87.204	84.631	82.285	80.117
46						93.695	90.300	87.459	84.929	82.621
47							93.825	90.494	87.703	85.216
48								93.950	90.681	87.939
49									94.069	90.860
50										94.184

95 PER CENT. RANKS

$j\backslash n$	SAMPLE SIZE									
	1	2	3	4	5	6	7	8	9	10
1	95.000	77.639	63.160	52.713	45.072	39.304	34.816	31.234	28.313	25.887
2		97.468	86.465	75.139	65.741	58.180	52.070	47.068	42.914	39.416
3			98.305	90.239	81.075	72.866	65.874	59.969	54.964	50.690
4				98.726	92.356	84.684	77.468	71.076	65.506	60.662
5					98.979	93.715	87.124	80.710	74.863	69.646
6						99.149	94.662	88.889	83.125	77.756
7							99.270	95.361	90.225	84.997
8								99.361	95.898	91.274
9									99.432	96.323
10										99.488

95 PER CENT. RANKS

$j\backslash n$	SAMPLE SIZE									
	11	12	13	14	15	16	17	18	19	20
1	23.840	22.092	20.582	19.264	18.104	17.075	16.157	15.332	14.587	13.911
2	36.436	33.868	31.634	29.673	27.940	26.396	25.012	23.766	22.637	21.611
3	47.009	43.811	41.010	38.539	36.344	34.383	32.619	31.026	29.580	28.262
4	56.437	52.733	49.465	46.566	43.978	41.657	39.564	37.668	35.943	34.366
5	65.019	60.914	57.262	54.000	51.075	48.440	46.055	43.888	41.912	40.103
6	72.875	68.476	64.520	60.928	57.744	54.835	52.192	49.783	47.580	45.558
7	80.042	75.470	71.295	67.497	64.043	60.899	58.029	55.404	52.997	50.782
8	86.492	81.898	77.604	73.641	70.001	66.663	63.599	60.784	58.194	55.803
9	92.118	87.715	83.434	79.393	75.627	72.140	68.917	65.940	63.188	60.641
10	96.668	92.813	88.733	84.728	80.913	77.331	73.989	70.880	67.991	65.307
11	99.535	96.954	93.395	89.595	85.834	82.223	78.809	75.604	72.605	69.805
12		99.573	97.195	93.890	90.334	86.789	83.364	80.105	77.028	74.135
13			99.606	97.400	94.315	90.975	87.623	84.366	81.250	78.293
14				99.634	97.577	94.685	91.535	88.357	85.253	82.269
15					99.659	97.732	95.010	92.030	89.009	86.045
16						99.680	97.868	95.297	92.471	89.592
17							99.699	97.989	95.553	92.865
18								99.715	98.097	95.783
19									99.730	98.193
20										99 744

95 PER CENT. RANKS

				SAMPLE SIZE					
$j \backslash n$	21	22	23	24	25	26	27	28	29	30
1	13.295	12.731	12.212	11.735	11.293	10.883	10.502	10.147	9.814	9.503
2	20.673	19.812	19.020	18.289	17.612	16.983	16.397	15.851	15.339	14.860
3	27.055	25.947	24.925	23.980	23.104	22.289	21.530	20.821	20.156	19.533
4	32.921	31.591	30.364	29.227	28.172	27.190	26.274	25.417	24.614	23.860
5	38.441	36.909	35.193	34.181	32.961	31.824	30.763	29.769	28.837	27.962
6	43.698	41.980	40.390	38.914	37.541	36.260	35.062	33.940	32.887	31.897
7	48.739	46.849	45.097	43.469	41.952	40.535	39.210	37.967	36.800	35.701
8	53.594	51.546	49.643	47.873	46.221	44.677	43.230	41.873	40.597	39.395
9	58.280	56.087	54.046	52.142	50.364	48.700	47.139	45.673	44.294	42.993
10	62.810	60.484	58.315	56.289	54.393	52.616	50.948	49.379	47.901	46.507
11	67.189	64.746	62.461	60.321	58.316	56.434	54.664	52.998	51.427	49.944
12	71.420	68.874	66.485	64.244	62.138	60.158	58.293	56.536	54.877	53.309
13	75.501	72.869	70.391	68.058	65.861	63.791	61.839	59.996	58.254	56.605
14	79.425	76.728	74.176	71.764	69.487	67.336	65.303	63.380	61.561	59.837
15	83.182	80.444	77.836	75.361	73.015	70.792	68.686	66.691	64.799	63.005
16	86.755	84.006	81.366	78.843	76.441	74.158	71.988	69.927	67.970	66.111
17	90.116	87.397	84.752	82.204	79.762	77.430	75.207	73.089	71.073	69.154
18	93.219	90.589	87.978	85.431	82.970	80.604	78.338	46.173	74.106	72.133
19	95.990	93.540	91.019	88.509	86.052	83.672	81.378	79.176	77.066	75.047
20	98.281	96.178	93.832	91.411	88.994	86.623	84.318	82.092	79.950	77.894
21	99.756	98.360	96.348	94.099	91.771	89.440	87.148	84.915	82.753	80.669
22		99.767	98.433	96.505	94.344	92.101	89.851	87.633	85.468	83.367
23			99.777	98.499	96.648	94.569	92.406	90.232	88.083	85.981
24				99.786	98.560	96.780	94.777	92.689	90.584	88.501
25					99.795	98.616	96.902	94.969	92.951	90.913
26						99.803	98.668	97.015	95.148	93.194
27							99.810	98.716	97.120	95.314
28								99.817	98.761	97.218
29									99.823	98.802
30										99.829

95 PER CENT. RANKS

$j \backslash n$	31	32	33	34	35	36	37	38	39	40
1	9.211	8.937	8.678	8.434	8.203	7.985	7.778	7.581	7.394	7.216
2	14.409	13.985	13.585	13.207	12.850	12.512	12.191	11.885	11.595	11.319
3	18.946	18.394	17.873	17.381	16.915	16.474	16.054	15.656	15.277	14.915
4	23.150	22.482	21.850	21.253	20.688	20.152	19.643	19.159	18.698	18.259
5	27.137	26.360	25.625	24.931	24.272	23.648	23.054	22.490	21.952	21.440
6	30.964	30.084	29.252	28.465	27.718	27.010	26.337	25.696	25.085	24.503
7	34.665	33.687	32.763	31.887	31.056	30.268	29.518	28.804	28.124	27.475
8	38.261	37.190	36.176	35.216	34.305	33.439	32.616	31.832	31.084	30.371
9	41.766	40.606	39.507	68.466	37.477	36.537	35.643	34.791	33.979	33.203
10	45.190	43.945	42.765	41.645	40.582	39.571	38.608	37.691	36.815	35.979
11	48.542	57.214	45.956	44.761	43.626	42.546	41.517	40.537	39.601	38.706
12	51.825	50.419	49.086	47.819	46.615	45.468	44.376	43.334	42.339	41.388
13	55.044	53.564	52.159	50.823	49.552	48.341	47.187	46.086	45.034	44.028
14	58.203	56.651	55.177	53.775	52.440	51.168	49.955	48.796	47.689	46.630
15	61.302	59.683	58.144	56.678	55.282	53.951	52.680	51.466	50.305	49.195
16	64.343	62.661	61.060	59.534	58.080	56.691	55.366	54.098	52.886	51.725
17	67.326	65.585	63.926	62.343	60.833	59.391	58.012	56.693	55.431	54.222
18	70.251	68.456	66.742	65.106	63.543	62.049	60.620	59.252	57.942	56.686
19	73.117	71.272	69.509	67.823	66.210	64.668	63.190	61.776	60.419	59.119
20	75.922	74.034	72.225	70.493	68.835	67.246	65.723	64.264	62.864	61.520
21	78.664	76.738	74.889	73.116	71.415	69.784	68.219	66.717	65.275	63.891
22	81.338	79.382	77.499	75.689	73.951	72.280	70.676	69.135	67.654	66.230
23	83.939	81.961	80.052	78.212	76.440	74.735	73.094	71.517	69.999	68.539
24	86.460	84.472	82.545	80.680	78.881	77.145	75.473	73.862	72.310	70.815
25	88.891	86.907	84.971	83.091	81.270	79.509	77.809	76.168	74.586	73.060
26	91.219	89.255	87.325	85.439	83.604	81.825	80.101	78.435	76.825	75.270
27	93.422	91.505	89.596	87.717	85.878	84.087	82.347	80.660	79.027	77.447
28	95.470	93.635	91.772	89.916	88.086	86.292	84.542	82.840	81.188	79.587
29	97.310	95.615	93.834	92.024	90.217	88.433	86.682	84.972	83.306	81.688
30	98.841	97.396	95.752	94.021	92.261	90.501	88.760	87.050	85.378	83.746
31	99.835	98.878	97.476	95.880	94.198	92.483	90.768	89.069	87.399	85.763
32		99.840	98.912	97.552	96.001	94.364	92.694	91.021	89.362	87.729
33			99.845	98.945	97.623	96.114	94.521	92.893	91.260	89.639
34				99.849	98.975	97.690	96.222	94.669	93.081	91.487
35					99.854	99.004	97.754	96.324	94.810	93.260
36						99.858	99.031	97.814	96.420	94.943
37							99.861	99.057	97.871	96.511
38								99.865	99.081	97.925
39									99.869	99.104
40										99.872

				SAMPLE SIZE					
$j \backslash n$	41	42	43	44	45	46	47	48	49	50
1	7.046	6.884	6.730	6.582	6.440	6.305	6.175	6.050	5.931	5.816
2	11.055	10.804	10.563	10.334	10.113	9.902	9.700	9.506	9.319	9.140
3	14.571	14.241	13.927	13.626	13.338	13.061	12.796	12.541	12.297	12.061
4	17.840	17.439	17.056	16.690	16.339	16.002	15.679	15.369	15.071	14.784
5	20.951	20.483	20.036	19.608	19.198	18.804	18.427	18.064	17.715	17.379
6	23.947	23.416	22.907	22.420	21.954	21.506	21.076	20.663	20.266	19.883
7	26.854	26.262	25.694	25.151	24.630	24.130	23.650	23.188	22.744	22.317
8	29.689	29.037	28.413	27.814	27.241	26.691	26.162	25.623	25.164	24.694
9	32.461	31.752	31.073	30.422	29.797	29.198	28.622	28.068	27.535	27.022
10	35.180	34.415	33.682	32.980	32.306	31.659	31.037	30.439	29.864	29.309
11	37.851	37.032	36.247	35.495	34.773	34.079	33.413	32.772	32.154	31.560
12	40.478	39.607	38.772	37.971	37.202	36.463	35.753	35.069	34.411	33.778
13	43.065	42.143	41.259	40.410	39.596	38.813	38.060	37.336	36.698	35.933
14	45.615	44.644	43.712	42.817	41.958	41.132	40.338	39.573	38.836	38.126
15	48.131	47.110	46.132	45.193	44.290	43.422	42.587	41.783	41.008	40.262
16	50.612	49.546	48.522	47.539	46.594	45.665	44.810	43.968	43.156	42.373
17	53.062	51.950	50.883	49.857	48.871	47.922	47.009	46.129	45.280	44.462
18	55.482	54.326	53.215	52.148	51.122	50.134	49.183	48.266	47.382	46.530
19	57.871	56.672	55.520	54.413	53.348	52.322	51.334	50.382	49.463	48.577
20	60.230	58.991	57.799	56.653	55.549	54.487	53.463	52.476	51.523	50.604
21	62.560	61.281	60.051	58.867	57.727	56.629	55.570	54.549	53.563	52.612
22	64.861	63.545	62.273	61.057	59.882	58.749	57.556	56.602	55.584	54.801
23	67.133	65.781	64.478	63.223	62.013	60.846	59.721	58.634	57.585	56.572
24	69.376	67.989	66.652	65.363	64.121	62.922	61.765	60.647	59.568	58.524
25	71.588	70.169	68.800	67.480	66.205	64.975	63.787	62.640	61.531	60.459
26	73.769	72.320	70.922	69.571	68.267	67.007	65.790	64.613	63.476	62.375
27	75.919	74.443	73.016	71.637	70.304	69.016	67.771	66.566	65.401	64.274
28	78.035	76.534	75.082	73.677	72.317	71.002	69.730	68.500	67.308	66.155
29	80.117	78.594	77.119	75.689	74.306	72.966	71.668	70.412	69.196	68.017
30	82.162	80.621	79.125	77.674	76.268	74.905	73.584	72.304	71.064	69.862
31	84.166	82.611	81.099	79.630	78.203	76.819	75.477	74.175	72.912	71.687
32	86.128	84.564	83.039	81.554	80.111	78.708	77.346	76.023	74.739	73.493
33	88.042	86.475	84.942	83.446	81.988	80.569	79.190	77.848	76.545	75.279
34	89.903	88.340	86.805	85.302	83.834	82.402	81.007	79.650	78.329	77.045
35	91.702	90.153	88.623	87.119	85.645	84.204	82.797	81.426	80.090	78.790
36	93.430	91.907	90.391	88.892	87.418	85.972	84.557	83.175	81.826	80.511
37	95.070	93.591	92.102	90.618	89.150	87.704	86.285	84.895	83.536	82.210
38	96.598	95.190	93.744	92.287	90.834	89.395	87.977	86.584	85.218	83.882
39	97.976	96.681	95.305	93.891	92.464	91.041	89.630	88.238	86.870	85.528
40	99.126	98.025	96.760	95.414	94.030	92.633	91.238	89.854	88.488	87.144
41	99.875	99.147	98.071	96.835	95.518	94.164	92.795	91.427	90.069	88.728
42		99.878	99.167	98.116	96.907	95.618	94.291	92.950	91.608	90.275
43			99.881	99.186	98.158	96.975	95.714	94.414	93.098	91.781
44				99.883	99.205	98.199	97.041	95.805	94.531	93.240
45					99.886	99.222	98.238	97.103	95.892	94.643
46						99.889	99.239	98.275	97.163	95.976
47							99.891	99.255	98.311	97.221
48								99.893	99.270	98.345
49									99.895	99.285
50										99.897

Appendix 7. Matrix algebra revision

The solution of the second-order matrix

$$\begin{vmatrix} a_1 & b_1 \\ a_2 & b_2 \end{vmatrix}$$

is $a_1 b_2 - a_2 b_1$.

The solution of the third-order matrix

$$\begin{vmatrix} a_1 & b_1 & c_1 \\ a_2 & b_2 & c_2 \\ a_3 & b_3 & c_3 \end{vmatrix}$$

is

$$a_1 \begin{vmatrix} b_2 & c_2 \\ b_3 & c_3 \end{vmatrix} - b_1 \begin{vmatrix} a_2 & c_2 \\ a_3 & c_3 \end{vmatrix} + c_1 \begin{vmatrix} a_2 & b_2 \\ a_3 & b_3 \end{vmatrix}$$

$$= a_1 b_2 c_3 + a_2 b_3 c_1 + a_3 b_1 c_2 - a_1 b_3 c_2 - a_2 b_1 c_3 - a_3 b_2 c_1$$

$a_1, a_2, \ldots, c_2, c_3$ are called the *elements* of the matrix.

The lower order matrix associated with a_1, i.e.

$$\begin{vmatrix} b_2 & c_2 \\ b_3 & c_3 \end{vmatrix}$$

is called the *cofactor* of a_1, denoted A_1. Similarly,

$$- \begin{vmatrix} a_2 & c_2 \\ a_3 & c_3 \end{vmatrix}$$

is the cofactor of b_1, denoted B_1, and so on. Thus the solution of the third-order matrix is

$$a_1 A_1 + b_1 B_1 + c_1 C_1$$

and so on for higher orders.

To get the cofactor signs correct, think of the element positions in the matrix as having positive and negative signs associated with them, as follows:

$$\begin{vmatrix} + & - & + & - & . & . & . \\ - & + & - & + & . & . & . \\ + & - & + & - & . & . & . \\ . & . & . & . \\ . & . & . & . \\ . & . & . & . \end{vmatrix}$$

388

Matrix multiplication

$$\begin{vmatrix} a_1 & b_1 \\ a_2 & b_2 \end{vmatrix} \times \begin{vmatrix} A_1 & B_1 \\ A_2 & B_2 \end{vmatrix}$$

$$= \begin{vmatrix} (a_1\,A_1 + b_1\,A_2) & (a_1\,B_1 + b_1\,B_2) \\ (a_2\,A_1 + b_2\,A_2) & (a_2\,B_1 + b_2\,B_2) \end{vmatrix}$$

Appendix 8 Selected list of organizations involved in reliability engineering

NORTH AMERICA

American Society for Quality Control (ASQC) (Reliability Society).
Institute of Electrical and Electronic Engineers (IEEE) (Reliability Division).
Institute of Environmental Sciences: Reliability and environmental testing.
Society of Reliability Engineers (SRE).
These societies and institutions all include reliability topics in their journals, and support conferences, courses, etc. on reliability. Several have overseas divisions.

USAF Rome Air Development Centre (RADC) (Reliability and Compatibility Division). RADC maintains and issues MIL-HDBK-217. The Reliability Analysis Centre (RAC) is operated on behalf of RADC by the Illinois Institute of Technology Research Institute (IITRI), and performs studies and produces reports for RADC. The Data and Analysis Center for Software (DACS) performs similar functions for software reliability. Both RAC and DACS provide useful newsletters.

Government–Industry Data Exchange Program (GIDEP). Data exchange programme on electronic parts reliability and problems.

UK/EUROPE

Institute of Quality Assurance (IQA) (UK).

Safety and Reliability Society (SRS) (UK).
National Quality Assurance Organizations in other countries.
National Centre for Systems Reliability (NCSR) (UK). NCSR is an offshoot of the UK Atomic Energy Authority, and provides a service on reliability of systems such as petrochemical and power generation plant.

Defence Quality Assurance Board (DQAB)(UK). DQAB operates the UK Ministry of Defence quality assurance organization, including the BS 9000 system for electronic components.

390

Eu Re Data. European data exchange system on reliability of components used in petrochem, generating plant, etc.

British Standards Institution. Reliability and quality standards, as well as design and safety standards.

Appendix 9 Reliability data systems

OUTPUTS

1. Listing of failures:

Failure Report No.	System.	Sub-system.	Assembly.	Part

Output selectable by system, sub-system, assembly, part. Also by time period and other appropriate feature, e.g. location, modification status.

2. Pareto analysis of top 10–20 failure modes, selectable as above. (Chapter 13). See Note 1.

3. Full printout of failure report data, selectable (e.g. all in top 10, by assembly number, etc.), as follows:

FR No. System	Sub-system	Assembly	Part	Date	Run time	Symptom/ Effect
Part No. Serial No.						

4. MTBF for system, sub-system, assembly. Show number of failures for each MTBF value. Selectable as above.

5. Trend analyses, selectable as above. See Chapter 3.

6. Probability/hazard plots and derivation of distribution parameters. Selected as above. See Chapter 3.

INPUTS

1. For each failure:

Failure Report Number	Date	Run time	Location	
	System	Sub-system	Assembly	Part
Part/drawing No. Serial No. Modification state				
Failure symptoms Repair action Repair time				

2. For all units
   Total run time. (regular update, say monthly. See note 2).

Notes:
1. The Pareto analysis should form the basis for the critical items list (page 171). Failure reporting for critical items should be amplified as necessary to aid investigation, and special reporting, e.g. by telex, could be used.
2. Calendar time could be used instead of run time when appropriate. Then trend analysis, probability plots, and MTBF would be calculated on a calendar time basis. This is acceptable if run time is quite closely correlated with calendar time and if run time data are not easily obtainable, e.g. if the equipment does not have run time indicators. Using calendar time can be easier and cheaper, since it is then not necessary to obtain the run time at each failure and total (fleet) run time. Only the startup date for each unit and the total number of units in use need be ascertained.
3. The system should be made simple to operate and to understand.
4. Input data should be sufficiently detailed to be meaningful to users. Coding of data such as causes of failure, either by the person filling in the job report or by people at the data centre, can lead to ambiguity and errors.
5. Data input directly into portable computers, rather than onto paper, can greatly improve the quality and speed of data collection and analysis.

INDEX

Roman numerals refer to definitions in Prelims

393